ITT Introduction to Calculus

ITT Introduction to Calculus

John C. Peterson

Chattanooga State Technical Community College

Delmar Publishers

I(T)P® An International Thomson Publishing Company

Albany • Bonn • Boston • Cincinnati • Detroit • London • Madrid • Melbourne
Mexico City • New York • Pacific Grove • Paris • San Francisco • Singapore • Tokyo
Toronto • Washington

NOTICE TO THE READER

Publisher does not warrant or guarantee any of the products described herein or perform any independent analysis in connection with any of the product information contained herein. Publisher does not assume, and expressly disclaims, any obligation to obtain and include information other than that provided to it by the manufacturer.

The reader is expressly warned to consider and adopt all safety precautions that might be indicated by the activities described herein and to avoid all potential hazards. By following the instructions contained herein, the reader willingly assumes all risks in connection with such instructions.

The publisher makes no representations or warranties of any kind, including but not limited to, the warranties of fitness for particular purpose or merchantability, nor are any such representations implied with respect to the material set forth herein, and the publisher takes no responsibility with respect to such material. The publisher shall not be liable for any special, consequential, or exemplary damages resulting, in whole or in part, from the readers' use of, or reliance upon, this material.

Delmar Staff
Publisher: Robert D. Lynch
Acquisitions Editor: Paul Shepherdson
Production Manager: Larry Main
Art and Design Coordinator: Nicole Reamer

Printed in the United States of America

For more information, contact:

Delmar Publishers Inc.
3 Columbia Circle, Box 15015
Albany, NY 12212-5015

International Thomson Publishing Europe
Berkshire House 168-173
High Holborn
London, WC1V 7AA
England

Thomas Nelson Australia
102 Dodds Street
South Melbourne, 3205
Victoria, Australia

Nelson Canada
1120 Birchmont Road
Scarborough, Ontario
Canada, M1K 5G4

International Thomson Editores
Campos Eliseos 385, Piso 7
Col Polanco
11560 Mexico D F Mexico

International Thomson Publishing GmbH
Konigswinterer Strasse 418
53227 Bonn
Germany

International Thomson Publishing Asia
221 Henderson Road
#05-10 Henderson Building
Singapore 0315

International Thomson Publishing—Japan
Hirakawacho Kyowa Building, 3F
2-2-1 Hirakawacho
Chiyoda-ku, Tokyo 102
Japan

5 6 7 8 9 10 XXX 02 01 00 99 98

ISBN: 0-8273-8476-9

Contents

CHAPTER

1

An Introduction to Calculus

In order to pave this parking lot, we must figure out how much asphalt is needed. In this chapter, we will see how calculus can be used to find this answer.

Courtesy of Michael A. Gallitelli, Metroland Photo Inc.

In the middle of the seventeenth century, scientists began to study speed, motion, and rates of change. From this study evolved a new branch of mathematics called calculus. For the next 250 years, most of the important developments in mathematics and science were connected with calculus. Two men, Isaac Newton (1642–1727) and Gottfried Wilhelm Leibniz (1646–1716) are given credit for its discovery.

The development of calculus was a result of attempts to answer several questions in geometry. One question dealt with the slope of the tangent line to a curve, which led to differential calculus. A second question, concerning the area enclosed by the graph of a function and the x-axis, led to integral calculus. The solutions to these questions, the tangent question, and the area question will be explored in Sections 1.1 and 1.2. In Chapters 2 and 4 we will begin a more detailed look at ways to solve these questions.

This chapter forms a foundation for your study of calculus. As you proceed through these next few chapters, you will be given opportunities to use calculus in such areas as heat, light, sound, electricity, and magnetism. Computer programs and calculators will help your study of the ideas behind this powerful field of mathematics.

1.1 THE TANGENT QUESTION

In earlier courses, you studied the slope of a straight line. In particular, we found that if $P(x_1, y_1)$ and $Q(x_2, y_2)$ are two points on a line, then the slope of the line is given by the formula

$$m = \frac{y_2 - y_1}{x_2 - x_1}$$

But, what about curves that are not straight lines?

Consider the curve in Figure 1.1. As you can see, this is not a straight line, but it is the curve of a function. Slope is an indication of the "steepness" of a line. The slope of a straight line is the same at every point on the line. The curve in Figure 1.1 gives the impression that it gets steeper as the values of x increase. We might expect that the slope of a nonlinear curve would be different at different points on the curve. We would like a way to measure the steepness, or slope, of a nonlinear curve at any particular point on that curve.

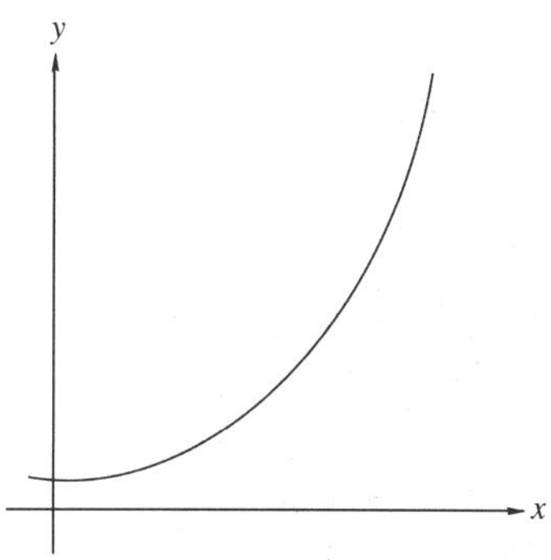

FIGURE 1.1

One way to think of the slope of a curve at some point is to draw the tangent line to the curve at that point. Look at the curve in Figure 1.2. We have drawn the tangent to this curve at point P. We have also drawn the tangent to the curve at point Q. We can tell at a glance that the slope of the tangent at Q is greater than the slope of the tangent at P. Thus, we can let the slope of the tangent to a curve at some point be used for the slope of the curve at that point.

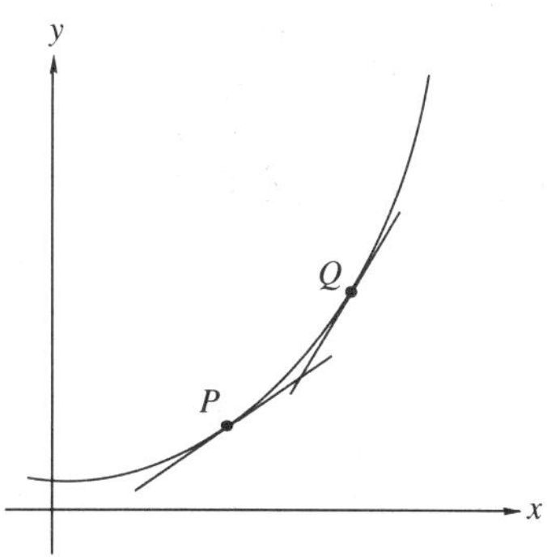

FIGURE 1.2

This is a helpful idea. Now, all we have to do is figure out some way to determine the slope of the tangent to a curve at any point. At present we do not have the background to determine the slope of a tangent line to a curve. We will substitute another idea until we have the necessary background.

Look at the graph of the function in Figure 1.3. This is the same curve we used in the previous example. As in Figure 1.2, P and Q are different points on the curve. P has the coordinates (x_1, y_1) and Q has the coordinates (x_2, y_2). The line that passes through points P and Q is called a **secant line**. The slope of the secant line through points P and Q is given by

$$m_{PQ} = \frac{y_2 - y_1}{x_2 - x_1}$$

Average Slope

This is an important idea. We have found that the slope of the secant line is a way to approximate the slope of the tangent line. What the secant line really gives is the **average rate of change** or **average slope** of the curve over the interval from (x_1, y_1) to (x_2, y_2). We will use a bar to indicate an average. Thus, the symbol for the average slope will be $\bar{m}$, the symbol for the average velocity $\bar{v}$, and so on. The average rate of change through any two points such as $P(x_1, y_1)$ and $Q(x_2, y_2)$ is the same as the

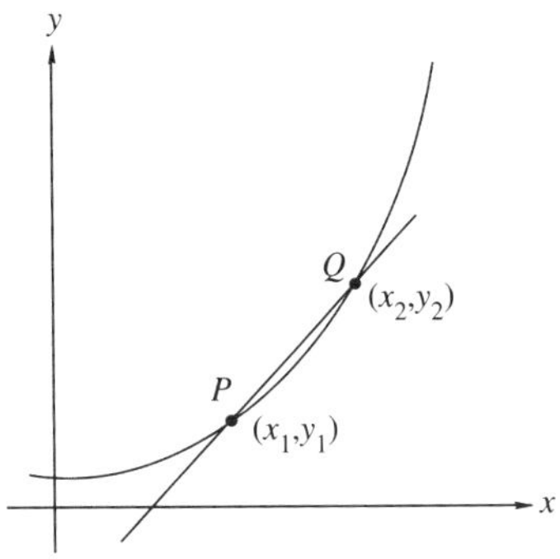

FIGURE 1.3

slope of the line through those two points, and is given by

$$\bar{m} = \frac{y_2 - y_1}{x_2 - x_1}$$

or, since $y_1 = f(x_1)$ and $y_2 = f(x_2)$,

$$\bar{m} = \frac{f(x_2) - f(x_1)}{x_2 - x_1}$$

EXAMPLE 1.1

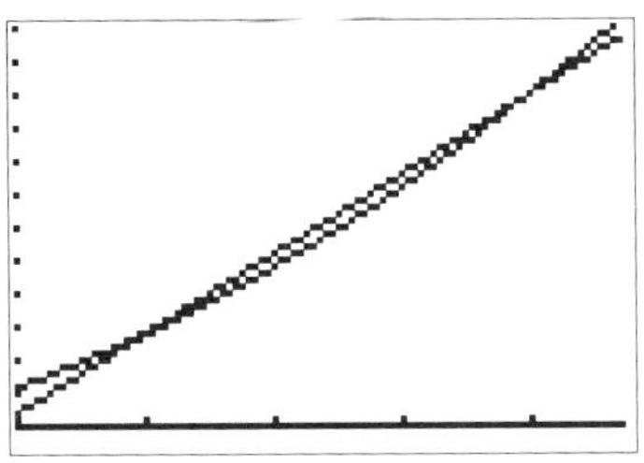

[2, 6.7, 1] x [−2, 60, 5]

FIGURE 1.4

Consider the function $f(x) = x^2 + 3x - 4$. Find the average slope $\bar{m}$ of the curve over the interval from $x = 3$ to $x = 6$.

Solution If we let $x_1 = 3$ and $x_2 = 6$, then, since $f(x) = x^2 + 3x - 4$, we have $f(3) = 14$ and $f(6) = 50$. Figure 1.4 shows a portion of the graph of $f(x) = x^2 + 3x - 4$ and the secant line through the points $(3, 14)$ and $(6, 50)$. Thus, the average slope of $\bar{m}$ is derived as follows:

$$\begin{aligned}\bar{m} &= \frac{f(x_2) - f(x_1)}{x_2 - x_1}\\ &= \frac{f(6) - f(3)}{6-3}\\ &= \frac{50 - 14}{6-3}\\ &= \frac{36}{3}\\ &= 12\end{aligned}$$

The average slope of $f(x) = x^2 + 3x - 4$ from $x = 3$ to $x = 6$ is 12.

EXAMPLE 1.2

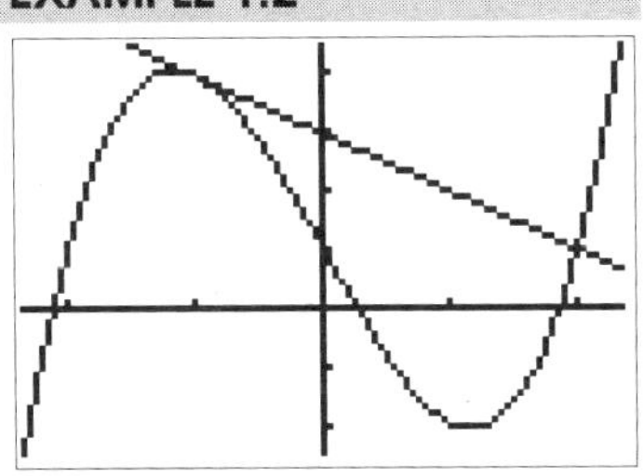

[−2.35, 2.35, 1] x [−2.5, 4.5, 1]

FIGURE 1.5

Find the average slope of the function $g(x) = x^3 - 4x + 1$ from $x = -1$ to $x = 2$.

Solution We begin by letting $x_1 = -1$ and $x_2 = 2$, and evaluating $g(-1)$ and $g(2)$.

$$\begin{aligned}g(-1) &= (-1)^3 - 4(-1) + 1\\ &= 4\end{aligned}$$

$$\begin{aligned}\text{and } g(2) &= 2^3 - 4(2) + 1\\ &= 1\end{aligned}$$

Figure 1.5 shows the graph of $g(x) = x^3 - 4x + 1$ over the interval $[-2.35, 2.35]$ and the secant line through the points $(-1, 4)$ and $(2, 1)$. Thus,

$$\bar{m} = \frac{g(2) - g(-1)}{2 - (-1)}$$

EXAMPLE 1.2 (Cont.)

$$= \frac{1-4}{2+1}$$
$$= \frac{-3}{3}$$
$$= -1$$

The average slope of $g(x) = x^3 - 4x + 1$ from $x = -1$ to $x = 2$ is -1.

EXAMPLE 1.3

Find the average slope of the curve $h(x) = 2x^2 + 5x - 6$ from $x = -4$ to $x = x_1$, where $x_1 \neq -4$.

Solution This is a more general version of the average slope problem. When $x = -4$, $h(x) = h(-4) = 6$ and when $x = x_1$, $h(x_1) = 2x_1^2 + 5x_1 - 6$. Figure 1.6 shows the graph of h and a line through the points $(-4, 6)$ and $(x_1, h(x_1))$. The average slope is

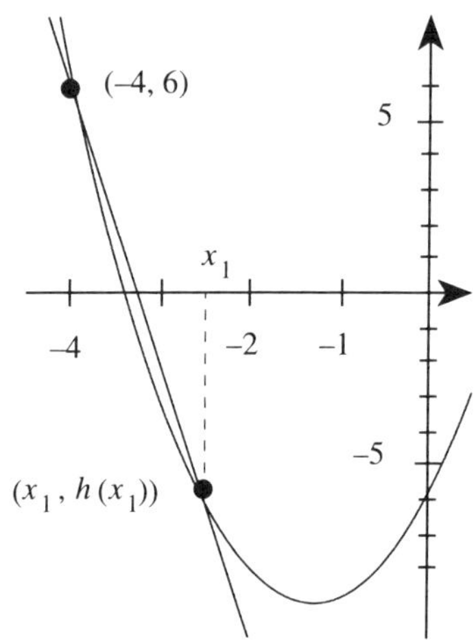

FIGURE 1.6

$$\bar{m} = \frac{h(x_1) - h(-4)}{x_1 - (-4)}$$
$$= \frac{(2x_1^2 + 5x_1 - 6) - 6}{x_1 + 4}$$
$$= \frac{2x_1^2 + 5x_1 - 12}{x_1 + 4}$$
$$= \frac{(2x_1 - 3)(x_1 + 4)}{x_1 + 4}$$
$$= 2x_1 - 3$$

since $x_1 \neq -4$.

This last example provides a general formula for finding the average slope from $x = -4$ to any other point on the curve. For example, if we want the average slope from -4 to -5, we can use this formula with $x_1 = -5$ and get $\bar{m} = 2x_1 - 3 = 2(-5) - 3 = -13$.

The idea for finding the average slope can be applied to finding any type of average change. This can best be shown by the following example.

Application

EXAMPLE 1.4

A tank is filled with water by opening a valve on an inlet pipe. The volume V in liters of water in the tank t min after the valve is opened is given by the formula $V(t) = 5t^2 + 4t$. **(a)** What is the average rate of increase in the volume during the second minute? **(b)** What is the average rate of increase in the volume during the next 30 seconds?

EXAMPLE 1.4 (Cont.)

Solution We want to find the average rate of change in the volume.

(a) When $t = 2$, $V(t) = V(2) = 28$ and when $t = 1$, $V(t) = V(1) = 9$. If R_{Vol} is the rate of change in the volume, then

$$\bar{R}_{\text{Vol}} = \frac{V(2) - V(1)}{2 - 1} = \frac{28 - 9}{2 - 1} = 19$$

The volume changed at the rate of 19 L/min from the first to the second minute.

(b) The next 30 seconds will be from $t = 2.0$ min to $t = 2.5$ min. When $t = 2.5$, we determine that $V(t) = V(2.5) = 41.25$. Thus,

$$\bar{R}_{\text{Vol}} = \frac{V(2.5) - V(2)}{2.5 - 2.0} = \frac{41.25 - 28}{2.5 - 2.0} = \frac{13.25}{0.5} = 26.5$$

Thus, the volume changed at the rate of 26.5 L/min from minutes 2 to 2.5. ▪

Using a Graphing Calculator

Many graphing calculators can be used to approximate the slope of a tangent line at a particular point on a curve. Example 1.5 will show how this can be done on the function used in Example 1.3.

EXAMPLE 1.5

Use a graphing calculator to find the slope of the tangent line to $h(x) = 2x^2 + 5x - 6$ at $x = -4$.

Solution Begin by graphing h. The result of graphing h on a TI-82 as $y_1 = 2x^2 + 5x - 6$ is shown in Figure 22.7a.

Next, press [TRACE] and move the cursor to $x = -4$ and then press [2nd] [DRAW] 5 [ENTER][1]. (Note that pressing [2nd] [DRAW] 5 accesses the "`Draw-Tangent`" option on the calculator.) The result is shown in Figure 1.7b. In the lower left-hand corner of the calculator screen shown in Figure 1.7b, you should see $dy/dx = -11$. The notation $dy/dx = -11$ indicates that the slope of this tangent line is -11. In Example 22.3, we found that the average slope of the tangent line to $h(x) = 2x^2 + 5x - 6$ near $x = -4$ was $\overline{m} = 2x_1 - 3$. If you let $x_1 = -4$, then $\overline{m} = 2(-4) - 3 = -11$, the value of dy/dx.

[1] On a TI-85, press [GRAPH] [MORE] [MATH] [MORE] [MORE] [TANLN] and then move the cursor to (or near) $x = -4$ and press [ENTER]. Setting `xMin=-6.3` and `xMax=6.3` will allow the cursor to stop at $x = -4$.

EXAMPLE 1.5 (Cont.)

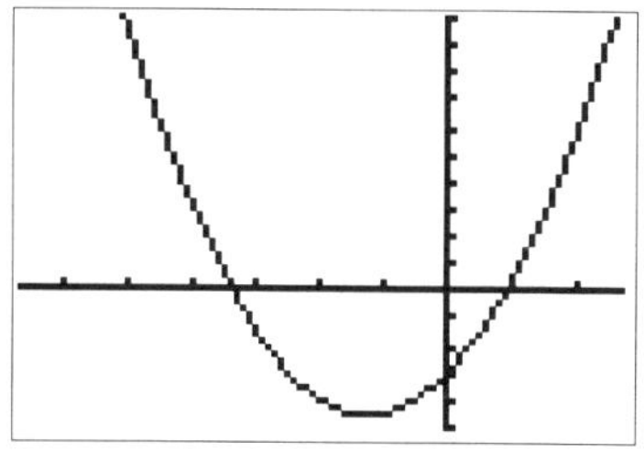

[−6.7, 2.7, 1] x [−10, 20, 2]

FIGURE 1.7a

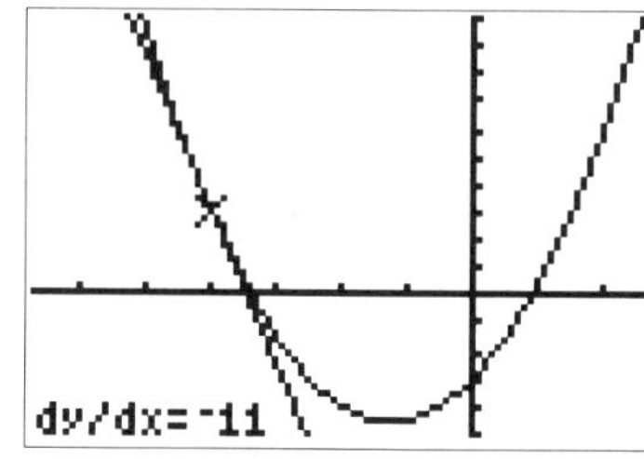

[−6.7, 2.7, 1] x [−10, 20, 2]

FIGURE 1.7b

Exercise Set 1.1

Find the average slope over the indicated intervals for the functions in Exercises 1–14.

1. $f(x) = x^2 + 5x$ from $x = 1$ to $x = 3$
2. $g(x) = 5x - 3$ from $x = 4$ to $x = 7$
3. $h(x) = 7x + 1$ from $x = -3$ to $x = -1$
4. $j(x) = 4x^2 - 5$ from $x = -2$ to $x = 0$
5. $k(x) = 3x^3 + 7x - 1$ from $x = -2$ to $x = -1$
6. $m(x) = 2x^4 - 5x^2 + x - 1$ from $x = 0$ to $x = 1$
7. $f(x) = \dfrac{5}{x}$ from $x = 2$ to $x = 5$
8. $g(x) = \dfrac{8}{x+2}$ from $x = -6$ to $x = -3$
9. $h(x) = x^2$ from $x = 1$ to $x = 0$
10. $j(x) = x^2 + 3x + 5$ from $x = -2$ to $x = b, b \neq -2$
11. $k(x) = x^2 + x - 7$ from $x = 1$ to $x = x_1, x_1 \neq 1$
12. $m(x) = x^3 + 4$ from $x = 1$ to $x = x_1, x_1 \neq 1$
13. $f(x) = x^2 + 1$ from $x = x_1$ to $x = x_1 + h, h \neq 0$
14. $g(x) = x^2 + 4x - 1$ from $x = x_1$ to $x = x_1 + h, h \neq 0$

Solve Exercises 15–22.

15. *Machine technology* What was the average rate of change in the volume for the tank in Example 1.4 **(a)** during the third minute, **(b)** during the fourth minute, that is, from the third to the fourth minute, and **(c)** from the second to the fourth minute?

16. *Physics* A stone is dropped into a pool of water and causes a ripple that travels in the shape of a circle out from the point of impact at a rate of 2 m/s. What is the average change in the area within this circle from the third to the fourth second?

17. Let $f(x) = 3x^2 - 5$. Find the average slope of the curve from $x_1 = 4$ to **(a)** $x_2 = 6$, **(b)** $x_3 = 5$, **(c)** $x_4 = 4.5$, **(d)** $x_5 = 4.25$, **(e)** $x_6 = 4.1$, and **(f)** $x_7 = 4.05$.

18. For the same function in Exercise 17, $f(x) = 3x^2 - 5$, find the average slope of the curve from $x_1 = 4$ to **(a)** $x_2 = 2$, **(b)** $x_3 = 3$, **(c)** $x_4 = 3.5$, **(d)** $x_5 = 3.75$, **(e)** $x_6 = 3.9$, and **(f)** $x_7 = 3.95$.

19. Write a computer program to determine the average slope between any two points on a curve. Use your program to check your results from Exercises 17 and 18.

20. *Medical technology* Consider the following sentence:

> The child's temperature has been rising for the last two hours, but not as rapidly since we gave the antibiotic an hour ago.

(a) With this statement in mind, sketch a graph of the child's temperature as a function of time.
(b) How did you (or how can you) use tangents to your graph to show that your graph is consistent with the statement?

21. *Transportation* On a 50-minute trip, a car travels for 15 min with an average velocity of 20 mph and then 35 min with an average velocity of 54 mph. Determine **(a)** the total distance traveled and **(b)** the average velocity for the entire trip.

22. *Agriculture* The graph in Figure 1.8 gives the number of $f(t)$ farms in a certain Iowa county t years after 1980.

(a) Calculate the average rate of change in the number of farms from 1980 to 1985.
(b) Calculate the average rate of change in the number of farms from 1990 to 1995.
(c) During what two-year period was the number of farms decreasing most rapidly?
(d) What was the rate of change in the number of farms for the two years you answered in (c)?

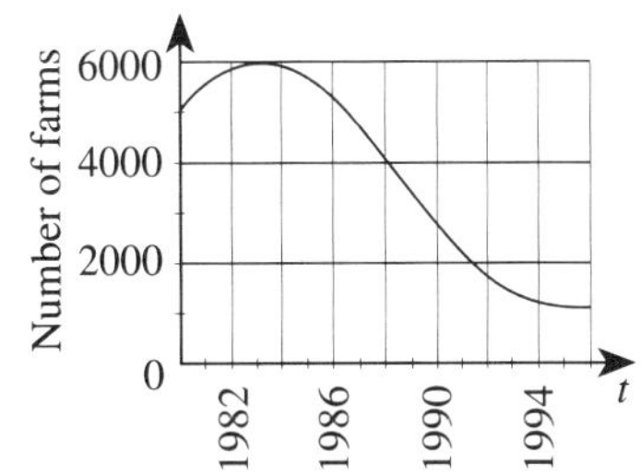

FIGURE 1.8

In Your Words

23. You are in an automobile that is traveling from your home to school. Explain how you can determine the average rate that the car is traveling.

24. What is the difference between a secant line and a tangent line?

1.2 THE AREA QUESTION

In Section 1.1, we examined one of the basic questions that led to the development of calculus. In this section, we will look at the other basic question—the area question.

The area question concerns the area between the graph of a function and the x-axis. There are some restrictions on this problem that can best be described by examining Figure 1.9. The function $f(x)$ should be nonnegative over a closed interval; that is, it should not fall below the x-axis in some closed interval. If the closed interval is $[a, b]$, then the desired area lies between the curve of $f(x)$, the x-axis, and the vertical lines $x = a$ and $x = b$. This is indicated by the shading in Figure 1.9.

We have used several different formulas for the area. We will use two of them to help determine the shaded area in Figure 1.9.

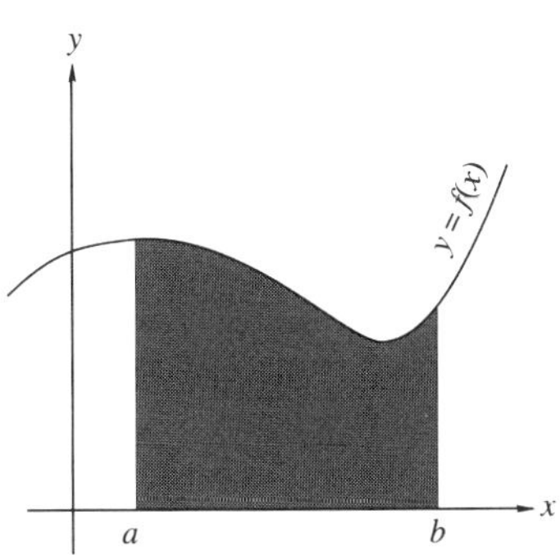

FIGURE 1.9

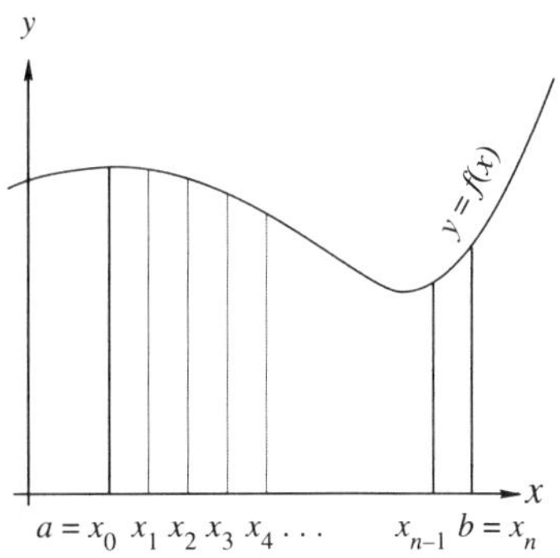

FIGURE 1.10

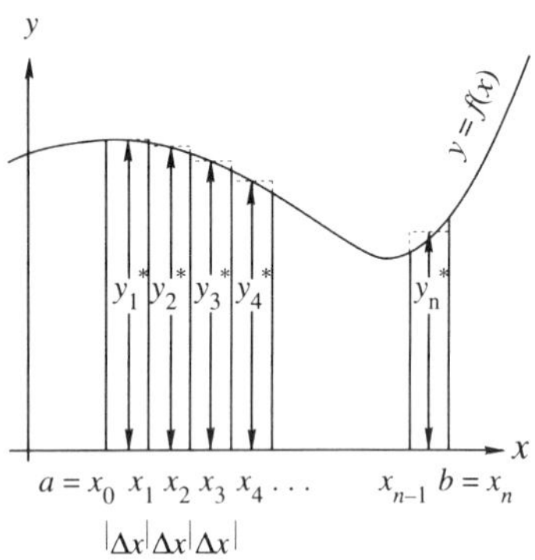

FIGURE 1.11

Rectangular Approximation

The first method we will use will approximate the area by adding the areas of several rectangles. To begin, we will divide the interval $[a, b]$ into n segments. While each segment can be a different length, we will make them all the same length. If we let $a = x_0$ and $b = x_n$, the first segment is $[x_0, x_1]$; the second segment $[x_1, x_2]$; the third segment $[x_2, x_3]$; and so on. The last segment is $[x_{n-1}, x_n]$. In general, the ith segment would be $[x_{i-1}, x_i]$. At each of these points, $x_0, x_1, x_2, \ldots, x_n$, we will erect a line perpendicular to the x-axis. This creates a group of n strips or bands, as shown in Figure 1.10.

We want to draw a line segment parallel to the x-axis across the top of each of these strips. Where we draw these segments is up to us. We would like to draw them so that we can get a close approximation to the area under the curve. What we will do is use an approach that should give a reasonable approximation.

Look at Figure 1.11. We have selected a point in the middle of each strip and drawn the line segment so that it went through the y-coordinate at that point. For example, in the first strip, the midpoint between x_0 and x_1 is the point we will call x_1^*. The height of the curve at x_1^* is $f(x_1^*)$, which we will call y_1^*. Now, since each segment is the same width, the width of each rectangle is the same. This width, which we will call Δx, is equal to $\dfrac{x_n - x_0}{n}$. This means that the area of the first rectangle is $(\Delta x)(y_1^*)$. In the same way, the area of the second rectangle would be $(\Delta x)(y_2^*)$ and the area of the last rectangle would be $(\Delta x)(y_n^*)$.

We can now get an idea of the area under this curve by adding all the areas of these rectangles. Thus, the area under the curve, A, can be written as

$$\begin{aligned} A &\approx (y_1^*)(\Delta x) + (y_2^*)(\Delta x) + (y_3^*)(\Delta x) + \cdots + (y_n^*)(\Delta) \\ &= (y_1^* + y_2^* + y_3^* + \cdots + y_n^*)(\Delta x) \\ &= \left(\sum_{i=1}^{n} y_i^*\right)\Delta x \end{aligned}$$

EXAMPLE 1.6

Suppose $f(x) = x^2 + 3$. Find an approximation of the area between this curve and the x-axis from $x = 1$ to $x = 6$.

Solution We will divide the interval $[1, 6]$ into 5 equal segments. (The number of segments is purely arbitrary and we selected 5 because it means that each segment has a length of 1, or $\Delta x = 1$.) The midpoints of the segments are at 1.5, 2.5, 3.5, 4.5, and 5.5 as can be seen in Figure 1.12. The value of the function at each of these points can be seen in the following table.

x^*	1.50	2.50	3.50	4.50	5.50
$f(x^*) = y^*$	5.25	9.25	15.25	23.25	33.25

EXAMPLE 1.6 (Cont.)

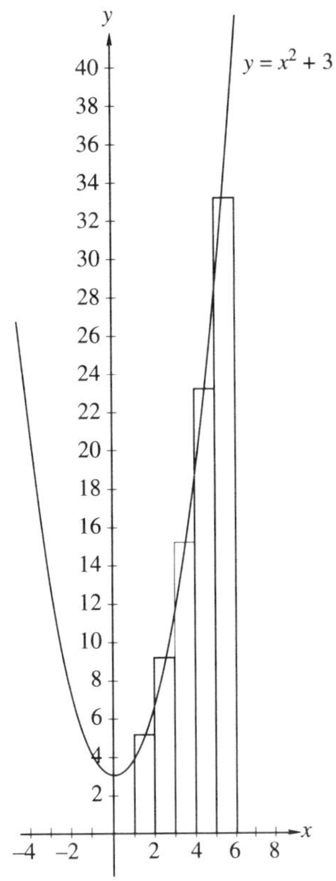

FIGURE 1.12

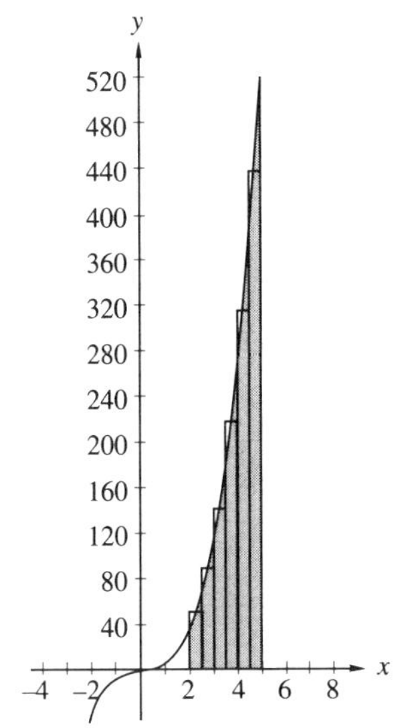

FIGURE 1.13

The area then is found using our new formula for area:

$$\begin{aligned} A &\approx (y_1^*)(\Delta x) + (y_2^*)(\Delta x) + (y_3^*)(\Delta x) + \cdots + (y_n^*)(\Delta x) \\ &\approx (y_1^* + y_2^* + y_3^* + \cdots + y_n^*)(\Delta x) \\ &= (5.25 + 9.25 + 15.25 + 23.25 + 33.25)(1) \\ &= 86.25 \end{aligned}$$

The area is approximately 86.25 square units.

We will work another example, only this time we will make the lengths of the segments something other than 1.

EXAMPLE 1.7

Approximate the area under the curve $g(x) = 4x^3 + 2x - 1$ from $x = 2$ to $x = 5$.

Solution The curve is shown in Figure 1.13 with the shaded region indicating the area we want to approximate. The total length of the interval is 3 units. We will divide this into 6 segments, so each segment will be 0.5 units long or $\Delta x = 0.5$. The following table gives the values of the function at the midpoints of each of these intervals.

EXAMPLE 1.7 (Cont.)

x^*	2.25	2.7500	3.2500	3.7500	4.2500	4.7500
y^*	49.0625	87.6875	142.8125	217.4375	314.5625	437.1875

Again, using our formula for approximating the area we get

$$\begin{aligned} A &\approx (y_1^* + y_2^* + y_3^* + \cdots + y_n^*)(\Delta x) \\ &= (49.0625 + 87.6875 + 142.8125 + 217.4375 \\ &\quad + 314.5625 + 437.1875)(0.5) \\ &= (1{,}248.75)(0.5) \\ &= 624.375 \end{aligned}$$

The approximate area under this curve from $x = 2$ to $x = 5$ is 624.375 square units.

Application

EXAMPLE 1.8

The following table gives the results of a series of drillings to determine the depth of the bedrock at a building site. These drillings were taken along a straight line down the middle of the lot where the building will be placed. In the table, x is the distance from the front of the parking lot and y is the corresponding depth. Both x and y are given in feet.

x	0	20	40	60	80	100	120	140	160
y	33	35	40	45	42	38	46	40	48

Approximate the area of this cross-section.

Solution The total length of the interval is 160 ft. This has been divided into 8 segments, with each segment 20 ft long, so $\Delta x = 20$. Since we do not know the function, we will have to approximate the depth at the midpoints of the intervals. We will do this by taking the average of the values at each end of an interval. Thus, the length at the midpoint of the first interval is $y_1^* = \dfrac{y_1 + y_2}{2} = \frac{33+35}{2} = 34$. The next table gives the approximate values at the midpoints of these intervals.

x^*	10	30	50	70	90	110	130	150
y^*	34	37.5	42.5	43.5	40	42	43	44

Once again, using our formula for approximating the area, we get

$$\begin{aligned} A &\approx (y_1^* + y_2^* + y_3^* + \cdots + y_8^*)\Delta x \\ &= (34 + 37.5 + 42.5 + 43.5 + 40 + 42 + 43 + 44)(20) \\ &= (326.5)(20) \\ &= 6{,}530\,\text{ft}^2 \end{aligned}$$

In Section 4.5, we will find another method to approximate the area of this cross-section, at which point we will get a more accurate approximation of $6{,}460\,\text{ft}^2$.

Application

EXAMPLE 1.9

Before the parking lot for the building in Example 1.8 is completed it will have to be paved. A photograph of the completed lot is shown in Figure 1.14a. In order to pave this parking lot, its area had to be determined. This area was then multiplied by the thickness of the asphalt to obtain the total volume of the asphalt that was needed.

In order to find the area, the contractor drew a base line through the "middle" of the parking lot and then drew lines perpendicular to the base line every 50 feet. Each of these lines were drawn until they reached the other side of the parking lot, as shown in Figure 1.14b. The following table shows the lengths labeled y (in feet) of each of these lines.

x	0	50	100	150	200	250	300	350	400	450	500	550	600
y	72	104	160	200	200	200	190	190	190	180	180	180	180

Use the data in the table to approximate the area of the parking lot.

Courtesy of Michael A. Gallitelli, Metroland Photo Inc.

FIGURE 1.14a

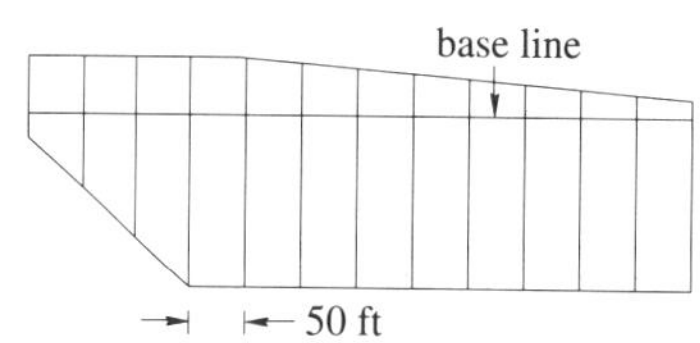

FIGURE 1.14b

Solution We are given that the length of each interval is 50 ft, and so $\Delta x = 50$. As in Example 1.8, we will approximate the lengths at the midpoints by taking the average of the values at each end of the intervals. The next table gives the approximate values at the midpoints of these intervals.

x^*	25	75	125	175	225	275	325	375	425	475	525	575
y^*	88	132	180	200	200	195	190	190	185	180	180	180

EXAMPLE 1.9 (Cont.)

As before, we use our formula for approximating the area and obtain

$$\begin{aligned} A &\approx (y_1^* + y_2^* + y_3^* + \cdots + y_{12}^*)\Delta x \\ &= (88 + 132 + 180 + 200 + 200 + 195 + 190 \\ &\quad + 190 + 185 + 180 + 180 + 180)(50) \\ &= (2{,}100)(50) \\ &= 105{,}000 \end{aligned}$$

Thus, we see that the area of this parking lot is approximately 105,000 ft^2. ▪

Exercise Set 1.2

In Exercises 1–10, find the area under the graph of the function from *a* to *b* using the method described in this section. The value of *n* in each problem indicates the number of segments into which you should divide the interval.

1. $f(x) = 3x + 2;\ a = 1,\ b = 7,\ n = 6$
2. $g(x) = 7 - 4x;\ a = -4,\ b = 1,\ n = 5$
3. $h(x) = x^2 + 1;\ a = 0,\ b = 3,\ n = 6$
4. $k(x) = 3x^2 - 2;\ a = 1,\ b = 3,\ n = 4$
5. $j(x) = 4x^2 + 3x - 5;\ a = 1,\ b = 5,\ n = 8$
6. $m(x) = x^3 + 2;\ a = -1,\ b = 2,\ n = 6$
7. $f(x) = 16 - x^2;\ a = -2,\ b = 4,\ n = 8$
8. $g(x) = 4x - x^3;\ a = 0,\ b = 2,\ n = 10$
9. $h(x) = 3x^2 - 2x + 7;\ a = -5,\ b = -1,\ n = 8$
10. $k(x) = x^4 - 3;\ a = 2,\ b = 4,\ n = 8$

In Exercises 11–12, find the approximate area under the curve defined by graphing the sets of experimental data.

11.

x	5	6	7	8	9	10	11
y	4.2	3.9	3.8	4.0	3.5	3.4	3.9

x	12	13
y	4.1	4.3

12.

x	1.00	1.25	1.50	1.75	2.00
y	16.32	16.48	16.73	16.42	16.38

x	2.25	2.50
y	16.29	16.25

Solve Exercises 13–18.

13. Write a computer program to use the rectangular method described in this section to approximate the area under a curve. Your program should be general enough that you can enter the function, and input the endpoints of the interval and the number of segments. Check your program by using it to solve Exercises 1–10.

14. Use the program you wrote in Exercise 13 to get more accurate approximations of the area under the curve in Exercises 9–10. Make the value of n **(a)** 15, **(b)** 24, **(c)** 48, and **(d)** 96.

15. *Business* The marginal profit for a certain type of coat is given by $P'(x) = 78 - 0.8x$ dollars per coat, where x is the number of coats produced and sold weekly. The profit for the first n coats that are produced and sold is determined by finding the area under the graph of P' from $a = 0$ to $b = n$. What is the profit for the first 40 coats produced and sold?

16. *Environmental science* The level of pollution in San Juan Pedro Bay, due to a sewage spill, is estimated to be $f(t) = \frac{1200t}{\sqrt{t^2+10}}$ parts per million, where t is the time in days since the spill occurred. Find the total amount of pollution during the first 5 days of the oil spill by using the method described in this section and 10 divisions.

17. *Electronics* The charge on a capacitor in millicoulombs can be estimated by finding the area under the graph of $f(t) = 0.6t^2 - 0.2t^3$ from $t = 1$ to $t = 3$. Estimate the charge on this capacitor by using the method described in this section and 8 divisions.

18. *Ecology* A town wants to drain and fill the swamp shown in Figure 1.15.
 (a) What is the surface area of the swamp?
 (b) If the swamp has an average depth of 6 ft, how many cubic yards of dirt will it take to fill the "hole" that is left after the swamp is drained?

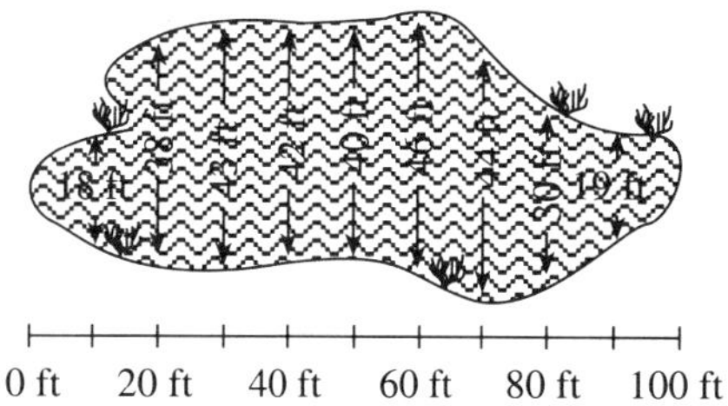

FIGURE 1.15

In Your Words

19. Describe how to use the procedures of this chapter to find the area of an irregular-shaped region.

20. What changes would you have to make in the procedures of this chapter if the rectangles were not all the same width?

21. Use the procedures described in Example 1.19 on an irregular-shaped region such as a parking lot or a lake to determine its area.

1.3 LIMITS: AN INTUITIVE APPROACH

In Sections 1.1 and 1.2, we have looked at ways in which we could attempt to solve two of the questions that led to the development of calculus. These two questions are known as the tangent question and the area question. So far we have laid a foundation for answering them. In this section, we will learn about a third part of the foundation—the limit.

To begin our introduction to the limit, we will return to our look at the tangent question. Consider the curve of the function $f(x) = x^2 - 4$. Suppose that we want to find the slope of the tangent to this curve at the point $(3, 5)$. In Section 1.1, we found out how to determine the slope of a secant line to this curve through the point $(3, 5)$. Let $(x, f(x))$ be any point, except the point $(3, 5)$, on the graph of $f(x) = x^2 - 4$. The secant line through these two points is shown in Figure 1.16. The slope of the secant line is given by

$$\bar{m} = \frac{f(x) - 5}{x - 3} = \frac{(x^2 - 4) - 5}{x - 3} = \frac{x^2 - 9}{x - 3}$$

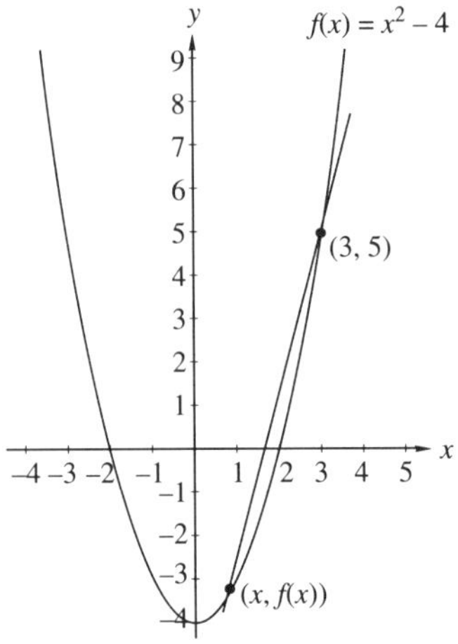

FIGURE 1.16

Since the point $(x, f(x))$ is different from $(3, 5)$, we know that $x \neq 3$ and so we can simplify our equation to

$$\bar{m} = \frac{(x+3)(x-3)}{x-3} = x+3$$

So far so good! Everything that we have done is just like what we did in Section 1.1. But now, we want to look at the problem a little closer. What happens as x gets closer to 3? The values of $\bar{m} = x+3$ get closer to 6. We put this idea into symbols by writing

$$\bar{m}_{\tan} = \lim_{x \to 3} \frac{x^2-9}{x-3} = \lim_{x \to 3}(x+3) = 6$$

Since we know that the slope of the tangent line at the point $(3, 5)$ is 6, we are able to determine that the equation of the tangent is

$$\begin{aligned} y - 5 &= 6(x-3) \\ y &= 6x - 13 \end{aligned}$$

This is the basic idea behind a limit. We will now begin to look at a more detailed explanation.

> **Limit of a Function**
>
> If a function $y = f(x)$ approaches the real number L as x approaches a particular value a, then we say that L is the **limit** of f as x approaches a. The notation for this concept is
>
> $$\lim_{x \to a} f(x) = L.$$

The $\lim\limits_{x \to a} f(x)$ may or may not exist for any function at any particular a. The following examples use numerical calculations to locate limits.

The Numerical Approach to a Limit

To get a better idea of a limit, we will return to Exercises 17–18 from Exercise Set 1.1.

In these two exercises $f(x) = 3x^2 - 5$. We will set up two tables of values. Each table will contain values for x and $\bar{m} = \dfrac{f(x) - f(4)}{x-4} = \dfrac{f(x) - 43}{x-4}$. In the first, we will put the values for Exercise 17, and in the second, the values for Exercise 18.

In Exercise 18, all of the values of x were larger than 4 with the first one being the largest, 6, and each one getting smaller. The last value was the smallest, 4.05. We will continue this process by selecting additional numbers larger than 4 (and smaller than 4.05), each one getting closer to 4.

x	6	5	4.5	4.25	4.10	4.05	4.01	4.001	4.0001
m	30	27	25.5	24.75	24.30	24.15	24.03	24.003	24.0003

The table of values for Exercise 18 contains values of x that are all smaller than 4 with the largest being 3.95. Again, we will continue the process. This time the numbers will all be smaller than 4, each getting increasingly closer to 4.

x	2	3	3.5	3.75	3.9	3.95	3.99	3.999	3.9999
m	18	21	22.5	23.75	23.70	23.85	23.97	23.997	23.9997

These tables give us a numerical method for approximating the $\lim_{x\to 4}(3x^2-5)$. As you can see from the two tables, when x is 4.0001, $m \approx 24.0004$ and when x is 3.9999, $m \approx 23.9995$. It appears that if we were to pick values of x that were closer to 4, we could then get a value of $f(x) = 3x^2 - 5$ that was closer to 24.

EXAMPLE 1.10

Use a numerical approach to find $\lim_{x\to 1}\frac{x^3-1}{x-1}$.

Solution As in the previous explanation, we will construct a table (using the program from Exercise 19, Exercise Set 1.1). This time, since we want to find the limits as x gets closer to 1, we will select values of x that keep getting closer to 1. Some of the values selected are less than 1, and some are larger than 1. The results are shown in the following table.

x	0.500	0.900	0.990	0.999	1.001	1.010	1.100	1.500
$\frac{x^3-1}{x-1}$	1.75	2.71	2.9701	2.997	3.003	3.0301	3.31	4.75

It appears from the table that $\lim_{x\to 1}\frac{x^3-1}{x-1} = 3$.

This result is confirmed by looking at the graph of $y = \frac{x^3-1}{x-1}$, as shown in Figure 1.17. The "hole" in the graph appears to be at $(1, 3)$. ▪

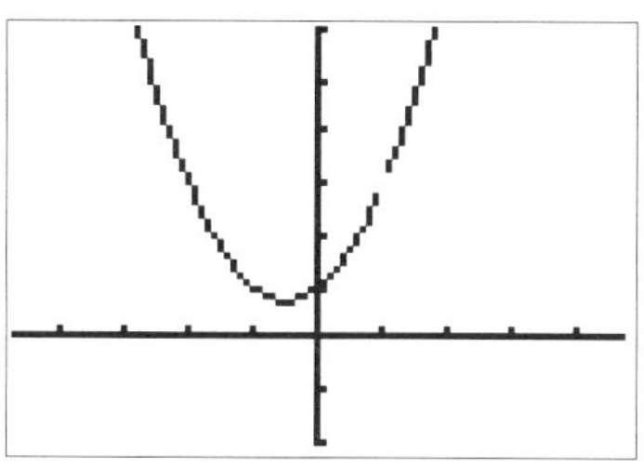

[−4.7, 4.7, 1] x [−2, 6, 1]

FIGURE 1.17

You may remember that one of the special factors from algebra was $x^3 - 1 = (x-1)(x^2+x+1)$. We can use this knowledge to see that as long as $x \neq 1$, $g(x) = \frac{x^3-1}{x-1} = x^2+x+1$. This makes it a lot easier to see that as x gets closer to 1, the value of the function gets closer to 3.

This next example will show that this type of simplification does not work for all limits.

EXAMPLE 1.11

Use a numerical approach to find $\lim_{x\to 0} \frac{\sin x}{x}$.

Solution Since we cannot factor an x out of $\sin x$, the simplification method we have been using will not work. But, if we carefully construct a table of values for x and $g(x) = \frac{\sin x}{x}$ for values of x near 0, we can get an idea of what this limit will be. The following table shows this. Notice that the values of x are in radians.

x (radians)	0.200	0.100	0.010	−0.010	−0.100	−0.200
$\sin x$	0.1987	0.0998	0.0099	−0.0099	−0.0998	−0.1987
$\frac{\sin x}{x}$	0.9933	0.9983	0.9999	0.9999	0.9983	0.9933

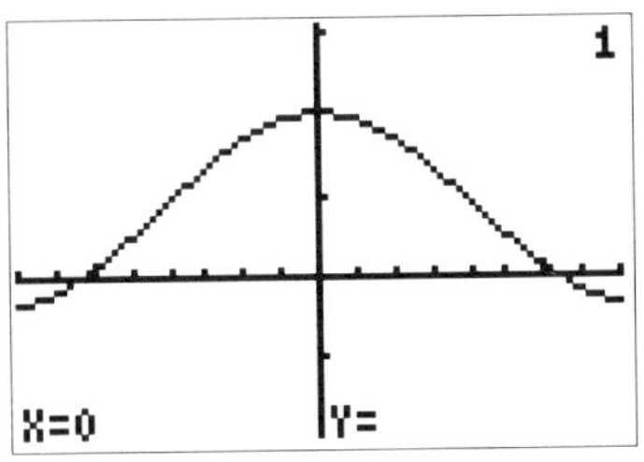

[−4, 4, 0.5] x [−1, 1.5, 1]

FIGURE 1.18

From this table of values it appears that

$$\lim_{x\to 0} \frac{\sin x}{x} = 1$$

The graph in Figure 1.18 seems to reinforce this numerical method that shows $\lim_{x\to 0} \frac{\sin x}{x} = 1$. Notice that if you TRACE this curve, there is no y-value when $x = 0$. That is because 0 is not in the domain of the function.

The Graphical Approach to a Limit

The numerical approach is often a good way to approximate a limit. There are some times, however, when a graphical approach will be easier. We will use the graphical approach in the next two examples.

EXAMPLE 1.12

Use a graphical approach to find $\lim_{x\to 1} h(x)$, where

$$h(x) = \begin{cases} 2x - 4 & \text{if } x \neq 1 \\ 5 & \text{if } x = 1 \end{cases}$$

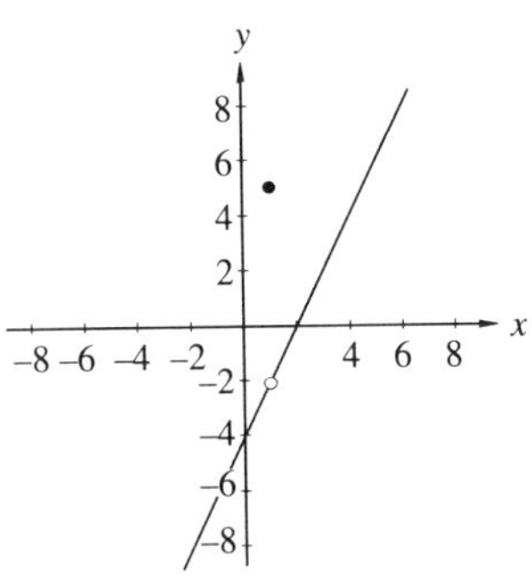

FIGURE 1.19

Solution The graph of this function is shown in Figure 1.19. As you can see, as the values of x get close to 1 the values of $h(x)$ get very close to −2. In fact, if we select x close enough to 1, we can get $h(x)$ as close to −2 as we want. We decide, based on our observation, that $\lim_{x\to 1} h(x) = -2$. Notice, however, that when $x = 1$, $h(x) = 5$. Thus, $\lim_{x\to 1} h(x) \neq h(1)$.

Note While Example 1.12 is somewhat artificial, you can see that the limit of a function at a given point is not always the same as the value of the function at that point.

The limit of a function at a certain point may not always exist. Until now, all of the functions we have examined have approached a certain number as the values of x got close to a particular value. The next example will show that this does not always happen.

EXAMPLE 1.13

Use a graphical approach to find $\lim_{x\to 0} k(x)$, where

$$k(x) = \begin{cases} 2 & \text{if } x < 0 \\ -4 & \text{if } x = 0 \\ 5 & \text{if } x > 0 \end{cases}$$

Solution The graph of this function is shown in Figure 1.20. As you can see, if the values of x are close to 0 and negative, the value of $k(x)$ is 2. On the other hand, for values of x that are close to 0 and positive, the value of $k(x)$ is 5. There is no specific number that the values of $k(x)$ are near when x is close to 0. We must conclude that the $\lim_{x\to 0} k(x)$ does not exist. The fact that the function is defined when $x = 0$, $k(0) = -4$, has no effect on whether the limit of the function exists at that point.

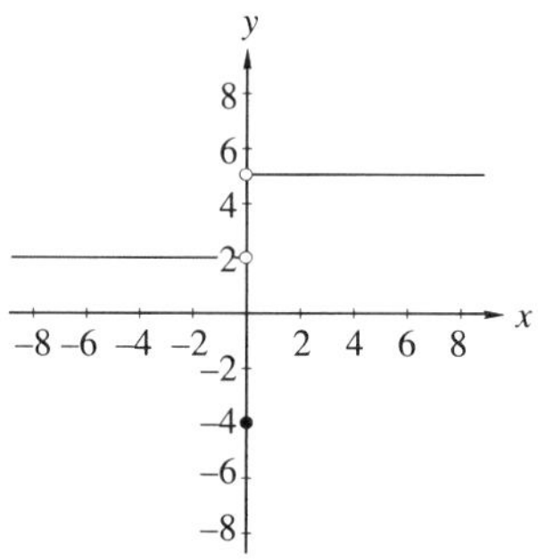

FIGURE 1.20

We will use the graphical approach to show two basic rules for limits:

Rules for Limits

Rule 1: $\lim_{x\to c} A = A$, where A and c are real numbers

Rule 2: $\lim_{x\to c} x = c$, where c is a real number

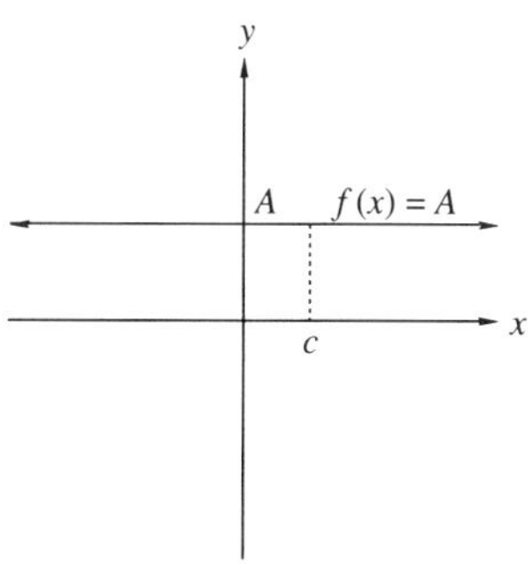

FIGURE 1.21

The graph for the first rule is shown in Figure 1.21. As you can see, the value of $f(x)$ is always A and so, as the values of x approach c, $f(x) = A$, thus we have $\lim_{x\to c} f(x) = A$.

The second rule is demonstrated in Figure 1.22. The graph of $f(x) = x$ is a straight line. For any value of c, as x gets close to c, the values of $f(x)$ are just as close to c, since $f(x) = x$. This shows that $\lim_{x\to c} f(x) = c$.

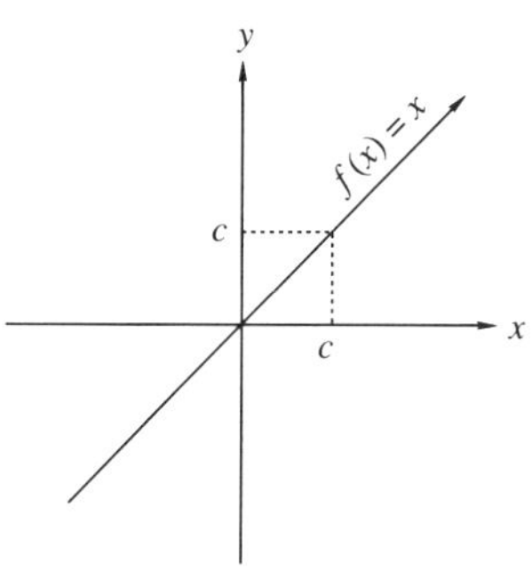

FIGURE 1.22

EXAMPLE 1.14

Evaluate each of the following limits: **(a)** $\lim_{x\to3} 6$, **(b)** $\lim_{x\to-2} 7$, **(c)** $\lim_{x\to-5} x$, and **(d)** $\lim_{x\to7} x$.

Solution The first two, (a) and (b), follow rule 1, $\lim_{x\to c} A = A$. In (a), $A = 6$ so, $\lim_{x\to3} 6 = 6$ and in (b), $A = 7$ and $\lim_{x\to-2} 7 = 7$.

The last two parts, (c) and (d), use rule 2, $\lim_{x\to c} x = c$. In (c) we have $\lim_{x\to-5} x = -5$ and in (d) $\lim_{x\to7} x = 7$. ▪

In this section, we have tried to give you a foundation for the idea of a limit. The following exercises are designed to further develop this foundation. In Section 1.4 we will take a more detailed look at the limit concept.

Exercise Set 1.3

In Exercises 1–8, complete each table and use it to determine the indicated limit.

1.

x	0.9	0.99	0.999	0.9999	1.0001	1.001	1.01	1.1
$f(x) = 3x$								

$\lim_{x\to1} 3x = ?$

2.

x	2.9	2.99	2.999	2.9999	3.0001	3.001	3.01	3.1
$g(x) = x - 4$								

$\lim_{x\to3} (x - 4) = ?$

3.

x	-1.1	-1.01	-1.001	-1.0001	-0.9999	-0.999	-0.99	-0.9
$h(x) = x^2 + 2$								

$\lim_{x\to-1} (x^2 + 2) = ?$

4.

x	-2.1	-2.01	-2.001	-2.0001	-1.9999	-1.999	-1.99	-1.9
$k(x) = \frac{x^2 - 4}{x + 2}$								

$\lim_{x\to-2} \frac{x^2 - 4}{x + 2} = ?$

5.

x	-0.1	-0.01	-0.001	-0.0001	0.0001	0.001	0.01	0.1
$f(x) = \frac{\tan x}{x}$								

$\lim_{x\to0} \frac{\tan x}{x} = ?$

6.

x	-0.1	-0.01	-0.001	-0.0001	0.0001	0.001	0.01	0.1
$g(x) = \frac{1 - \cos x}{x^2}$								

$\lim_{x\to0} \frac{1 - \cos x}{x^2} = ?$

7.

x	0.9	0.99	0.999	0.9999	1.0001	1.001	1.01	1.1
$h(x) = \dfrac{x}{x-1}$								

$\lim\limits_{x \to 1} \dfrac{x}{x-1} = ?$

8.

x	−0.1	−0.01	−0.001	−0.0001	0.0001	0.001	0.01	0.1
$k(x) = \dfrac{1}{x^2}$								

$\lim\limits_{x \to 0} \dfrac{1}{x^2} = ?$

Solve Exercises 9 and 10.

9. For the function f in the given figure, find the following limits or state that the limit does not exist. Give a reason for your answers. **(a)** $\lim\limits_{x \to 0} f(x)$ **(b)** $\lim\limits_{x \to 2} f(x)$ **(c)** $\lim\limits_{x \to 3} f(x)$

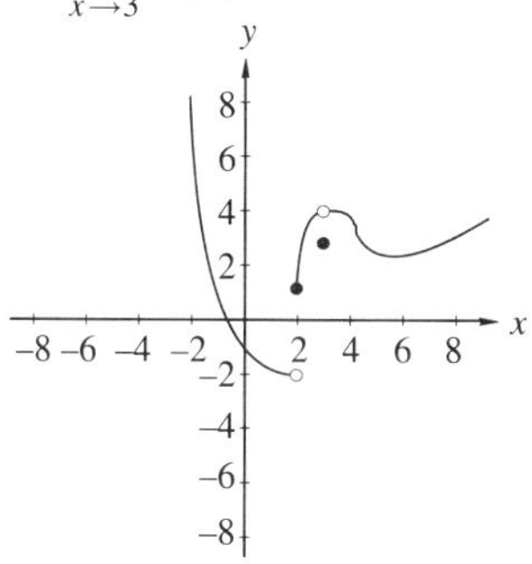

10. For the function g in the given figure, find the following limits or state that the limit does not exist. Give reasons for your answers. **(a)** $\lim\limits_{x \to -2} g(x)$ **(b)** $\lim\limits_{x \to 0} g(x)$ **(c)** $\lim\limits_{x \to 3} g(x)$

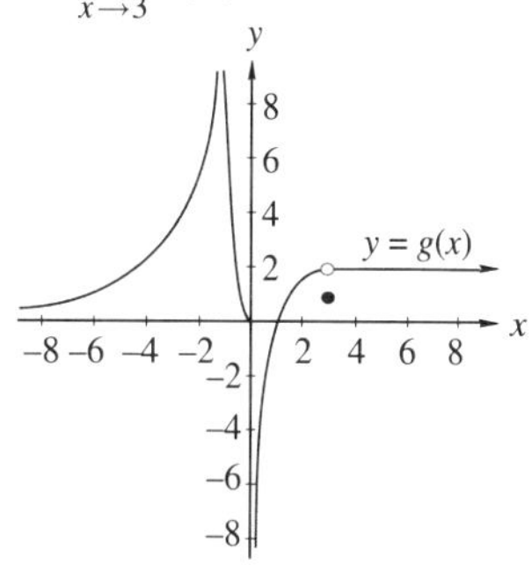

In Exercises 11–18, sketch the graphs of the functions and find out if $\lim\limits_{x \to c}$ exists at the given value of c or state that the limit does not exist.

11. $f(x) = \begin{cases} 3x - 2 & \text{if } x \neq 1 \\ 0 & \text{if } x = 1 \end{cases} \qquad c = 1$

12. $g(x) = \begin{cases} x + 1 & \text{if } x < 0 \\ x^2 + 1 & \text{if } x \geq 0 \end{cases} \qquad c = 0$

13. $h(x) = \begin{cases} 5 - 2x & \text{if } x < 0 \\ 0 & \text{if } x = 0 \\ 4x + 5 & \text{if } x > 0 \end{cases} \qquad c = 0$

14. $k(x) = \begin{cases} x + 1 & \text{if } x < 1 \\ \text{not defined} & \text{if } x = 1 \\ 2x & \text{if } x > 1 \end{cases} \qquad c = 1$

15. $f(x) = \begin{cases} 1 - 3x & \text{if } x < 0 \\ 3x + 1 & \text{if } x \geq 0 \end{cases} \qquad c = 0$

16. $g(x) = \begin{cases} 3x + 1 & \text{if } x \leq -1 \\ 1 - 3x & \text{if } x > -1 \end{cases} \qquad c = -1$

17. $h(x) = \begin{cases} 3x + 1 & \text{if } x < 0 \\ 4 & \text{if } x = 0 \\ 1 - 3x & \text{if } x > 0 \end{cases} \qquad c = 0$

18. $k(x) = \begin{cases} 3x + 1 & \text{if } x < 0 \\ 2 & \text{if } x = 0 \\ x + 1 & \text{if } x > 0 \end{cases} \qquad c = 0$

Evaluate the limits in Exercises 19–24.

19. $\lim_{x\to 2}(-15)$
20. $\lim_{x\to 0} x$
21. $\lim_{x\to -4} x$
22. $\lim_{x\to -7} 19$
23. $\lim_{x\to 8} x$
24. $\lim_{x\to -5}(-9)$

In Exercises 25–28, use a graphing calculator or graphing software to guess whether each of the following limits exist. If the limit does exist, estimate its value.

25. $\lim_{x\to 0}\dfrac{2x^2}{x^2+\sin x}$
26. $\lim_{x\to 0}\dfrac{5x^2}{2x^2+\sin^2 x}$
27. $\lim_{x\to 0}\dfrac{2x^2}{x^2+\cos x-1}$
28. $\lim_{x\to 0}\dfrac{3x^2}{x^2+\tan x}$

In Your Words

29. Explain the notation $\lim_{x\to a} f(x) = L$.
30. Suppose that $y = g(x)$ is a function and $\lim_{x\to 2} g(x) = 4$ and $g(2) = -3$.
    (a) Sketch a possible graph for g.
    (b) Use your graph to explain what is happening to g for values of x close to $x = 2$.
31. Suppose that the function $y = h(x)$ is not defined at $x = 4$. Is it possible for $\lim_{x\to 4} h(x)$ to exist? Explain your answer.

1.4 ONE-SIDED LIMITS

In using both the numerical and the graphical approaches to finding $\lim_{x\to c} f(x)$, we studied the behavior of the function f on both sides of c. There are times when it is necessary to investigate the limit on just one side of c. When this is done, we are investigating one-sided limits.

There are two types of **one-sided limits**. One, the **left-hand limit**, is written as $\lim_{x\to c^-} f(x)$, and the other, the **right-hand limit**, is written as $\lim_{x\to c^+} f(x)$.

To get a better idea of one-sided limits, we will return to an example we worked earlier. In Example 1.13, we wanted to find $\lim_{x\to 0} k(x)$, where

$$k(x) = \begin{cases} 2 & \text{if } x < 0 \\ -4 & \text{if } x = 0 \\ 5 & \text{if } x > 0 \end{cases}$$

The graph of the function is shown in Figure 1.23.

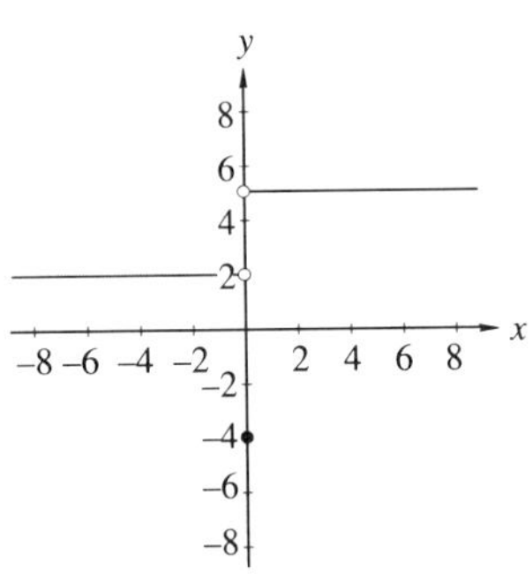

FIGURE 1.23

We know that $\lim_{x\to 0} k(x)$ does not exist, because when x is positive, the value of $k(x)$ is 5, and when x is negative, the value of $k(x)$ is 2. What we have really said is that on the right-hand side of 0 the limit of $k(x)$ is 5 and on the left-hand side, the limit is 2. These are examples of one-sided limits.

In general, the **one-sided limits** of a function $f(x)$ at a point c are the **left-hand limit**, $\lim_{x\to c^-} f(x)$, and the **right-hand limit**, $\lim_{x\to c^+} f(x)$. In the function in Example 1.13, $\lim_{x\to 0^-} k(x) = 2$ and $\lim_{x\to 0^+} k(x) = 5$.

EXAMPLE 1.15

Determine the one-sided limits of $f(x)$ at $x = 3$, if

$$f(x) = \begin{cases} 2x+1 & \text{if } x < 1 \\ 4-x & \text{if } 1 \le x \le 3 \\ x & \text{if } x > 3 \end{cases}$$

Solution The way in which $f(x)$ behaves depends on whether $x < 3$ or $x > 3$, as shown in Figure 1.24. We have different definitions for $f(x)$ for each one-sided limit. When $x < 3$ (but close to 3), f is defined by $f(x) = 4 - x$, and so the left-hand limit is

$$\lim_{x \to 3^-} f(x) = \lim_{x \to 3^-} (4 - x) = 1$$

But, when $x > 3$, f is defined by $f(x) = x$, and so the right-hand limit is

$$\lim_{x \to 3^+} f(x) = \lim_{x \to 3^+} x = 3$$

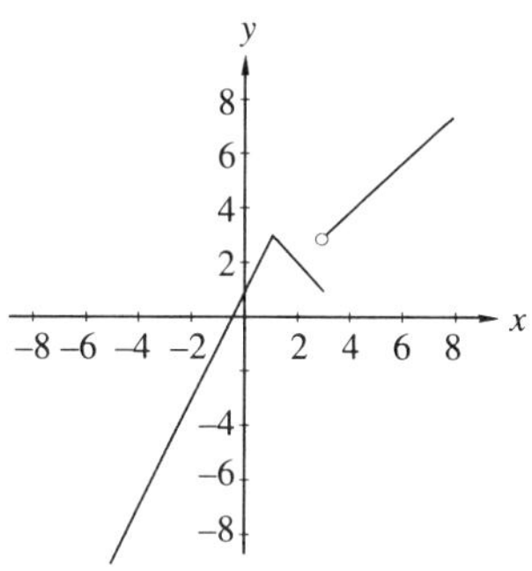

FIGURE 1.24

One-sided limits provide a way to determine whether or not the limit of a function exists at a point. The limit of a function $f(x)$ will exist at a point c, if and only if the right-hand limit of $f(x)$ at c and the left-hand limit of $f(x)$ at c both equal the same real number. Symbolically, this relationship is written as follows.

Relationship Between One-Sided and Two-Sided Limits

$\lim_{x \to c} f(x) = L$, if and only if $\lim_{x \to c^-} = L$ and $\lim_{x \to c^+} f(x) = L$, where L is a real number.

EXAMPLE 1.16

Determine the one-sided limits of $f(x)$ at $x = 1$, where

$$f(x) = \begin{cases} 2x+1 & \text{if } x < 1 \\ 4-x & \text{if } 1 \le x \le 3 \\ x & \text{if } x > 3 \end{cases}$$

Solution This is the same function we studied in Example 1.15. We will take the one-sided limits at $x = 1$. Here, when $x < 1$, f behaves as if $f(x) = 2x + 1$, and we get

$$\lim_{x \to 1^-} f(x) = \lim_{x \to 1^-} (2x + 1) = 3$$

When $x > 1$ (but close to 1), we use $f(x) = 4 - x$, with the result that

$$\lim_{x \to 1^+} f(x) = \lim_{x \to 1^+} (4 - x) = 3$$

Since $\lim_{x \to 1^-} f(x) = \lim_{x \to 1^+} f(x) = 3$, we can say that $\lim_{x \to 1} f(x) = 3$.

We just specified that in order for the limit to exist, both of the one-sided limits must equal the same real number. The next example will show why we must make sure that it is a real number.

EXAMPLE 1.17

Find the one-sided limits of $g(x) = \frac{1}{x^2}$, where $x = 0$.

Solution We will use a table of values and a graph to give us an indication of these limits. The following table and the graph in Figure 1.25 seem to indicate that as $x \to 0$, from both the left and the right, $g(x)$ increases without bound or without limit.

x	± 1	± 0.5	± 0.1	± 0.01	± 0.001	± 0.0001
$g(x)$	1	4	100	10,000	1,000,000	100,000,000

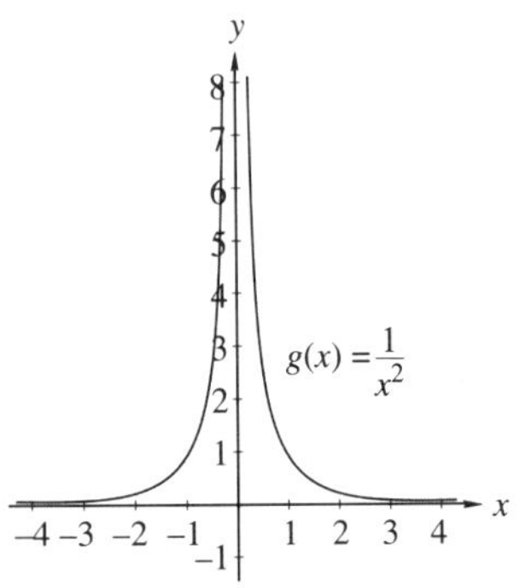

FIGURE 1.25

So, there is no limit for $g(x)$ at $x = 0$. But, because both one-sided limits increase without bound, we say that $g(x)$ becomes positively infinite as x approaches 0. Symbolically, we write

$$\lim_{x \to 0} \frac{1}{x^2} = \infty$$

Caution

Be careful! The fact that we used an equal sign in this last limit does not mean that the limit exists. In fact, just the opposite is true. The infinity symbol ∞ here means that the limit does not exist and shows this because the one-sided limits increase without bound.

A somewhat different result is shown in the next example, where we look at $\frac{1}{x}$ as it gets close to 0.

EXAMPLE 1.18

Find the one-sided limits of $h(x) = \frac{1}{x}$ as $x \to 0$.

Solution Again, we will use a table of values and a graph to demonstrate our answer. The table follows and the graph of $h(x)$ is in Figure 1.26.

x	-1	-0.1	-0.01	-0.001	0.001	0.01	0.1	1
$h(x)$	-1	-10	-100	$-1{,}000$	$1{,}000$	100	10	1

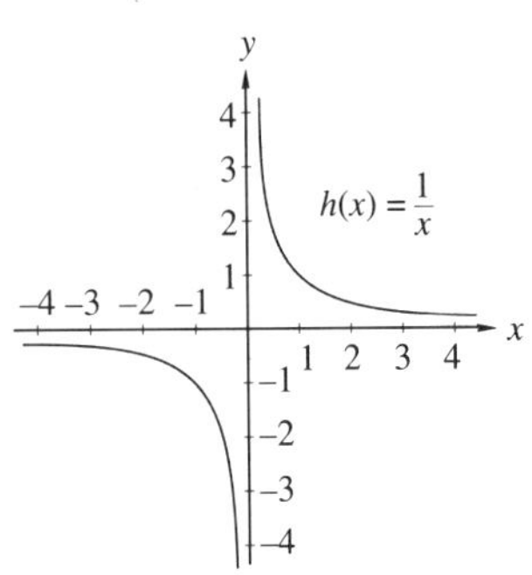

FIGURE 1.26

As we can see, $\lim_{x \to 0+} \frac{1}{x} = \infty$. But, what happens when x is negative? In this case, the values of $h(x)$ become negatively infinite, so

$$\lim_{x \to 0-} \frac{1}{x} = -\infty$$

Thus we see two reasons that $\lim_{x \to 0} \frac{1}{x}$ does not exist. One reason is that the left-hand and right-hand limits are not equal; the other is that infinity, either positive or negative, is not a real number.

Limits at Infinity

What happens to $h(x)$ as x gets increasingly larger? The following table gives some indication of what happens when the values of x get larger.

x	100	1,000	10,000	100,000	1,000,000
$h(x) = \frac{1}{x}$	0.01	0.001	0.0001	0.00001	0.000001

This table and the graph in Figure 1.26 indicate that

$$\lim_{x \to \infty} \frac{1}{x} = 0$$

That is, as x gets larger without bound, the value of $\frac{1}{x}$ approaches 0. Similarly, as x gets smaller without bound, the value of $\frac{1}{x}$ approaches 0. This is written as

$$\lim_{x \to -\infty} \frac{1}{x} = 0$$

These two results are summarized in the following box.

Limits at Infinity

$$\lim_{x \to \infty} \frac{1}{x} = 0$$

and

$$\lim_{x \to -\infty} \frac{1}{x} = 0$$

Exercise Set 1.4

Solve Exercises 1 and 2.

1. For the function f given in the figure below, find the following limits or state that the limit does not exist.

   (a) $\lim_{x \to 2^-} f(x)$ **(c)** $\lim_{x \to 3^-} f(x)$

   (b) $\lim_{x \to 2^+} f(x)$ **(d)** $\lim_{x \to 3^+} f(x)$

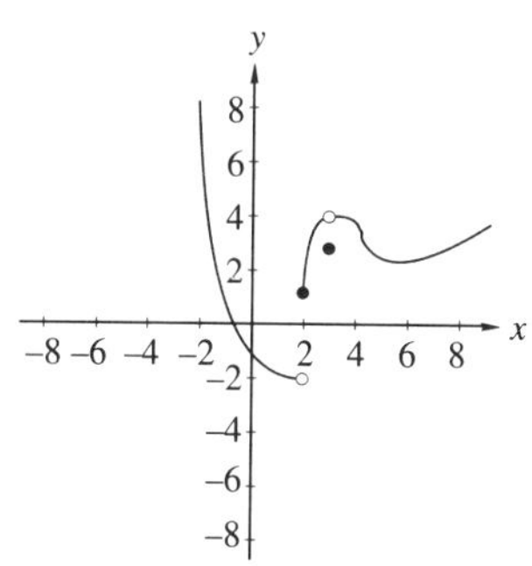

2. For the function g given in the figure below, find the following limits or state that the limit does not exist.

(a) $\lim_{x \to -1^-} g(x)$ **(c)** $\lim_{x \to 0^-} g(x)$ **(e)** $\lim_{x \to \infty} g(x)$

(b) $\lim_{x \to -1^+} g(x)$ **(d)** $\lim_{x \to 0^+} g(x)$ **(f)** $\lim_{x \to -\infty} g(x)$

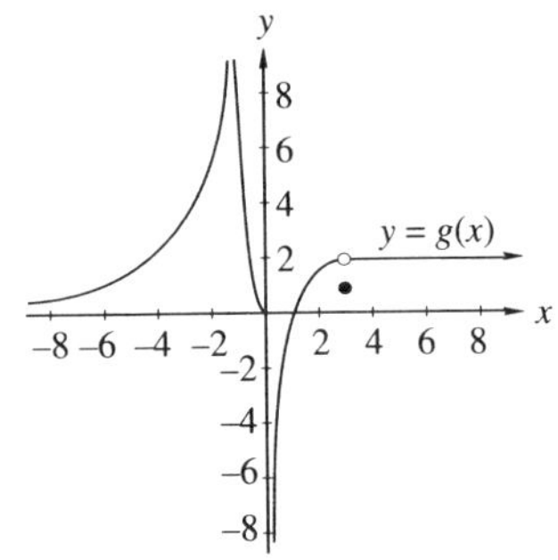

In Exercises 3–18, use a table or a graph to determine the one-sided limits.

3. $\lim_{x \to 5^+} (x-2)$

4. $\lim_{x \to 2^-} (x^2-1)$

5. $\lim_{x \to 4^+} \frac{x}{x-4}$

6. $\lim_{x \to 3^-} \frac{2x}{x-3}$

7. $\lim_{x \to -2^+} \frac{x^2-4}{x+2}$

8. $\lim_{x \to -5^-} \frac{x^2+25}{x+5}$

9. $\lim_{x \to 1^-} \frac{|x-1|}{x-1}$

10. $\lim_{x \to 1^+} \frac{|x-1|}{x-1}$

11. $\lim_{x \to 2^-} \frac{|x-2|}{x-2}$

12. $\lim_{x \to 2^+} \frac{|x-2|}{x-2}$

13. $\lim_{x \to 2^+} \sqrt{x-2}$

14. $\lim_{x \to -4^+} \sqrt{x+4}$

15. $\lim_{x \to \infty} \frac{9}{3x}$

16. $\lim_{x \to -\infty} \frac{-8}{4x}$

17. $\lim_{x \to -\infty} \frac{1}{2x+1}$

18. $\lim_{x \to \infty} \frac{-5}{9x-4}$

In Exercises 19–24, use a graphing calculator or graphing software to guess whether each of the following limits exists. If the limit does exist, estimate its value.

19. $\lim_{x \to \infty} \frac{2x^2}{x^2+\sin x}$

20. $\lim_{x \to \infty} \frac{5x^2}{2x^2+\sin^2 x}$

21. $\lim_{x \to 0^-} \frac{2x^2}{x^2-\cos x}$

22. $\lim_{x \to 0^+} \frac{2x^2}{x^2-\tan x}$

23. $\lim_{x \to \infty} \frac{5^x-5}{2^x-2}$

24. $\lim_{x \to -\infty} \frac{5^x-5}{2^x-2}$

Solve Exercises 25–28.

25. *Medical technology* The concentration of a drug in a patient's bloodstream t hours after it was injected is given by

$$C(t) = \frac{0.25t}{t^2+5}$$

Find each of the following. **(a)** $C(0.5)$, **(b)** $C(2)$, **(c)** $\lim_{t \to \infty} C(t)$

26. *Business* It has been determined that after t weeks of training, a certain student in a word processing class can keyboard at the rate of $W(t) = 65 + \frac{70t^2}{t^2+15}$ words per minute. What are the most words per minute that we can expect this student to keyboard, that is, what is $\lim_{t \to \infty} \left(65 + \frac{70t^2}{t^2+15}\right)$?

27. *Business* The cost, in dollars, of an overseas telephone call is given by the function

$$C(t) = \begin{cases} 4.75 & \text{if } 0 < t \le 3 \\ 0.65t + 2.80 & \text{if } t > 3 \end{cases}$$

where t is the length of the call in minutes. Determine each of the following:

(a) $\lim_{t \to 3^-} C(t)$

(b) $\lim_{x \to 3^+} C(t)$

(c) $\lim_{x \to 3} C(t)$

28. *Finance* A certain state income tax schedule can be given by the function

$$T(x) = \begin{cases} 0.04x & \text{if } 0 < x \leq 12{,}500 \\ 0.03x + 125 & \text{if } 12{,}500 < x \leq 35{,}000 \\ 0.02x + 465 & \text{if } x > 35{,}000 \end{cases}$$

where x is the taxable income in dollars, $0 < x$, and $T(x)$ is in dollars. Determine each of the following:

(a) $\lim_{x \to 12{,}500^-} T(x)$
(b) $\lim_{x \to 12{,}500^+} T(x)$
(c) $\lim_{x \to 12{,}500} T(x)$
(d) $\lim_{x \to 35{,}000^-} T(x)$
(e) $\lim_{x \to 35{,}000^+} T(x)$
(f) $\lim_{x \to 35{,}000} T(x)$

In Your Words

29. **(a)** Describe the differences among $\lim_{x \to a^-} f(x)$, $\lim_{x \to a^+} f(x)$, and $\lim_{x \to a} f(x)$.
(b) Are all three of the limits in (a) ever the same? If so, what does that mean?
(c) Can two of the limits in (a) ever be the same (and the third one different)? If so, what does that mean?
(d) Can all three of the limits in (a) be different? If so, what does that mean?

30. Explain the difference between $\lim_{x \to c} f(x) = \infty$ and $\lim_{x \to \infty} f(x) = c$.

1.5 ALGEBRAIC TECHNIQUES FOR FINDING LIMITS

Until now, we have found limits by using graphs or using tables of values. In many cases it would be a lot easier to use our knowledge of algebra to determine a limit. The following rules of limits will help us find a limit with much less difficulty.

In Section 1.3, we had the following two rules for limits:

> **Rules 1 and 2 for Limits**
>
> Rule 1: $\lim_{x \to c} A = A$, where A and c are real numbers
>
> Rule 2: $\lim_{x \to c} x = c$, where c is a real number

For all of the following rules, we will assume that f and g are two functions and that $\lim_{x \to c} f(x) = L$ and $\lim_{x \to c} g(x) = M$, where L and M are both real numbers.

> **Rule 3: Limit of a Sum or Difference** $\lim_{x \to c}[f(x) \pm g(x)] = \lim_{x \to c} f(x) \pm \lim_{x \to c} g(x)$
> $= L \pm M$

Thus, in order to find the limit of the sum (or difference) of two functions, you can find the sum (or difference) of their limits.

EXAMPLE 1.19

Evaluate $\lim_{x\to 9}(x+6)$.

Solution

$$\begin{aligned}\lim_{x\to 9}(x+6) &= \lim_{x\to 9} x + \lim_{x\to 9} 6 && \text{(rule 3)}\\ &= 9+6 && \text{(rules 1 and 2)}\\ &= 15\end{aligned}$$

EXAMPLE 1.20

Evaluate $\lim_{x\to 2}(3-x)$.

Solution

$$\begin{aligned}\lim_{x\to 2}(3-x) &= \lim_{x\to 2} 3 - \lim_{x\to 2} x && \text{(rule 3)}\\ &= 3-2 && \text{(rules 1 and 2)}\\ &= 1\end{aligned}$$

Rule 4: Limit of a Product

$$\begin{aligned}\lim_{x\to c}[f(x)g(x)] &= \left[\lim_{x\to c} f(x)\right]\left[\lim_{x\to c} g(x)\right]\\ &= LM\end{aligned}$$

This rule states that the limit of the product of two functions is the product of their limits.

EXAMPLE 1.21

Evaluate $\lim_{x\to -3} 7x$.

Solution

$$\begin{aligned}\lim_{x\to -3} 7x &= \left(\lim_{x\to -3} 7\right)\left(\lim_{x\to -3} x\right) && \text{(rule 4)}\\ &= (7)(-3) && \text{(rules 1 and 2)}\\ &= -21\end{aligned}$$

EXAMPLE 1.22

Evaluate $\lim_{x\to 4} 6x^2$.

Solution

$$\lim_{x\to 4} 6x^2 = \left(\lim_{x\to 4} 6x\right)\left(\lim_{x\to 4} x\right) \qquad \text{(rule 4)}$$

EXAMPLE 1.22 (Cont.)

$$= 6\left(\lim_{x\to 4} x\right)\left(\lim_{x\to 4} x\right) \qquad \text{(rules 1 and 4)}$$

$$= 6 \cdot 4 \cdot 4 \qquad \text{(rule 2)}$$

$$= 96$$

Rule 5: Limit of a Quotient

$$\lim_{x\to c} \frac{f(x)}{g(x)} = \frac{\lim_{x\to c} f(x)}{\lim_{x\to c} g(x)} = \frac{L}{M}, \quad \text{if } M \neq 0$$

If the limit of the denominator is not 0, then the limit of the quotient of two functions is the quotient of their limits.

EXAMPLE 1.23

Evaluate $\lim_{x\to 5} \frac{2x+4}{x-1}$.

Solution

$$\lim_{x\to 5} \frac{2x+4}{x-1} = \frac{\lim_{x\to 5}(2x+4)}{\lim_{x\to 5}(x-1)} \qquad \text{(rule 5)}$$

$$= \frac{14}{4} = \frac{7}{2}$$

Rule 6: Limits of $[f(x)]^n$ and $\sqrt[n]{f(x)}$.

If n is a positive integer,

$$\lim_{x\to c} [f(x)]^n = \left[\lim_{x\to c} f(x)\right]^n = L^n$$

and

$$\lim_{x\to c} \sqrt[n]{f(x)} = \sqrt[n]{\lim_{x\to c} f(x)} = \sqrt[n]{L}$$

EXAMPLE 1.24

Evaluate $\lim_{x\to -2} (2x-1)^3$.

Solution

$$\lim_{x\to -2} (2x-1)^3 = \left[\lim_{x\to -2} (2x-1)\right]^3 \qquad \text{(rule 6)}$$

$$= (-5)^3 = -125$$

EXAMPLE 1.25

Evaluate $\lim_{x\to 3} \sqrt[6]{(3x-1)^2}$.

Solution

$$\begin{aligned}\lim_{x\to 3} \sqrt[6]{(3x-1)^2} &= \sqrt[6]{\lim_{x\to 3}(3x-1)^2} && \text{(rule 6)}\\ &= \sqrt[6]{[\lim_{x\to 3}(3x-1)]^2} && \text{(rule 6)}\\ &= \sqrt[6]{(8)^2} = \sqrt[6]{64} = 2\end{aligned}$$

EXAMPLE 1.26

Evaluate $\lim_{x\to 2} \frac{x^2-5x+6}{x^2-4}$.

Solution Checking the denominator, we see that $\lim_{x\to 2}(x^2-4)=0$. This means that we cannot use rule 5. But, this does not mean the limit does not exist. If we factor both the numerator and denominator, we see that

$$\frac{x^2-5x+6}{x^2-4} = \frac{(x-2)(x-3)}{(x-2)(x+2)}$$

We are interested in the limits as x approaches 2 and not when x has the value of 2. Thus, the factor $x-2$ is not 0 for these values and we can cancel this common factor. We then get

$$\begin{aligned}\lim_{x\to 2}\frac{x^2-5x+6}{x^2-4} &= \lim_{x\to 2}\frac{(x-2)(x-3)}{(x-2)(x+2)}\\ &= \lim_{x\to 2}\frac{x-3}{x+2}\\ &= \frac{2-3}{2+2} = \frac{-1}{4}\end{aligned}$$

Caution

In Example 1.26, substituting 2 for x in the expression $\frac{x^2-5x+6}{x^2-4}$ resulted in $\frac{0}{0}$. Whenever substitution results in $\frac{0}{0}$, we must do more work to determine whether a limit exists.

Until now, we have applied the six rules for limits only for the limit of a function as x approaches a specific point. As the next examples show, we often have to be more careful when we find limits at infinity.

EXAMPLE 1.27

Evaluate $\lim\limits_{x\to\infty} \frac{1}{x^n}$, $n > 0$.

Solution

$$\lim_{x\to\infty} \frac{1}{x^n} = \left(\lim_{x\to\infty} \frac{1}{x}\right)^n = 0^n = 0$$

Notice from this example that as $x \to \infty$, $\frac{1}{x^2}$, $\frac{1}{x^3}$, $\frac{1}{x^4}$, etc., all have limits of 0. Using this with rule 3, we see that

$$\lim_{x\to\infty} \frac{c}{x^n} = \left(\lim_{x\to\infty} c\right)\left(\lim_{x\to\infty} \frac{1}{x^n}\right) = c \cdot 0 = 0$$

Thus, we have the following.

Limits at Infinity

If c is a constant and n is a positive rational number, then

$$\lim_{x\to\infty} \frac{c}{x^n} = 0$$

We will now show some techniques for finding limits at infinity.

EXAMPLE 1.28

Evaluate $\lim\limits_{x\to\infty} \frac{4x^3+2x^2+5}{5x^3+x}$.

Solution We will first divide both the numerator and denominator by the largest power of x in the denominator, in this case x^3. This will make each term into a constant or a term with a variable in the denominator and allow us to use the properties for limits at infinity.

$$\begin{aligned}
\lim_{x\to\infty} \frac{4x^3+2x^2+5}{5x^3+x} &= \lim_{x\to\infty} \frac{\dfrac{4x^3+2x^2+5}{x^3}}{\dfrac{5x^3+x}{x^3}} \\
&= \lim_{x\to\infty} \frac{\dfrac{4x^3}{x^3}+\dfrac{2x^2}{x^3}+\dfrac{5}{x^3}}{\dfrac{5x^3}{x^3}+\dfrac{x}{x^3}} \\
&= \lim_{c\to\infty} \frac{4+\dfrac{2}{x}+\dfrac{5}{x^3}}{5+\dfrac{1}{x^2}}
\end{aligned}$$

EXAMPLE 1.28 (Cont.)

Now, according to the properties for limits at infinity, all the terms with an x in the denominator have a limit of 0. So, we now have

$$\lim_{x\to\infty}\frac{4+\frac{2}{x}+\frac{5}{x^3}}{5+\frac{1}{x^2}}=\frac{4+0+0}{5+0}=\frac{4}{5}$$

Hint When you want to find a limit at infinity, divide both the numerator and denominator by the largest power of the variable in the denominator.

EXAMPLE 1.29

Evaluate $\lim_{x\to\infty}\frac{7x^5-2x^2+1}{4x^6+5x^3+x-5}$.

Solution As in Example 1.28, we will divide both the numerator and denominator by the largest power of x in the denominator, in this case x^6.

$$\begin{aligned}\lim_{x\to\infty}\frac{7x^5-2x^2+1}{4x^6+5x^3+x-5} &= \lim_{x\to\infty}\frac{\frac{7x^5-2x^2+1}{x^6}}{\frac{4x^6+5x^3+x-5}{x^6}}\\ &= \lim_{x\to\infty}\frac{\frac{7x^5}{x^6}-\frac{2x^2}{x^6}+\frac{1}{x^6}}{\frac{4x^6}{x^6}+\frac{5x^3}{x^6}+\frac{x}{x^6}-\frac{5}{x^6}}\\ &= \lim_{x\to\infty}\frac{\frac{7}{x}-\frac{2}{x^4}+\frac{1}{x^6}}{4+\frac{5}{x^3}+\frac{1}{x^5}-\frac{5}{x^6}}\\ &= \frac{0-0+0}{4+0+0-0}=\frac{0}{4}=0\end{aligned}$$

Thus, $\lim_{x\to\infty}\frac{7x^5-2x^2+1}{4x^6+5x^3+x-5}=0$.

Exercise Set 1.5

In Exercises 1–34, evaluate the given limit using algebraic techniques.

1. $\lim_{x\to 5} 8$
2. $\lim_{x\to -4} x$
3. $\lim_{x\to 8}(x-3)$
4. $\lim_{x\to -7}(x+4)$
5. $\lim_{x\to -6}(x+7)$
6. $\lim_{x\to 6}(9-x)$
7. $\lim_{x\to 2} 3x$
8. $\lim_{t\to -3} 5t$
9. $\lim_{x\to 6} \frac{2}{3}x+5$
10. $\lim_{x\to 4} \left(\frac{3}{2}x-2\right)$
11. $\lim_{p\to 3} \dfrac{p^2+6}{p}$
12. $\lim_{x\to -2} \dfrac{x^2-6}{x}$
13. $\lim_{s\to 5} \dfrac{3s^2+5}{2s}$
14. $\lim_{x\to 6} \dfrac{2x^2-3}{3-x}$
15. $\lim_{x\to 2}(x+1)^3$
16. $\lim_{y\to 5}(y-1)^3$
17. $\lim_{x\to -1}(2x^2-1)^2$
18. $\lim_{t\to 4}(3t^2+2)^2$
19. $\lim_{x\to 3}\sqrt{x+3}$
20. $\lim_{x\to -6}\sqrt{6-7x}$
21. $\lim_{x\to 2} \dfrac{x^2-2x}{x-2}$
22. $\lim_{w\to 3} \dfrac{w^2-1}{w+1}$
23. $\lim_{x\to -1} \dfrac{x^2-1}{x+1}$
24. $\lim_{y\to 3} \dfrac{y^2-7y+12}{y-3}$
25. $\lim_{x\to -4} \dfrac{x^2+2x-8}{x^2+5x+4}$
26. $\lim_{x\to \frac{1}{2}} \dfrac{2x^2+5x-3}{4x^2-2x}$
27. $\lim_{t\to -7} \dfrac{2t^2+15t+7}{t^2+5t-14}$
28. $\lim_{x\to 0} \dfrac{(x+3)^2-9}{x}$
29. $\lim_{h\to 0} \dfrac{(4+h)^2-4^2}{h}$
30. $\lim_{x\to c} \dfrac{x^3-c^3}{x-c}$
31. $\lim_{x\to \infty} \dfrac{2x^2+4}{5x^2+3x+2}$
32. $\lim_{x\to \infty} \dfrac{6x^3+9x+1}{2x^3+4x+8}$
33. $\lim_{x\to \infty} \dfrac{7x^4+5x^2}{x^5+2}$
34. $\lim_{x\to \infty} \dfrac{4x^5+3x}{2x^4+1}$

In Your Words

35. Suppose that $\lim_{x\to 1} f(x)=0$ and $\lim_{x\to 1} g(x)=\infty$.
    (a) Give an example of two functions f and g so that $\lim_{x\to 1}[f(x)+g(x)]=\infty$.
    (b) Give an example of two functions f and g so that $\lim_{x\to 1}[f(x)\cdot g(x)]=0$.
    (c) Give an example of two functions f and g so that $\lim_{x\to 1}[f(x)\cdot g(x)]=5$.
36. Suppose that neither $\lim_{x\to 2} f(x)$ and $\lim_{x\to 2} g(x)$ exists. Give an example of two functions f and g so that $\lim_{x\to 2}[f(x)+g(x)]$ does exist or explain why it is not possible.

1.6 CONTINUITY

In Sections 1.4 and 1.5, we have talked about two of the important ideas leading up to calculus. In this section, we will look at the last new idea we need before we begin calculus—**continuity at a point**.

Let's consider the two functions

$$f(x)=x$$

and

$$g(x)=\begin{cases} x & \text{if } x\neq 2 \\ 3 & \text{if } x=2 \end{cases}$$

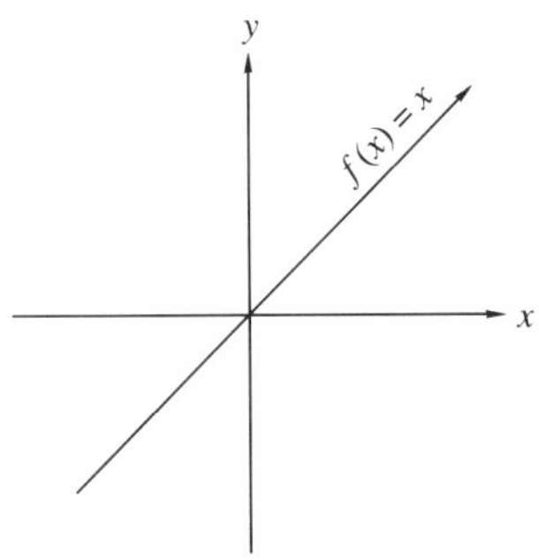

FIGURE 1.27

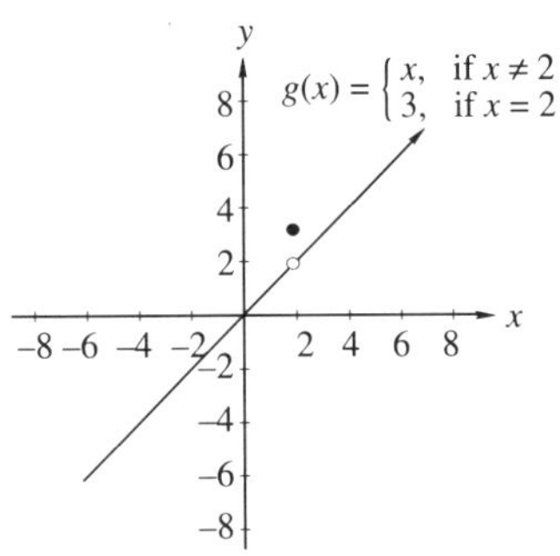

FIGURE 1.28

The graphs of these functions appear in Figures 1.27 and 1.28, respectively. Looking at either the graphs or the algebraic definitions, you can see that the two functions are exactly the same, except at one point. At that point, when $x = 2$, there is a "break" in the graph of g, but no "break" in the graph of f. The function g is an example of a function that is not continuous, while f demonstrates a continuous function.

The break in g is used to determine that the function is not continuous. One way of thinking of a continuousfunction is that it is a function you can draw without lifting your pencil from the paper. Look again at the graphs of f and g. You could draw f without lifting your pencil. You could not draw g without lifting your pencil.

Continuity at a Point

As we saw above, we can describe a function as either continuous or noncontinuous by looking at how it is drawn. We can use the idea of limits to provide an algebraic definition of continuity. Look again at the definitions of f and g. What is $\lim_{x\to 2} f(x)$? What are $f(2)$ and $g(2)$?

As you can see, $\lim_{x\to 2} f(x) = f(2) = 2$. We know that f is a continuous function at $x = 2$. But $\lim_{x\to 2} g(x) = 2 \neq g(2) = 3$, thus g is not a continuous function, or is **discontinuous** at $x = 2$. This provides us with a more formal definition:

Continuity at a Point

A function f is **continuous** at $x = c$, if and only if the following three conditions are satisfied:

(a) $f(x)$ is defined at $x = c$,
(b) $\lim_{x\to c} f(x)$ exists, and
(c) $\lim_{x\to c} f(x) = f(c)$.

A function that is not continuous at $x = c$ is **discontinuous** at that point.

Note A function is discontinuous at $x = c$ if any one of these three conditions are not satisfied.

EXAMPLE 1.30

Discuss the continuity of the function f at $x = 4$, where

$$f(x) = \begin{cases} \dfrac{x^2 - 16}{x - 4} & \text{if } x \neq 4 \\ 8 & \text{if } x = 4 \end{cases}$$

Solution In order to determine if f is continuous, we need to check the three conditions in the definition: **(a)** The function is defined at $x = 4$, since $f(4) = 8$, **(b)** $\lim_{x\to 4} \dfrac{x^2 - 16}{x - 4} = \lim_{x\to 4} \dfrac{(x+4)(x-4)}{x - 4} = 8$, and **(c)** $\lim_{x\to 4} \dfrac{x^2 - 16}{x - 4} = 8 = f(4)$.

The three conditions are satisfied, so this function is continuous at $x = 4$. ▪

EXAMPLE 1.31

Discuss the continuity of each of the following functions at the indicated points:

(a) $g(x) = \dfrac{1}{x}$ at $x = 0$

(b) $h(x) = \begin{cases} x - 1 & \text{if } x < 2 \\ x + 1 & \text{if } x \geq 2 \end{cases}$ at $x = 2$

(c) $j(x) = \begin{cases} x^2 + 2 & \text{if } x \neq 0 \\ 0 & \text{if } x = 0 \end{cases}$ at $x = 0$

Solutions None of these are continuous at the designated points.

(a) $g(x)$ is not defined when $x = 0$, so it does not satisfy the first condition.

(b) $h(x)$ is defined at $x = 2$, but we can see that $\lim_{x\to 2^-} h(x) = 1$ and $\lim_{x\to 2^+} h(x) = 3$, so the $\lim_{x\to 2} h(x)$ does not exist and the second condition is not satisfied.

(c) $j(x)$ is defined at 0, $j(0) = 0$. $\lim_{x\to 0} j(x) = 2$. The third condition is not satisfied, since $\lim_{x\to 0} j(x) = 2 \neq 0 = j(0)$.

The graphs in these three functions are shown in Figures 1.29a, 1.29b, and 1.29c, respectively. ▪

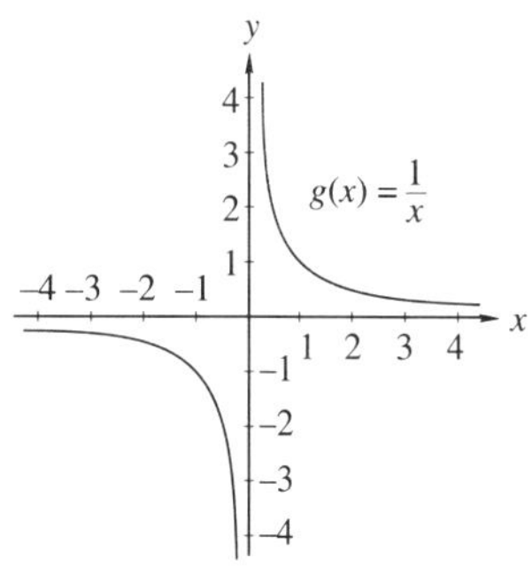

FIGURE 1.29a

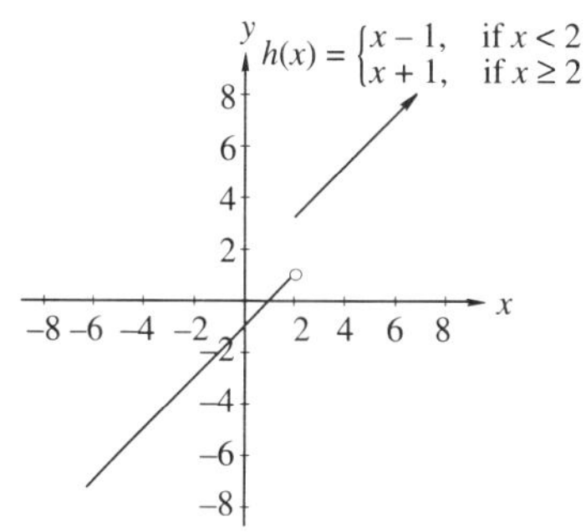

FIGURE 1.29b

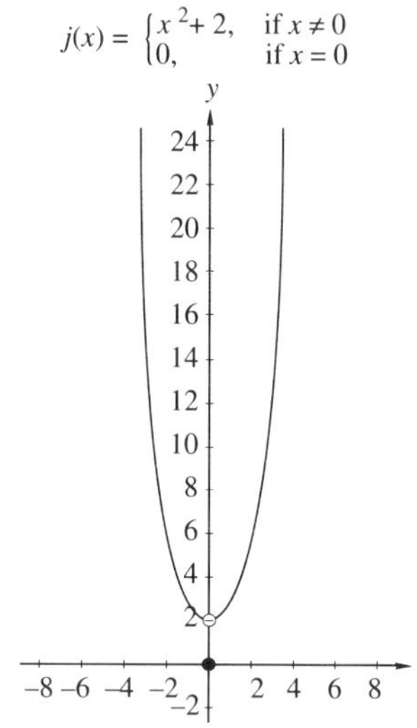

FIGURE 1.29c

Continuity on Intervals

When we talk about functions, we usually consider them over an interval. An **interval** is a set of all points on the number line between a and b. Because the endpoints are not always included in the intervals, notation has been developed to make these distinctions. We will use the following notations and terms:

Interval Notation	
	$(a,b): a < x < b$
	$[a,b): a \leq x < b$
	$(a,b]: a < x \leq b$
	$[a,b]: a \leq x \leq b$

The interval (a,b) is called an **open interval** and does not include either endpoint. The interval $[a,b]$ is a **closed interval** and includes both endpoints. The other two intervals, $[a,b)$ and $(a,b]$, are **half-open intervals**. Because $+\infty$ and $-\infty$ are not numbers, any interval that includes either of these is open on that end. For example $a < x < \infty$, or simply $a < x$, is the interval, (a,∞) and $-\infty < x \leq b$, or simply $x \leq b$, is the interval $(-\infty, b]$.

The definition of continuity we just gave is for a function continuous at a point. We are usually more interested in the continuity of a function over an interval.

Continuity Over an Interval

We say that a function is **continuous on an interval** if it is continuous at each point in the interval.

If you are concerned with the continuity of a function on an interval that includes an endpoint of the interval, then you need to examine the one-sided limits at that endpoint.

EXAMPLE 1.32

The function $k(x) = \sqrt{x-2}$ is continuous on the interval $[2, \infty)$.
The function $m(x) = \sqrt{3-x}$ is continuous on the interval $(-\infty, 3]$.
The function $f(x) = \sqrt[3]{x}$ is continuous on $(-\infty, \infty)$.

Note All polynomial functions are continuous at every point or on the interval $(-\infty, \infty)$.

Functions that are continuous at all points, such as a polynomial, are said to be **continuous everywhere**, or continuous, and are referred to as **continuous functions**. Rational functions are continuous over their domains.

EXAMPLE 1.33

Find any points of discontinuity for each of the following functions:

(a) $f(x) = \dfrac{x^3 - 3x^2 + 17x - 5}{x^2 - 1}$

(b) $g(x) = \begin{cases} x^2 & \text{if } x < 2 \\ x^2 + 5 & \text{if } x \geq 2 \end{cases}$

(c) $h(x) = \begin{cases} 6 & \text{if } x < 3 \\ x + 3 & \text{if } x > 3 \end{cases}$

Solutions

(a) Possible points of discontinuity occur when the denominator is 0. The denominator is 0 when x is ± 1, so f is not defined at these two points. By condition (a), if the function is not defined at ± 1, it is not continuous at those two points. These are the only points of discontinuity for f.

(b) Possible points of discontinuity occur when $x = 2$, since this is the only place where a break may occur. When $x < 2$, $\lim_{x \to 2^-} g(x) = \lim_{x \to 2^-} x^2 = 4$ and when $x > 2$, $\lim_{x \to 2^+} g(x) = \lim_{x \to 2^+} (x^2 + 5) = 9$. The limit does not exist at $x = 2$, so this is a point of discontinuity.

(c) The function h is not defined at $x = 3$, so the function is not continuous at this point. It is continuous for all other values of x.

Properties of Continuous Functions

The formal definition that we gave for a continuous function was based on a limit. The rules that apply to limits help us develop three properties that apply to continuous functions.

> **Properties of Continuous Functions**
>
> Property 1: If f and g are both continuous at a point c, then so are $f+g, f-g$, and $f \cdot g$.
>
> Property 2: If f and g are both continuous at c and $g(c) \neq 0$, then $f/g = \dfrac{f}{g}$ is also continuous at c.
>
> Property 3: If f is continuous at $g(c)$ and g is continuous at c, then $f \circ g = f(g(x))$ is continuous at c.

We used property 2 in the solution of Example 1.33a. There, we let f be the polynomial $x^3 - 3x^2 + 17x - 5$ and g the polynomial $x^2 - 1$. Since f and g are both polynomials they are continuous everywhere, so f/g is continuous where $g \neq 0$, namely when $x \neq \pm 1$.

This concludes this section on continuity and completes your foundation for calculus. Limits and continuity will both provide necessary tools as we continue to develop our knowledge of calculus.

Exercise Set 1.6

In Exercises 1–10, determine where the function is continuous at c. If the function is not continuous, give a reason.

1. $f(x) = \begin{cases} 3x+1 & \text{if } x \le 4 \\ x^2 - 3 & \text{if } x > 4 \end{cases} \qquad c = 4$

2. $g(x) = \begin{cases} 2x-1 & \text{if } x \le 1 \\ 2x & \text{if } x > 1 \end{cases} \qquad c = 1$

3. $h(x) = \begin{cases} 3x+1 & \text{if } x < 1 \\ 3 & \text{if } x = 1 \\ 4x & \text{if } x > 1 \end{cases} \qquad c = 1$

4. $j(x) = \begin{cases} 5x-1 & \text{if } x < 1 \\ 4 & \text{if } x = 1 \\ 4x & \text{if } x > 1 \end{cases} \qquad c = 1$

5. $k(x) = \begin{cases} x^2+2 & \text{if } x > 2 \\ 3x & \text{if } x < 2 \end{cases} \qquad c = 2$

6. $f(x) = \begin{cases} x^3+1 & \text{if } x > 2 \\ 0 & \text{if } x = 2 \\ x^2+5 & \text{if } x < 2 \end{cases} \qquad c = 2$

7. $g(x) = \begin{cases} x^2 & \text{if } x < -2 \\ 3 & \text{if } x = -2 \\ -4x+2 & \text{if } x > -2 \end{cases} \qquad c = -2$

8. $h(x) = \begin{cases} x^2 & \text{if } x \le -2 \\ 2x & \text{if } x > -2 \end{cases} \qquad c = -2$

9. $j(x) = \begin{cases} x^3-4 & \text{if } x < 2 \\ 5 & \text{if } x = 2 \\ x^2 & \text{if } x > 2 \end{cases} \qquad c = 2$

10. $k(x) = \dfrac{x^3 - 1}{x - 1} \qquad c = 1$

In Exercises 11–20, find all points of discontinuity.

11. $f(x) = 2x^2 - 5$

12. $g(x) = \dfrac{4}{x - 5}$

13. $h(x) = \dfrac{x^2 + 3x + 2}{x + 1}$

14. $j(x) = 7t - 4$

15. $k(x) = \dfrac{x + 1}{x^2 + x - 6}$

16. $f(x) = \dfrac{x^2 + x - 6}{x^2 + 5x + 6}$

17. $g(x) = \dfrac{x}{x^2 + 1}$

18. $h(x) = \dfrac{x}{x}$

19. $j(x) = \begin{cases} x^2 + 2 & \text{if } x < 2 \\ \sqrt{x - 2} & \text{if } x \geq 2 \end{cases}$

20. $k(x) = \begin{cases} \dfrac{1}{x} & \text{if } x \neq 5 \\ 3 & \text{if } x = 5 \end{cases}$

In Exercises 21–30, determine the intervals over which the given function is continuous.

21. $f(x) = \dfrac{x}{x + 3}$

22. $g(x) = \dfrac{x}{x^2 - 1}$

23. $h(x) = \dfrac{x + 2}{x^2 - 4}$

24. $j(x) = \dfrac{x^2 + 3x - 4}{x + 4}$

25. $k(x) = \sqrt{x + 3}$

26. $f(x) = \sqrt{5 - x}$

27. $g(x) = \sqrt{x^2 - 9}$

28. $h(x) = \sqrt{x^2 + 4x + 4}$

29. $j(x) = \dfrac{1}{x}$

30. $k(x) = \dfrac{1}{\sqrt{x}}$

Solve Exercises 31–34.

31. *Physics* The acceleration due to gravity, g, varies with the height above the surface of the Earth. The acceleration varies in a different way below the Earth's surface. It has been determined that, as a function of r, the distance from the center of the Earth, g is described by

$$g(r) = \begin{cases} \dfrac{GMr}{R^3} & \text{for } r < R \\ \dfrac{GM}{r^2} & \text{for } r \geq R \end{cases}$$

where $R \approx 6.371 \times 10^3$ km is the radius of the Earth, $M \approx 5.975 \times 10^{24}$] kg is the mass of the Earth, and $G \approx 6.672 \times 10^{-11}$ N·m^2/kg^2 is the gravitational constant.

(a) Is g a continuous function of r? Explain your answer.

(b) Sketch a graph of g near $r = R$.

32. *Business* The cost, in dollars, of an overseas telephone call is given by the function

$$C(t) = \begin{cases} 8.50 & \text{if } 0 < t \leq 3 \\ 0.85t + 5.95 & \text{if } t > 3 \end{cases}$$

where t is the length of the call in minutes. Is C a continuous function at $t = 3$?

33. *Finance* A recent federal income tax schedule can be given by the function

$$T(x) = \begin{cases} 0.15x & \text{if } 0 < x \leq 23{,}900 \\ 0.28x - 3{,}107 & \text{if } 23{,}900 < x \leq 61{,}650 \\ 0.33x - 6{,}189.50 & \text{if } 61{,}650 < x \leq 123{,}790 \end{cases}$$

where x is the taxable income in dollars, $0 < x \leq 123{,}790$, and $T(x)$ is in dollars. Determine each of the following:

(a) $\lim_{x \to 23{,}900-} T(x)$
(b) $\lim_{x \to 23{,}900+} T(x)$
(c) $\lim_{x \to 23{,}900} T(x)$
(d) Is T continuous at $x = 23{,}900$?
(e) $\lim_{x \to 61{,}650-} T(x)$
(f) $\lim_{x \to 61{,}650+} T(x)$
(g) $\lim_{x \to 61{,}650} T(x)$
(h) Is T continuous at $x = 61{,}650$?

34. *Business* The local gas company uses the following function for computing their customers' monthly gas bills:

$$C(t) = \begin{cases} 0.47x + 2.95 & \text{if } 0 < x \leq 24 \\ 0.89x - 7.13 & \text{if } x > 24 \end{cases}$$

where x is the number of thermal units (therms) used by the customer and $C(x)$ is the cost in dollars. Determine each of the following:

(a) $\lim_{x \to 24-} C(x)$
(b) $\lim_{x \to 24+} C(x)$
(c) $\lim_{x \to 24} C(x)$
(d) Is C continuous at $x = 24$?

In Your Words

35. The definition of continuity at a point has three conditions that must be satisfied if the function is to be continuous at a specific point.
(a) List each of these conditions.
(b) For each condition, give an example of a function which is not continuous because it does not satisfy that condition but does satisfy any previous conditions.

36. Suppose that f and g are functions and that neither is continuous as $x = 5$. Explain how $f + g$ can be continuous at $x = 5$.

CHAPTER 1 REVIEW

Important Terms and Concepts

Area under a graph
Average slope
Continuity
  At a point
  Over an interval
Continuous function
Interval notation
Limits
  At infinity
  Left-hand
  One-sided
  Right-hand
Properties of continuous functions
Rules for limits

Review Exercises

Find the average slope over the indicated intervals for the given functions in Exercises 1–6.

1. $f(x)=x^2-7$ from $x=0$ to $x=2$
2. $g(x)=4x+5$ from $x=-3$ to $x=-1$
3. $h(x)=2x^2+1$ from $x=-3$ to $x=-2$
4. $j(x)=\dfrac{8}{x+4}$ from $x=0$ to $x=\frac{1}{2}$
5. $k(x)=x^2+2x$ from $x=-1$ to $x=b$
6. $m(x)=x^2-5x$ from $x=6$ to $x=x_1$

In Exercises 7–10, find the areas under the graph of the given function over the indicated interval. The value of *n* indicates the number of segments into which each interval should be divided.

7. $f(x)=-(4x+7)$ over $[-5,-2], n=6$
8. $g(x)=2x^2+5$ over $[0,4], n=8$
9. $h(x)=3x^2+2x-1$ over $[2,4], n=8$
10. $j(x)=5-2x^2$ over $[-1,1], n=8$

In Exercise 11–20, use either the graphical or algebraic approach to determine the limit of the given function at the indicated point, or state that the limit does not exist.

11. $\lim\limits_{x\to 1}(3x^2+2x-5)$
12. $\lim\limits_{x\to -2}\dfrac{3x^2-12}{x+2}$
13. $\lim\limits_{x\to 0}\dfrac{3x^2-12}{x-2}$
14. $\lim\limits_{x\to -3}\dfrac{x^2-9}{3x+9}$
15. $\lim\limits_{x\to\infty}\dfrac{6x^2+3}{2x^2}$
16. $\lim\limits_{x\to\infty}\dfrac{9x^3+2x-1}{3x^2+x+1}$
17. $\lim\limits_{x\to 5^+}\sqrt{x-5}$
18. $\lim\limits_{x\to -4^-}\dfrac{x+4}{x^2-16}$
19. $\lim\limits_{x\to\infty}\dfrac{x+4}{2x^2+1}$
20. $\lim\limits_{x\to 1}f(x)$
    if $f(x)=\begin{cases}x^2-1 & \text{if } x<1\\ 1-x & \text{if } x>1\end{cases}$

In Exercises 21–24, determine the points of discontinuity for the given function.

21. $f(x)=\dfrac{x}{x-5}$
22. $g(x)=\dfrac{3}{x^3}$
23. $h(x)=\begin{cases}x+5 & \text{if } x<-3\\ 5-x & \text{if } x\ge -3\end{cases}$
24. $g(x)=\begin{cases}\dfrac{x}{x+1} & \text{if } x<1\\ \dfrac{1}{3-x} & \text{if } x\ge 1\end{cases}$

In Exercises 25–28, determine the intervals where the given function is continuous.

25. $f(x)=\dfrac{5}{x+7}$
26. $g(x)=\dfrac{x^2-4}{x^2+5x+6}$
27. $h(x)=\dfrac{x+5}{\sqrt{x^2-25}}$
28. $j(x)=\sqrt{\dfrac{3-x}{3+x}}$

CHAPTER 1 TEST

Solve Exercises 1 and 2.

1. Find the average slope of $f(x)=2x^3-3$ from $x=0$ to $x=1$.
2. Find the area under the graph of $g(x)=3x^2-2x$ over the interval $[1,5]$, when it is divided into $n=4$ subintervals.

In Exercises 3–6, determine the limit of the given function or state that the limit does not exist.

3. $\lim_{x \to 2} (5x^2 - 7)$

4. $\lim_{x \to 2} \left(\frac{x^2 - 4}{3x - 6} \right)$

5. $\lim_{x \to 1^+} f(x)$ where $f(x) = \begin{cases} x^3 - 1 & \text{if } x < 1 \\ 4 & \text{if } x = 1 \\ 1 + x & \text{if } x > 1 \end{cases}$

6. $\lim_{x \to \infty} \frac{5x^3 - 4x + 1}{7x^3 - 2}$

Solve Exercises 7 and 8.

7. Determine the points of discontinuity for $h(x) = \begin{cases} x + 7 & \text{if } x < 2 \\ \frac{1}{x^2 - 9} & \text{if } x \geq 2. \end{cases}$

8. Determine the intervals where $j(x) = \frac{x^2 - 9}{x^2 - x - 12}$ is continuous.

CHAPTER

2

The Derivative

When working with lenses, optical technicians need to be able to determine the line perpendicular to the surface of the lens at the point where the light enters the lens. Section 2.5 explains how to determine the equations of these normal lines.

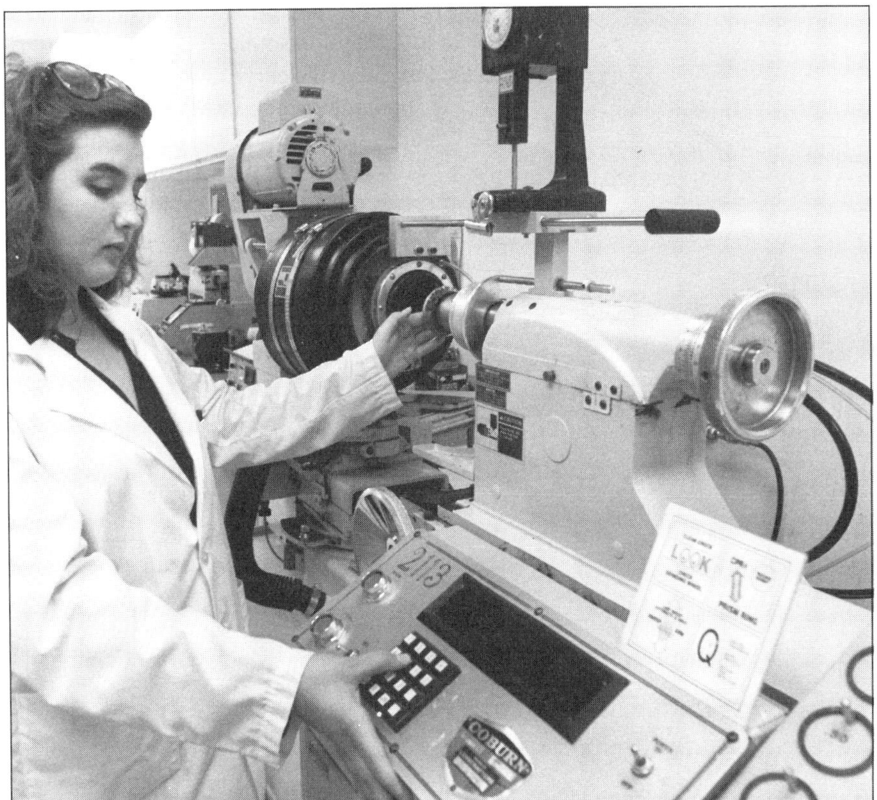

In Chapter 1, we looked at one of the questions that led to the development of calculus—the tangent question. The answer to that question depends on the use of limits. As you pursue the answer to the tangent question, you will develop an understanding of a part of calculus called the derivative. In this chapter, you will not only develop an understanding of a derivative but you will develop techniques for finding derivatives of functions.

2.1 THE TANGENT QUESTION AND THE DERIVATIVE

We will begin with the graph of the function we examined in Section 1.1. The graph is shown again in Figure 2.1. The curve has two points, P and Q. P has the coordinates (x_1, y_1) and Q the coordinates (x_2, y_2). The slope of the secant line through P and Q is

$$m_{PQ} = \frac{y_2 - y_1}{x_2 - x_1}$$

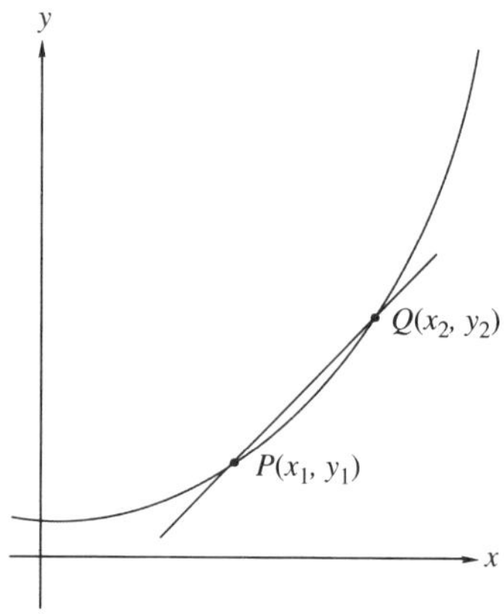

FIGURE 2.1

But, we are not really interested in the slope of a secant line. What we want to determine is the slope of the line that is tangent to this curve at the point P. To do this we are going to look at the various secant lines that we get by leaving P fixed. Figure 2.2 contains the same curve as in Figure 2.1, again with the points P and Q. A series of points Q_1, Q_2, Q_3, and Q_4 have been selected with each of these points lying between P and Q, each closer to P than the previous point. As each point gets closer to P, the slope of the secant line through that point and P approaches the slope of the tangent to the curve at P.

A similar thing happens if points Q^{I}, Q^{II}, Q^{III}, and Q^{IV} are on the curve but on the opposite side of P, as shown in Figure 2.3. As each point gets closer to P, the slope of the secant line through that point and P approaches the slope of the tangent line through P. Both of these examples lead us to the idea that the slope of a curve at a point P is the slope of the tangent to the curve at P.

It looks as if both of these examples involve the idea of a limit. Let's look at a more general case for the slope of these secant lines. Again, we will examine a function $y = f(x)$ as shown by the curve in Figure 2.4. We want to find an expression for the slope of the curve, or the slope of the tangent to the curve, at the point P.

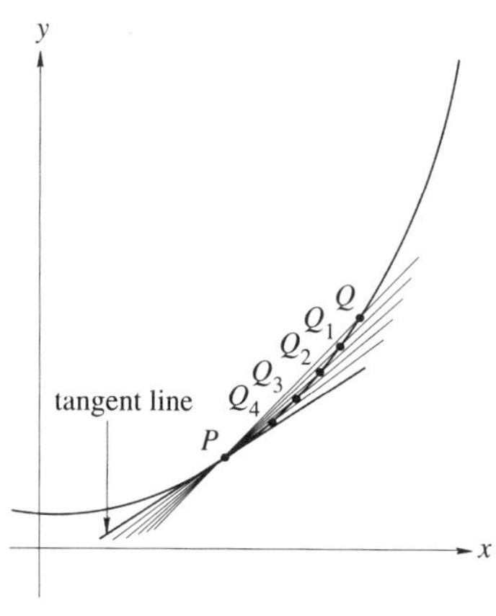

FIGURE 2.2

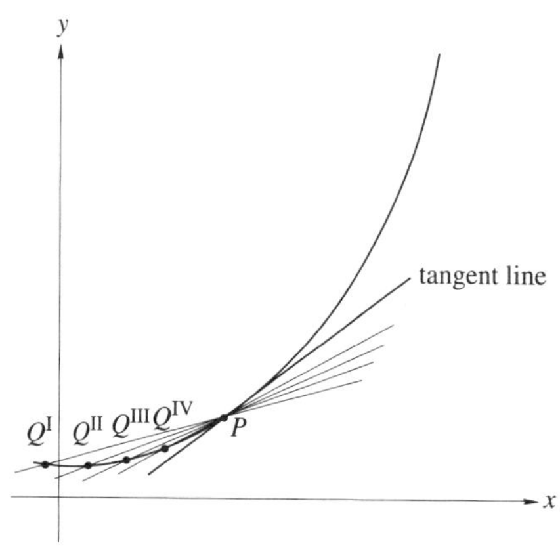

FIGURE 2.3

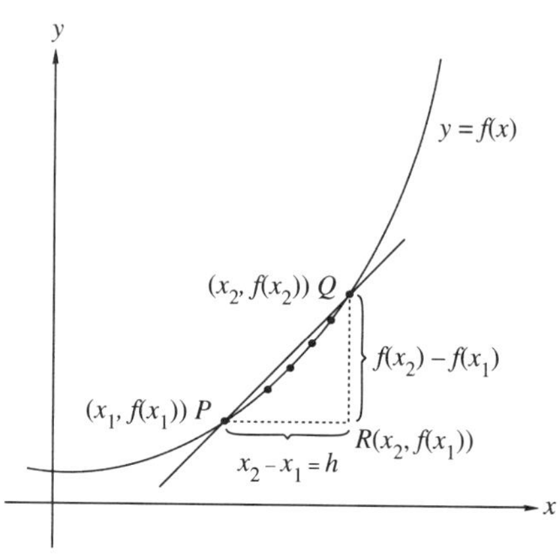

FIGURE 2.4

The point P has the coordinates $(x_1, y_1) = (x_1, f(x_1))$. We will let Q be any other point on the curve and let its coordinates be $(x_2, f(x_2))$. The slope of the secant line through P and Q is

$$m_{PQ} = \frac{f(x_2) - f(x_1)}{x_2 - x_1}$$

Now draw a line through Q parallel to the y-axis and a line through P parallel to the x-axis. These two lines intersect at the point R, which has the coordinates $(x_2, f(x_1))$. The distance from P to R is $x_2 - x_1$. If we let $h = x_2 - x_1$, then $x_2 = x_1 + h$. This allows us to write the slope of the secant line through P and Q as

$$\begin{aligned} m_{PQ} &= \frac{f(x_1+h) - f(x_1)}{(x_1+h) - x_1} \\ &= \frac{f(x_1+h) - f(x_1)}{h} \end{aligned}$$

We know that h is not 0 because if $x_1 = x_2$, either we would not have two distinct points or we would not have a function.

As Q is picked closer to P, x_2 approaches x_1. This means that h is getting closer to 0. The slope of the tangent line will be the limit

$$\lim_{h \to 0} \frac{f(x_1+h) - f(x_1)}{h}$$

This formula gives us the slope of the curve at the point $(x_1, f(x_1))$.

Slope of a Curve

The slope of a curve $y = f(x)$ at the point $(x_1, f(x_1))$ is

$$\lim_{h \to 0} \frac{f(x_1+h) - f(x_1)}{h}$$

provided that this limit exists.

EXAMPLE 2.1

Find the slope of $f(x) = x^2$ at the point $(3, 9)$.

Solution

$$\begin{aligned} \lim_{h \to 0} \frac{f(3+h) - f(3)}{h} &= \lim_{h \to 0} \frac{(3+h)^2 - 3^2}{h} \\ &= \lim_{h \to 0} \frac{9 + 6h + h^2 - 9}{h} \\ &= \lim_{h \to 0} \frac{6h + h^2}{h} \\ &= \lim_{h \to 0} \frac{h(6+h)}{h} \\ &= \lim_{h \to 0} (6+h) = 6 \end{aligned}$$

So, the slope of the tangent line to $y = x^2$ at $(3, 9)$ is 6.

The process that we have been describing will determine the slope of a tangent to a curve or the slope of the curve at a point. This process can be used to describe how any function is changing at a particular point. In other words, we have been using the method for finding the average change in a function and then the limit of that average change as the denominator approaches 0. This gives the instantaneous change of the function at that point. We will see this by taking a different look at Example 1.4.

Application

EXAMPLE 2.2

A tank is filled with water by opening an inlet pipe. The volume V in liters of water in the tank t minutes after the valve is opened is given by $V(t) = 5t^2 + 4t$. What is the instantaneous rate of increase in the volume at the first minute?

Solution We want to find the instantaneous rate of change R in the volume when $t = 1$. We can do this by using the formula for the rate of change $R_{\text{Vol}} = \dfrac{V(t_2) - V(t_1)}{t_2 - t_1}$ and letting $t_1 = 1$ and $t_2 = t_1 + h = 1 + h$, then finding the limit when h approaches 0.

$$\begin{aligned}
R_{\text{Vol}} &= \lim_{h \to 0} \frac{V(1+h) - V(1)}{h} \\
&= \lim_{h \to 0} \frac{[5(1+h)^2 + 4(1+h)] - [5(1)^2 + 4(1)]}{(1+h) - 1} \\
&= \lim_{h \to 0} \frac{[5(1+2h+h^2) + 4(1+h)] - [5+4]}{h} \\
&= \lim_{h \to 0} \frac{(5 + 10h + 5h^2 + 4 + 4h) - (9)}{h} \\
&= \lim_{h \to 0} \frac{9 + 14h + 5h^2 - 9}{h} \\
&= \lim_{h \to 0} \frac{14h + 5h^2}{h} \\
&= \lim_{h \to 0} (14 + 5h) = 14
\end{aligned}$$

When $t = 1$, the volume of the tank is increasing at the rate of 14 L/min. ▪

The Derivative

All of this discussion about instantaneous change of a function leads us to the following definition of the derivative.

Derivative of a Function f at a Number c

If $y = f(x)$ defines a function f and if the point c is in the domain of f, then the **derivative of f at c**, written as $f'(c)$, is defined as

$$f'(c) = \lim_{h \to 0} \frac{f(c+h) - f(c)}{h}$$

provided that this limit exists.

We used the notation $f'(c)$ to indicate the derivative of the function at the point c. In general, the derivative of the function is the following.

Derivative of a Function

$$f'(x) = \lim_{h \to 0} \frac{f(x+h) - f(x)}{h}$$

Note The notation $f'(x)$ is not a unique notation for the derivative. Other notations include $D_x y$, $\frac{dy}{dx}$, $\frac{d}{dx}[f(x)]$, y', $Df(x)$, $\frac{df}{dx}$, and $D_x[f(x)]$. There are times when a meaning is clearer if one of these notations is used instead of $f'(x)$.

Caution $\frac{dy}{dx}$, $\frac{df}{dx}$, and $\frac{d}{dx}$ are not fractions but are symbols for a derivative.

Four-Step Method

We can calculate the derivative of a function f at the point c by using the following four steps.

Four Steps for Calculating $f'(c)$

Step 1: Find $f(c+h)$.

Step 2: Subtract $f(c)$ from $f(c+h)$.

Step 3: Divide the result in Step 2 by h to get the difference quotient.

Step 4: Find the limit, if it exists, of the difference quotient in Step 3, as h approaches 0.

EXAMPLE 2.3

Find the derivative of $f(x) = x^3$ at $x = 3$ by using the four-step method.

Solution

Step 1: $f(3+h) = (3+h)^3 = 27 + 27h + 9h^2 + h^3$

Step 2: Since $f(3) = 3^3 = 27$, $f(3+h) - f(3) = [27 + 27h + 9h^2 + h^3] - 27$
$= 27h + 9h^2 + h^3$

Step 3: $\frac{f(3+h) - f(3)}{h} = \frac{27h + 9h^2 + h^3}{h} = 27 + 9h + h^2$

Step 4: $\lim_{h \to 0}(27 + 9h + h^2) = 27.$

So, the derivative of $f(x) = x^3$ at $x = 3$ is $f'(3) = 27$.

We can also use the four-step method to find the derivative of a function at any point x, rather than at a particular point.

EXAMPLE 2.4

Find the derivative of $g(x) = 3x^2 + x$ using the four-step method.

Solution

Step 1: $g(x+h) = 3(x+h)^2 + (x+h)$
$$= 3(x^2 + 2xh + h^2) + (x+h)$$
$$= 3x^2 + 6xh + 3h^2 + x + h$$

Step 2: $g(x+h) - g(x) = (3x^2 + 6xh + 3h^2 + x + h) - (3x^2 + x)$
$$= 6xh + 3h^2 + h$$

Step 3: $\dfrac{g(x+h) - g(x)}{h} = \dfrac{6xh + 3h^2 + h}{h} = 6x + 3h + 1\text{u}$

Step 4: $\lim\limits_{h \to 0} \dfrac{g(x+h) - g(x)}{h} = \lim\limits_{h \to 0} (6x + 3h + 1) = 6x + 1$

So, the derivative of $g(x) = 3x^2 + x$ is $g'(x) = 6x + 1$.

Once you have the derivative of a function, you can evaluate it at any particular point.

EXAMPLE 2.5

If $g(x) = 3x^2 + x$, what are **(a)** $g'(2)$, **(b)** $g'(0)$, and **(c)** $g'(-3)$?

Solutions In Example 2.4, we learned that the derivative of g is $g'(x) = 6x + 1$. As a result, we have

(a) $g'(2) = 6(2) + 1 = 13$
(b) $g'(0) = 6(0) + 1 = 1$
(c) $g'(-3) = 6(-3) + 1 = -17$

EXAMPLE 2.6

Use the four-step method to find the derivative of $y = \dfrac{1}{x+2}$.

Solution

Step 1: $y = f(x)$, so $f(x+h) = \dfrac{1}{x+h+2}$

Step 2: $f(x+h) - f(x) = \dfrac{1}{x+h+2} - \dfrac{1}{x+2}$.

In order to conduct the subtraction, we need to first find a common denominator and then subtract. The common denominator is $(x+2)(x+h+2)$.

$$\frac{1}{x+h+2} - \frac{1}{x+2} = \frac{1}{x+h+2}\left(\frac{x+2}{x+2}\right) - \frac{1}{x+2}\left(\frac{x+h+2}{x+h+2}\right)$$

EXAMPLE 2.6 (Cont.)

$$= \frac{x+2}{(x+h+2)(x+2)} - \frac{x+h+2}{(x+h+2)(x+2)}$$
$$= \frac{(x+2)-(x+h+2)}{(x+h+2)(x+2)}$$
$$= \frac{-h}{(x+h+2)(x+2)}$$

Step 3: $\dfrac{f(x+h)-f(x)}{h} = \dfrac{\dfrac{-h}{(x+h+2)(x+2)}}{h}$
$$= \frac{-h}{h(x+h+2)(x+2)}$$
$$= \frac{-1}{(x+h+2)(x+2)}$$

Step 4: $\lim\limits_{h\to 0} \dfrac{-1}{(x+h+2)(x+2)} = \dfrac{-1}{(x+2)^2}$

So, $y' = \dfrac{-1}{(x+2)^2}$.

EXAMPLE 2.7

Use the four-step method to find the derivative of $f(x) = \sqrt{x-3}$ at $x = 4$.

Solution We will first find the general expression for $f'(x)$ and then determine $f'(4)$.

Step 1: $f(x+h) = \sqrt{x+h-3}$

Step 2: $f(x+h) - f(x) = \sqrt{x+h-3} - \sqrt{x-3}$

Step 3: $\dfrac{f(x+h)-f(x)}{h} = \dfrac{\sqrt{x+h-3}-\sqrt{x-3}}{h}$

In order to simplify this expression, we will rationalize the numerator by multiplying both numerator and denominator by $\sqrt{x+h-3}+\sqrt{x-3}$.

$$\frac{\sqrt{x+h-3}-\sqrt{x-3}}{h} = \frac{(\sqrt{x+h-3}-\sqrt{x-3})(\sqrt{x+h-3}+\sqrt{x-3})}{h(\sqrt{x+h-3}+\sqrt{x-3})}$$
$$= \frac{(x+h-3)-(x-3)}{h(\sqrt{x+h-3}+\sqrt{x-3})}$$
$$= \frac{h}{h(\sqrt{x+h-3}+\sqrt{x-3})}$$
$$= \frac{1}{\sqrt{x+h-3}+\sqrt{x-3}}$$

EXAMPLE 2.7 (Cont.)

Step 4: $$\lim_{h\to 0} \frac{1}{\sqrt{x+h-3}+\sqrt{x-3}} = \frac{1}{\sqrt{x-3}+\sqrt{x-3}} = \frac{1}{2\sqrt{x-3}}$$

Now that we know $f'(x) = \dfrac{1}{2\sqrt{x-3}}$, we can find $f'(4) = \dfrac{1}{2\sqrt{4-3}} = \frac{1}{2}$. Thus, the derivative of $f(x) = \sqrt{x-3}$ at $x = 4$ is $\frac{1}{2}$. ■

Exercise Set 2.1

In Exercises 1–24, use the definition of the derivative (that is, use the four-step method) to differentiate the given function.

1. $f(x) = 3x+2$
2. $g(x) = 5x-7$
3. $f(x) = 4-3x$
4. $g(t) = 6-4t$
5. $y = 3x^2$
6. $y = -5t^2$
7. $y = 2x^2+5$
8. $j(x) = 3x^2-x+2$
9. $k(x) = 3-4x^2$
10. $g(x) = 2x-7x^2$
11. $y = -4x^3$
12. $y = x^2-2x^3$
13. $y = t-3t^3$
14. $f(x) = \dfrac{2x+1}{3}$
15. $f(x) = \dfrac{1}{x+1}$
16. $g(x) = \dfrac{5}{2x-3}$
17. $j(x) = \dfrac{4}{1-x^2}$
18. $f(x) = \dfrac{9}{x^2+2x}$
19. $s = 16t^2-6t+3$
20. $q = t^3-4t^2+5t-2$
21. $y = \sqrt{x^2+4}$
22. $y = \sqrt{5-x}$
23. $y = \sqrt{2x-x^2}$
24. $y = \sqrt{5x-4x^3}$

In Exercises 25–36, evaluate the derivative at each of the indicated points.

25. $f(x) = 3x+7$; $f'(2)$, $f'(9)$
26. $g(x) = x^2-2$; $g'(2)$, $g'(9)$
27. $s(t) = 16t^2+2t$; $s'(3)$, $s'(2)$, $s'(0)$
28. $s(t) = 2t^2-1$; $s'(1)$, $s'(5)$
29. $q(t) = t^3-4t^2$; $q'(0)$, $q'(3)$
30. $j(x) = \dfrac{1}{\sqrt{x}}$; $j'(4)$, $j'(9)$
31. $k(x) = \dfrac{1}{x^2-4}$; $k'(5)$, $k'(-5)$
32. $f(x) = \dfrac{4}{x^2+1}$; $f'(0)$, $f'(3)$, $f'(-5)$
33. $j(t) = \sqrt{5t-t^2}$; $j'(1)$, $j'(4)$
34. $g(x) = \dfrac{1}{\sqrt{x+4}}$; $g'(0)$, $g'(5)$, $g'(-2)$
35. $f(x) = \dfrac{5}{\sqrt{25-x^2}}$; $f'(0)$, $f'(3)$, $f'(-4)$
36. $h(x) = \sqrt{x^2+4x+9}$; $h'(-4)$, $h'(0)$, $h'(4)$

In Your Words

37. Without looking in the book, write the four steps used to calculate $f'(c)$.
38. Explain the relationship between the slope of a curve at some point x_0 and the derivative of the function that graphs that curve at x_0.

2.2 DERIVATIVES OF POLYNOMIALS

At this point, the definition by the four-step method is our only way to take derivatives. The four-step method is a very important process and a method that we will use again and again. But because of its length, it is time consuming and there are many opportunities to make mistakes. In this section, we are going to use the definition to develop some formulas for taking the derivative of any polynomial. In the remainder of the chapter, we will develop some rules for taking more involved derivatives.

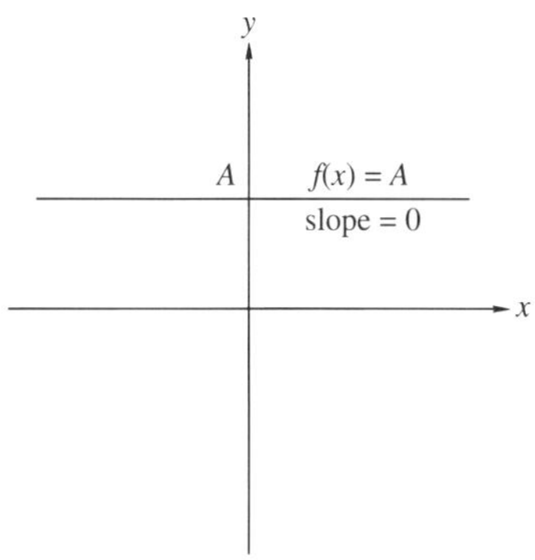

FIGURE 2.5

Constant Function

This first polynomial we will study is the constant function, $f(x) = k$. A look at the graph of this function in Figure 2.5 will show us what to expect. The slope of a constant function is 0, therefore, since the derivative is a measure of the slope of the tangent to the curve, we should expect the derivative to be 0. Let's use the four-step method to confirm our theory.

Step 1: $f(x+h) = k$

Step 2: $f(x+h) - f(x) = k - k = 0$

Step 3: $\dfrac{f(x+h) - f(x)}{h} = \dfrac{0}{h} = 0$

Step 4: $\lim\limits_{h\to 0} \dfrac{f(x+h) - f(x)}{h} = \lim\limits_{h\to 0} 0 = 0$

Thus, we have determined the following derivative of a constant function.

> **Rule 1: Derivative of $f(x) = k$**
>
> For the function $f(x) = k$, where k is a constant, the derivative is $f'(x) = 0$.

Linear Function

The next polynomial is the linear function $f(x) = mx + b$. Notice that we have written this function in the slope-intercept form of the equation. We know that the slope of this line is m. The derivative is supposed to give the slope of a tangent to the curve. Since this curve is a line, and its tangent is the same line, we expect that its derivative is also m. We use the four-step method:

Step 1: $f(x+h) = m(x+h) + b = mx + mh + b$

Step 2: $f(x+h) - f(x) = (mx + mh + b) - (mx + b) = mh$

Step 3: $\dfrac{f(x+h) - f(x)}{h} = \dfrac{mh}{h} = m$

Step 4: $\lim\limits_{h\to 0} \dfrac{f(x+h) - f(x)}{h} = \lim\limits_{h\to 0} m = m$

Again, we confirm our expectation and have shown the following for the derivative of a straight line.

Rule 2: Derivative of $f(x) = mx + b$

For the linear function $f(x) = mx + b$, the derivative is $f'(x) = m$.

EXAMPLE 2.8

Find the derivative of **(a)** $f(x) = -8$ and **(b)** $g(x) = 9x - 4$.

Solution

(a) $f(x) = -8$ is a constant function, so $f'(x) = 0$.

(b) $g(x) = 9x - 4$ is a linear function, so $g'(x) = 9$.

Polynomials of the Form $f(x) = x^n$

We will now consider higher-degree polynomials. This will be done in three steps. The first step will consider two specific forms of x^n, where n is 2 and 3. The second step will generalize this for any real number n. The third step will consider the sums and differences of these terms.

In order to develop the rule for the derivative of x^n, we will determine derivatives for x^2 and x^3. We begin with $f(x) = x^2$.

Step 1: $f(x+h) = (x+h)^2 = x^2 + 2xh + h^2$

Step 2: $f(x+h) - f(x) = (x^2 + 2xh + h^2) - x^2$

$$= 2xh + h^2$$

Step 3: $\dfrac{f(x+h) - f(x)}{h} = \dfrac{2xh + h^2}{h} = 2x + h$

Step 4: $\lim\limits_{h\to 0} \dfrac{f(x+h) - f(x)}{h} = \lim\limits_{h\to 0}(2x + h) = 2x$

So, the derivative of $f(x) = x^2$ **is** $f'(x) = 2x$.

Next, we find the derivative of $g(x) = x^3$.

Step 1: $g(x+h) = (x+h)^3 = x^3 + 3x^2h + 3xh^2 + h^3$

Step 2: $g(x+h) - g(x) = (x^3 + 3x^2h + 3xh^2 + h^3) - x^3$

$$= 3x^2h + 3xh^2 + h^3$$

Step 3: $\dfrac{g(x+h) - g(x)}{h} = \dfrac{3x^2h + 3xh^2 + h^3}{h} = 3x^2 + 3xh + h^2$

Step 4: $\lim\limits_{h\to 0} \dfrac{g(x+h) - g(x)}{h} = \lim\limits_{h\to 0}(3x^2 + 3xh + h^2) = 3x^2$

Thus, the derivative of $g(x) = x^3$ **is** $g'(x) = 3x^2$.

If we continued this process, we could show that the derivative of x^4 is $4x^3$, the derivative of x^5 is $5x^4$, the derivative of x^6 is $6x^5$, and so on. We could also show that the derivative of x^{-7} is $-7x^{-8}$, the derivative of $x^{5/2}$ is $\frac{5}{2}x^{3/2}$, and the derivative of $x^{-7/3}$ is $-\frac{7}{3}x^{-10/3}$.

What we have demonstrated is how to find the derivative of x^n, when n is a rational number. Thus, we now have

> **Rule 3: Derivative of $f(x) = x^n$**
>
> The derivative of any function $f(x) = x^n$, n a rational number, is $f'(x) = nx^{n-1}$.

EXAMPLE 2.9

Find the derivative of **(a)** $f(x) = x^7$, **(b)** $g(x) = x^{-3}$, **(c)** $h(x) = x^{2/3}$, and **(d)** $j(x) = x^{-5/2}$.

Solution

(a) $f(x) = x^7$ and so $f'(x) = 7x^{7-1} = 7x^6$

(b) $g(x) = x^{-3}$ and $g'(x) = -3x^{-3-1} = -3x^{-4}$

(c) $h(x) = x^{2/3}$ and $h'(x) = \frac{2}{3}x^{\frac{2}{3}-1} = \frac{2}{3}x^{-\frac{1}{3}}$

(d) $j(x) = x^{-5/2}$ and $j'(x) = -\frac{5}{2}x^{-\frac{5}{2}-1} = -\frac{5}{2}x^{-7/2}$ ▪

The next example demonstrates the usefulness of fractional exponents—both positive and negative.

EXAMPLE 2.10

Find the derivatives of **(a)** $y = x^2\sqrt{x}$ and **(b)** $f(x) = \frac{1}{x^{5/2}}$.

Solution

(a) $y = x^2\sqrt{x}$. To find the derivative of y, or $D_x y$, we must first rewrite $x^2\sqrt{x}$ as $x^2x^{1/2} = x^{5/2}$. So, $y = x^{5/2}$ and

$$D_x y = \frac{5}{2}x^{3/2} = \frac{5}{2}x\sqrt{x}$$

(b) $f(x) = \frac{1}{x^{5/2}}$. To apply rule 3, first write $f(x)$ as $f(x) = x^{-5/2}$. In Example 2.9(d), we showed that $f'(x) = -\frac{5}{2}x^{-7/2}$. ▪

Caution

Be careful. Do not just differentiate the denominator. If you do, you get

$$D_x\left(\frac{1}{x^{5/2}}\right) \neq \frac{1}{\frac{5}{2}x^{3/2}} = \frac{2}{5}x^{-3/2}$$

which is not the correct answer.

Three General Formulas

Suppose we have a function $F(x) = kf(x)$, where k is a constant and $f'(x)$ exists. [This might be a function such as $G(x) = 9x^3$, where $k = 9$ and $g(x) = x^3$.] We will use the four-step method to find the derivative of $F(x)$.

Step 1: $F(x+h) = kf(x+h)$

Step 2: $F(x+h) - F(x) = kf(x+h) - kf(x) = k[f(x+h) - f(x)]$

Step 3: $$\frac{F(x+h)-F(x)}{h} = \frac{k[f(x+h)-f(x)]}{h} = k\left[\frac{f(x+h)-f(x)}{h}\right]$$

Step 4: $$\lim_{h\to 0}\frac{F(x+h)-F(h)}{h} = \lim_{h\to 0} k\left[\frac{f(x+h)-f(x)}{h}\right] = k\lim_{h\to 0}\frac{f(x+h)-f(x)}{h} = kf'(x)$$

> **Rule 4: Derivative of $F(x) = kf(x)$**
>
> If $F(x) = kf(x)$, k is a constant and $f'(x)$ exists, then
> $$F'(x) = kf'(x)$$

EXAMPLE 2.11

Find the derivative of **(a)** $f(x) = 9x^3$, **(b)** $g(x) = 7x^{-5}$, and **(c)** $h(x) = \frac{2}{3}x^{5/2}$.

Solutions

(a) $f(x) = 9x^3$ so $f'(x) = 9(3x^2) = 27x^2$

(b) $g(x) = 7x^{-5}$ and $g'(x) = 7(-5x^{-6}) = -35x^{-6}$

(c) $h(x) = \frac{2}{3}x^{5/2}$ and $h'(x) = \frac{2}{3}(\frac{5}{2}x^{3/2}) = \frac{5}{3}x^{3/2}$ ▪

The next two formulas have to do with the sum and difference of two functions. These last two rules will allow us to find derivatives of higher-degree polynomials.

If f and g are two functions that can be differentiated, then what is the derivative of their sum or difference? It turns out that the derivative of a sum (or difference) is the sum (or difference) of the derivative of each term. These are stated formally in the following box, which contains the last two rules for this section.

Rules 5 and 6: Derivative of $F(x) = f(x) \pm g(x)$

Rule 5: If $F(x) = f(x) + g(x)$ and $f'(x)$ and $g'(x)$ exist, then

$$F'(x) = f'(x) + g'(x)$$

Rule 6: If $F(x) = f(x) - g(x)$ and $f'(x)$ and $g'(x)$ exist, then

$$F'(x) = f'(x) - g'(x)$$

EXAMPLE 2.12

Find the derivative of $f(x) = 7x^5 - 4x^3 + 2x - 8$.

Solution According to the last two rules, the derivative of a sum (or difference) is the sum (or difference) of the individual derivatives. We will differentiate this term by term:

$$\begin{aligned} f'(x) &= 7(5x^4) - 4(3x^2) + 2(1) - 0 \\ &= 35x^4 - 12x^2 + 2 \end{aligned}$$

Exercise Set 2.2

In Exercises 1–37, find the derivative of the function by using the six rules given in this section.

1. $f(x) = 19$
2. $g(x) = -4$
3. $h(x) = 7x - 5$
4. $j(x) = -3x + 7$
5. $k(x) = 9x^2$
6. $m(x) = 15x^3$
7. $f(x) = \frac{1}{3}x^{15}$
8. $g(x) = \frac{2}{5}x^{10}$
9. $h(x) = 5x^{-2}$
10. $j(x) = -3x^{-4}$
11. $k(x) = -\frac{2}{3}x^{3/2}$
12. $j(x) = \dfrac{4\sqrt{x}}{3}$
13. $h(x) = 4\sqrt{3x}$
14. $m(x) = \frac{3}{4}x^{-5/3}$
15. $g(x) = \dfrac{3}{\sqrt[3]{x}}$
16. $f(x) = 4x^3 - 5x^2 + 2$
17. $g(x) = 9x^2 + 3x - 4$
18. $n(x) = 7x^7 - 5x^5$
19. $h(x) = \frac{1}{3}x^3 + \frac{1}{2}x^2 - 5x + \frac{2}{3}x^{-2} + 5$
20. $j(x) = \frac{3}{4}x^4 - \frac{2}{3}x^3 + \frac{5}{4}x^2 - \frac{1}{2}x + 2x^{-3} + x^{-4} + 7$
21. $f(x) = \frac{4}{5}x^5 - \frac{3}{2}x^3 - 7x^0 + \frac{1}{2}x^{-2} - \frac{2}{3}x^{-3}$
22. $k(x) = \sqrt{2}x^5 - \sqrt{3}x^3 + 3x + \pi$
23. $\ell(x) = 3\sqrt{3}x^4 + 3\sqrt{5x} + \sqrt{4x^4}$
24. $m(x) = \sqrt{7}x^6 - \sqrt{4}x^3 + 5\sqrt{3}x^2 - \sqrt{7}$
25. $s(t) = 16t^2 - 32t + 5$
26. $q(t) = 4.3t^3 - 2.7t + 3.0$
27. $\alpha(t) = 30 - 4.0t^2 + 2t^{1/2}$
28. $s(t) = \frac{5}{3}t^4 + 7\sqrt{2}t^3 - 8t^{2/3} + 7t^{-1/2}$
29. $q(t) = \sqrt{6t} - 4t\sqrt{3t} + \sqrt[3]{t^2}$
30. $f(x) = 3x\sqrt{x}$
31. $g(x) = 5x^2\sqrt[3]{x}$
32. $j(x) = \dfrac{3x^3}{\sqrt[3]{x}}$
33. $h(x) = \dfrac{4}{x^2}$

34. $j(x) = 7x^2 + 2x\sqrt{x} + \dfrac{5}{x^3}$

35. $K(x) = \dfrac{4\sqrt[3]{x}}{\sqrt{x}}$

36. $M(x) = \dfrac{3\sqrt{x^3}}{\sqrt{3x}}$

37. $f(x) = \dfrac{4\sqrt{3x} + 3x^2 - 7\sqrt{x^3} + 2x^{-1}}{\sqrt{x}}$

Solve Exercises 38–44.

38. Derive rule 6.

39. Find all values of x where $f'(x) = 0$, if $f(x) = x^3 + 3x^2$.

40. Find all values of t where $\theta'(t) = 0$, if $\theta(t) = 3t^3 - t$.

41. Find all values of x where $g'(x) = 0$, if $g(x) = 2x^3 + x^2 - 4x$.

42. If $\omega(t) = 4.0t^2 - 2.0\sqrt{t}$, find the values of t for which $\omega'(t) = 0$.

43. If $j(t) = 4t^3 + \dfrac{1}{t}$, for what values of t is $j'(t) = 0$?

44. If $h(x) = 4 + \dfrac{1}{x} - \dfrac{8}{x^3}$, for what values of x is $h'(x) = 0$?

Solve Exercises 45–56.

45. *Demography* The population of a certain city t months from now can be estimated by the formula $P(t) = 146{,}000 + 64t - 0.04t^2$. Find

(a) The population of the city six months from now

(b) The rate of change in the city's population six months from now

46. *Business* A manufacturer estimates that the revenue in dollars from the sale of c cellular telephone is $R(c) = 120c - c^2$. Find

(a) The revenue from the sale of 30 telephones

(b) The rate of change in the revenue when $c = 30$

47. *Physics* A ball is rolled down an incline. The distance in feet of the ball from the starting point after t seconds is given by $s(t) = 15t + 7t^2$. Find

(a) The distance the ball has traveled when $t = 4$ s

(b) The velocity (rate of change) of the ball when $t = 4$ s

48. *Petroleum engineering* When an oil tank is drained for cleaning, the volume, V, of oil in gallons left in the tank t minutes after the drain valve is opened is given by the formula $V(t) = 75{,}000 - 3{,}000t + 30t^2$.

(a) How much oil is left in the tank 20 min after the drain is opened?

(b) What is the average rate that the oil drains during the first 20 min?

(c) What is the rate of change that the oil is draining when $t = 20$?

49. *Demography* A certain city's population t years from now can be estimated by using the function $P(t) = 125{,}000 + 1{,}500t - 36\sqrt{t}$. Find

(a) The city's population six years from now

(b) The rate at which the city's population is growing six years from now

(c) The city's population ten years from now

(d) The rate at which the city's population is growing ten years from now

50. *Business* The total cost, C, in dollars, of producing n units of a certain product is given by $C(n) = 5{,}000 + 25n - 0.4n^2$, where $0 \le n \le 60$.

(a) Determine the rate of change in the cost when $n = 10$.

(b) Determine the rate of change in the cost when $n = 30$.

(c) Determine the rate of change in the cost when $n = 50$.

51. *Electronics* The power, P, in watts, delivered to a $25\,\Omega$ resistor is $P = 25i^2$. Find the instantaneous rate of change of P with respect of i when $i = 2.4$ A.

52. *Material science* For certain temperatures, the tensile strength, S, in pounds, of a piece of material as a function of its temperature is described by $S(T) = 520 - 0.000085T^2$. Find the rate of change of S at the instant that $T = 135°$F.

53. *Business* A manufacturer has determined that the revenue in dollars from the sale of n cellular telephones is given by $R(n) = 78n - 0.025n^2$. The cost of producing n telephones is $C(n) = 9{,}800 + 22.5n$.

(a) If the profit, P, is the difference between the revenue and the cost, what is the profit function?

(b) Find the marginal profit function, P'.

(c) Find $P'(50)$.

54. *Electronics* The output voltage, V, in volts, of a circuit increases gradually with time t, in seconds, according to the function $V(t) = 30.0 + 0.04t^2$. Find an expression for the instantaneous rate of change of the voltage with time.

55. *Physics* The altitude, h, of a rocket fired from an airplane flying over the ocean can be described by the function $h(t) = -t^4 + 12t^3 - 16t^2 + 605$, where t is in seconds and the altitude is in feet above sea level. The velocity of the rocket at time t is $v(t) = h'(t)$.

(a) From what altitude was the rocket launched?

(b) At what time does the rocket fall into the ocean?

(c) What is the rocket's altitude and velocity at $t = 2$ s?

(d) What is the rocket's altitude and velocity at $t = 4$ s?

(e) What is the rocket's altitude and velocity at $t = 6$ s?

(f) What is the rocket's altitude and velocity at $t = 10$ s?

(g) What is the rocket's velocity when it falls into the ocean?

56. *Navigation* Upon reversing the engines, the distance traveled in meters by a supertanker in a straight line before coming to a stop is given by $s(t) = 84.5t - \frac{1}{6}t^3$, where t is in minutes.

(a) If the velocity of the supertanker at time t is $v(t) = s'(t)$, find an equation for the velocity.

(b) What was the velocity of the tanker when the engines were put in reverse?

(c) How long does it take for the supertanker to stop? That is, when is $v(t) = 0$?

(d) How far does the supertanker travel from the time the engines are reversed until it comes to a stop?

In Your Words

57. Describe how to find the derivative of $f(x) = kx^n$.

58. Describe how to find the derivative of the sum of two functions.

2.3 DERIVATIVES OF PRODUCT AND QUOTIENTS

In this section, we will learn general formulas for finding the derivatives of functions that are products or quotients.

The Product Rule

As before, we will assume that we have two functions, $f(x)$ and $g(x)$, and that their derivatives $f'(x)$ and $g'(x)$ exist. If we let F be the product of f and g, then $F(x) = f(x)g(x)$. We want to find $F'(x)$. Again, we will use the four-step method. The first two steps will be quite straightforward. In the third step, we will have to employ some more complicated algebra, which will be explained at that time.

Step 1: $F(x+h) = f(x+h)g(x+h)$

Step 2: $F(x+h) - F(x) = f(x+h)g(x+h) - f(x)g(x)$

Step 3: $\dfrac{F(x+h) - F(x)}{h} = \dfrac{f(x+h)g(x+h) - f(x)g(x)}{h}$

This does not look too promising. But, what we will do is add a fancy version of 0 to the numerator. We will add $-f(x+h)g(x)+f(x+h)g(x)$ to the numerator. Since this is equal to 0, the value of the numerator is not changed and we get

$$\begin{aligned}&\frac{f(x+h)g(x+h)-f(x+h)g(x)+f(x+h)g(x)-f(x)g(x)}{h}\\&=f(x+h)\frac{g(x+h)-g(x)}{h}+\frac{f(x+h)-f(x)}{h}g(x)\end{aligned}$$

Now we will go to Step 4 and take the limit of the quantity. Remember that the limit of a product is the product of the limits.

Step 4: $$\begin{aligned}\lim_{h\to 0}\frac{F(x+h)-F(x)}{h}&=\lim_{h\to 0}f(x+h)\lim_{h\to 0}\frac{g(x+h)-g(x)}{h}\\&\quad+\lim_{h\to 0}\frac{f(x+h)-f(x)}{h}\lim_{h\to 0}g(x)\\&=f(x)g'(x)+f'(x)g(x)\end{aligned}$$

This gives us the product rule for derivatives.

Product Rule for Derivatives

If $F(x)=f(x)g(x)$ and $f'(x)$ and $g'(x)$ exist, then

$$F'(x)=f(x)g'(x)+f'(x)g(x)$$

In words, this can be stated as "the derivative of the product of two functions, whose derivatives exist, is equal to the first function times the derivative of the second plus the second function times the derivative of the first."

Note Note carefully that the derivative of a product does not behave as nicely as the limit of a product. **The derivative of a product is not the product of the derivatives.**

EXAMPLE 2.13

Find the derivative of $F(x)=(x^2+x)(x^3-1)$.

Solution Here the function $F(x)$ is the product of $f(x)=x^2+x$ and $g(x)=x^3-1$. We calculate the derivatives of each of these two functions as $f'(x)=2x+1$ and $g'(x)=3x^2$. According to the new rule

$$\begin{aligned}F'(x)&=f(x)g'(x)+f'(x)g(x)\\&=(x^2+x)(3x^2)+(2x+1)(x^3-1)\\&=3x^4+3x^3+2x^4+x^3-2x-1\\&=5x^4+4x^3-2x-1\end{aligned}$$

■

Caution

Once again we stress that $F'(x) \neq f'(x)g'(x)$. In Example 2.13 we found that if $F(x) = (x^2+x)(x^3-1)$, then $F'(x) = 5x^4+4x^3-2x-1$. But, $f'(x)g'(x) = (2x+1)(3x^2) = 6x^3+3x^2 \neq F'(x)$.

EXAMPLE 2.14

Find $G'(x)$, if $G(x) = (x^3+2x^2+x)(x^2-5x)$.

Solution We could multiply these two polynomials and then take the derivative of that product, but we will use the product rule instead. The time will come when we will have to use the product rule, so we might as well get some practice. Also, the product rule makes the task somewhat easier. We will let

$$f(x) = x^3+2x^2+x \text{ and } g(x) = x^2-5x$$

so

$$\begin{aligned} f'(x) &= 3x^2+4x+1 \text{ and } g'(x) = 2x-5 \\ G'(x) &= f(x)g'(x) + f'(x)g(x) \\ &= (x^3+2x^2+x)(2x-5) + (3x^2+4x+1)(x^2-5x) \\ &= 5x^4-12x^3-27x^2-10x \end{aligned}$$

■

The Quotient Rule

Suppose now that $F(x) = \dfrac{f(x)}{g(x)}$, where $g(x) \neq 0$, and $f'(x)$ and $g'(x)$ both exist. What is $F'(x)$? There are a couple of ways to develop this formula. One way uses the four-step method. We are going to use a slightly different approach that depends on the assumption that $F'(x)$ exists.

If $F(x) = \dfrac{f(x)}{g(x)}$, then $F(x)g(x) = f(x)$. Now, using the product rule on the second equation, we see that

$$F(x)g'(x) + F'(x)g(x) = f'(x)$$

Solving this for $F'(x)$ we get

$$F'(x) = \frac{f'(x) - F(x)g'(x)}{g(x)}$$

But, $F(x) = \dfrac{f(x)}{g(x)}$, and so

$$F'(x) = \frac{f'(x) - \dfrac{f(x)}{g(x)}g'(x)}{g(x)} = \frac{g(x)f'(x) - f(x)g'(x)}{[g(x)]^2}$$

And so we have the second, and last, rule of this section, the quotient rule.

Quotient Rule for Derivatives

If $F(x) = \dfrac{f(x)}{g(x)}$, $g(x) \neq 0$ and both $f'(x)$ and $g'(x)$ exist, then

$$F'(x) = \frac{g(x)f'(x) - f(x)g'(x)}{[g(x)]^2}$$

EXAMPLE 2.15

If $F(x) = \dfrac{x^2+1}{3x-2}$, find $F'(x)$.

Solution According to the quotient rule,

$$F'(x) = \frac{g(x)f'(x) - f(x)g'(x)}{[g(x)]^2}$$

We will let $f(x) = x^2 + 1$ and $g(x) = 3x - 2$, so $f'(x) = 2x$ and $g'(x) = 3$. Thus,

$$\begin{aligned} F'(x) &= \frac{(3x-2)(2x) - (x^2+1)(3)}{(3x-2)^2} \\ &= \frac{(6x^2-4x) - (3x^2+3)}{(3x-2)^2} \\ &= \frac{3x^2-4x-3}{(3x-2)^2} \end{aligned}$$

Note Note that the derivative of the quotient of two functions is **not** the quotient of the derivatives. Thus, if $F(x) = \dfrac{f(x)}{g(x)}$, then $F'(x) \neq \dfrac{f'(x)}{g'(x)}$. In Example 2.15, $\dfrac{f'(x)}{g'(x)} = \dfrac{2x}{3} \neq F'(x)$.

EXAMPLE 2.16

If $y = \dfrac{1}{x^3}$, find y'.

Solution Here y can be considered a quotient with $f(x) = 1$ and $g(x) = x^3$. This means that $f'(x) = 0$ and $g'(x) = 3x^2$. Using the quotient rule, we get

$$\begin{aligned} y' &= \frac{x^3(0) - (1)(3x^2)}{(x^3)^2} \\ &= \frac{0 - 3x^2}{x^6} = \frac{-3}{x^4} \end{aligned}$$

EXAMPLE 2.16 (Cont.)

We could have used rule 3 to solve this by rewriting $y = \frac{1}{x^3} = x^{-3}$. Then we would have gotten $y' = -3x^{-4} = \frac{-3}{x^4}$.

EXAMPLE 2.17

If $f(x) = \frac{4x^2 - 8\sqrt{x}}{x^3 + \sqrt{x^5}}$, find $f'(x)$.

Solution We begin by writing the radicals with fractional exponents. Thus, we get

$$f(x) = \frac{4x^2 - 8x^{1/2}}{x^3 + x^{5/2}}$$

We now use the quotient rule:

$$\begin{aligned} f'(x) &= \frac{(x^3 + x^{5/2})(8x - 4x^{-1/2}) - (4x^2 - 8x^{1/2})(3x^2 + \frac{5}{2}x^{3/2})}{(x^3 + x^{5/2})^2} \\ &= \frac{(8x^4 - 4x^{5/2} + 8x^{7/2} - 4x^2) - (12x^4 + 10x^{7/2} - 24x^{5/2} - 20x^2)}{(x^3 + x^{5/2})^2} \\ &= \frac{-4x^4 - 2x^{7/2} + 20x^{5/2} + 16x^2}{(x^3 + x^{5/2})^2} \end{aligned}$$

Normal Lines

Until now we have concentrated on finding the slope of a line that is tangent to a curve at a particular point. We are also interested in the line that is perpendicular to the curve at a given point. This is called the **normal line**. If the slope of the tangent line is m_1 and the slope of the normal line is m_2, then, since the lines are perpendicular, we know that $m_1 = \frac{-1}{m_2}$.

EXAMPLE 2.18

Find an equation of the tangent and normal lines to $y = \frac{1}{x^3}$ at the point $\left(2, \frac{1}{8}\right)$.

Solution In Example 2.16, we found that the derivative of this function is $y' = \frac{-3}{x^4}$. The derivative is used to find the slope of a line tangent to y. Evaluating y' at the point $\left(2, \frac{1}{8}\right)$, we find that the line tangent to y at this point will have slope $m = \frac{-3}{2^4} = \frac{-3}{16}$. We now know that the slope is $-\frac{3}{16}$ at $\left(2, \frac{1}{8}\right)$, so the equation of the tangent line is

$$\begin{aligned} y - y_0 &= m(x - x_0) \\ y - \frac{1}{8} &= -\frac{3}{16}(x - 2) \end{aligned}$$

EXAMPLE 2.18 (Cont.)

$$\text{or} \qquad y = -\frac{3}{16}x + \frac{1}{2}$$

The slope of the normal line is $-\dfrac{1}{-3/16} = \frac{16}{3}$. The equation of the normal line to $y = \dfrac{1}{x^3}$ at the point $\left(2, \frac{1}{8}\right)$ is

$$y - \frac{1}{8} = \frac{16}{3}(x-2)$$

$$\text{or} \qquad y = \frac{16}{3}x - \frac{253}{24}$$

The next example will require the use of the quotient rule to find the slope of a normal line to a curve.

EXAMPLE 2.19

Determine the equation for the normal line to $f(x) = \dfrac{x^2+3x-1}{x+4}$ at $(3,2)$.

Solution We begin by finding the slope of the tangent line at this point. To do this, we find the derivative of f. Using the quotient rule,

$$f'(x) = \frac{(x+4)(2x+3) - (x^2+3x-1)(1)}{(x+4)^2}$$

To find the slope of f at $(3,2)$, we evaluate f' at $x = 3$. Notice that it is not necessary to simplify the derivative in order to evaluate it.

$$f'(3) = \frac{(7)(9) - (17)(1)}{7^2} = \frac{46}{49}$$

Since the slope of the tangent line is $\frac{46}{49}$, the slope of the normal line is $-\frac{1}{46/49} = -\frac{49}{46}$. We now know the slope of the normal line to f at $(3,2)$, and so the equation of that normal line is

$$y - 2 = -\frac{49}{46}(x-3)$$

$$\text{or} \qquad y = -\frac{49}{46}x + \frac{239}{46}$$

EXAMPLE 2.20

Find the equation of the tangent and normal lines to the curve

$$y = \frac{(x+1)(x^2+2x+5)}{3-x}$$

at the point $(1,8)$.

Solution We have one point on the line, $(1,8)$. What we need to find is the slope. To do this, we will find the derivative y' and evaluate it at the point $(1,8)$. While y

EXAMPLE 2.20 (Cont.)

is in the form $\dfrac{f(x)}{g(x)}$, the function f is a product of two functions. We will use the quotient rule, but indicate the derivatives that need to be taken by using the notation D_x. This way we get

$$y' = \frac{(3-x)D_x[(x+1)(x^2+2x+5)] - (x+1)(x^2+2x+5)D_x(3-x)}{(3-x)^2}$$

We use the product rule to find

$$D_x[(x+1)(x^2+2x+5)] = (x+1)(2x+2) + (1)(x^2+2x+5)$$

We also have $D_x(3-x) = -1$, so

$$y' = \frac{(3-x)[(x+1)(2x+2)+(x^2+2x+5)] - (x+1)(x^2+2x+5)(-1)}{(3-x)^2}$$

To find the slope of y at $(1,8)$, we need to evaluate y' when $x=1$. It is not necessary to simplify the derivative to do this. When $x=1$,

$$y' = \frac{(2)[(2)(4)+8] - (2)(8)(-1)}{2^2} = \frac{2(16)+16}{4} = \frac{48}{4} = 12$$

We now know that the slope of the tangent line is 12 and a point on the line is $(1,8)$, so an equation of the line is

$$y-8 = 12(x-1)$$

The slope of the normal line is $\dfrac{-1}{12}$, so an equation of that line is $y-8 = \dfrac{-1}{12}(x-1)$.

Exercise Set 2.3

In Exercises 1–38, differentiate the function.

1. $f(x) = (3x+1)(2x-7)$
2. $g(x) = (6x-2)(5-4x)$
3. $h(x) = (2x^2+x-1)(3x-5)$
4. $k(x) = (4x^3-1)(x^2-7x)$
5. $j(x) = (4-3x)(6x^2+6x-4)$
6. $s(t) = (6-4t)(3t^2-7t)$
7. $q(t) = 4(t^2-3t)^2$
8. $r(p) = (4p-3p^2)(2p-4)$
9. $f(w) = (3w^3-4w^2+2w-5)(w^2-w^{-1})$
10. $g(r) = (2r^4-4r^2+2r)(r^2+2r-1)$
11. $f(x) = \dfrac{4x-1}{2x+3}$
12. $h(y) = (y^3-1)(y^2-1)(y-1)$
13. $g(x) = \dfrac{9x^2+2}{4x-1}$
14. $h(x) = 4x^2 - \dfrac{4}{x^2}$
15. $j(s) = 6s^2 - \dfrac{1}{6s^2}$
16. $K(s) = \dfrac{3-4s}{5s-2s^2}$
17. $f(t) = 4t^3 - \dfrac{2t}{t-2}$
18. $j(x) = \dfrac{1+2x}{1-2x}$
19. $H(x) = \dfrac{x^3-1}{x-1}$

20. $k(x) = \dfrac{x^4+4}{3x}$

21. $f(\phi) = \dfrac{\phi^2}{3\phi^2-1}$

22. $L(t) = \dfrac{t^2+3t+2}{t^2-4t-4}$

23. $h(x) = \dfrac{3x^3-x+1}{x^3-3x-1}$

24. $m(x) = \dfrac{x^2+5x+6}{x^2-5x+6}$

25. $f(t) = \dfrac{3t^2-t-1}{\sqrt[3]{t}}$

26. $F(x) = \dfrac{2x^{3/2}-4x^{1/2}}{4x^{1/2}-2}$

27. $g(w) = 12w^2 + \dfrac{w-1}{w+1}$

28. $j(t) = \dfrac{t^2+1}{2t+1} + \dfrac{t-1}{2t+1}$

29. $h(x) = \dfrac{1}{x^2+1}$

30. $k(x) = \dfrac{5}{x^3-1}$

31. $H(x) = \dfrac{4-x^3}{(2-x^2)(3x-x^3)}$

32. $L(x) = \dfrac{(2x+1)(3x^2-4x)}{7x-1}$

33. $n(s) = \dfrac{2s^3}{(s^2-1)(s-1)}$

34. $m(z) = \dfrac{(z-5)(z^2+7)}{z^2+3z}$

35. $y = \dfrac{(2x-3)(x^2-4x+1)}{3x^3+1}$

36. $y = \dfrac{5x}{5-x}(25-x^2)$

37. $y = \dfrac{(t^2-1)}{(2t+1)} \cdot \dfrac{(t-1)}{(2t+1)}$

38. $y = \left(\dfrac{1-x}{x}\right)(1-x^2)$

Solve Exercises 39 and 40.

39. Find the slope of the curve $y = (3x^2+2x-1) \times (x^3-x+1)$ at $(1,4)$.

40. Find the slope of the curve $y = \dfrac{x^3}{x^2+1}$ at $(-1, -\frac{1}{2})$.

In Exercises 41–46, find the equation of the tangent and normal lines to the curves at the given points.

41. $y = \dfrac{8}{x+1}$ at $(3,2)$

42. $y = \dfrac{4x+1}{5x}$ at $(1,1)$

43. $y = \dfrac{5x+4}{2x+1}$ at $(1,3)$

44. $y = (x^2-3x-6)(2x-6)$ at $(-1,16)$

45. $y = \dfrac{9x+2}{x^2(x+4)}$ at $(-2,-2)$

46. $y = (x^3+5x^2-1)(x^2-2x)$ at $(1,-5)$

Solve Exercises 47–52.

47. *Demography* It is estimated that t years from now the population of a certain city, in thousands of people, will be $P(t) = (0.8t-6)(0.5t+9)+87$. How fast will the population be growing in 5 years?

48. *Business* The profit from the sale of n items of a certain product is given by $P(n) = (4-0.2n)(2.6n+7)$, where $P(n)$ is in hundreds of dollars and $1 \leq n \leq 20$. If the marginal profit is $P'(n)$, find the marginal profit when $n = 7$.

49. *Electronics* In a voltage divider circuit the output, V, in volts, depends on the value of a variable resistor, R, in ohms, according to the function $V = \dfrac{110R}{R+60}$. Find $V' = \dfrac{dV}{dR}$.

50. *Electronics* The impedance, Z, in ohms, of a Wein bridge depends on the resistance, R, in ohms, of one arm of the bridge so that $Z = \frac{(1+8R)^2+8R}{8(1+8R)}$. Find $Z' = \frac{dZ}{dR}$.

51. *Medical technology* It is estimated by a medical research team that the population of a bacterial culture after t hours is approximately $N(t) = \frac{t^2-2t+1}{3\sqrt{t}+2}$, where $N(t)$ is in thousands and $0 \le t \le 12$. **(a)** Find $N'(t)$. **(b)** Find the rate of growth after 5 hours.

52. *Optics* For thin lenses the object distance, s, is given by $s = \frac{s'f}{s'-f}$, where f is the focal length and s' is the image distance. If f is a constant, find the rate of change of s with respect to s', that is, find $\frac{ds}{ds'}$. (Note that in this standard optics formula, s' does not indicate the derivative of s.)

In Your Words

53. Without looking in the text, explain how to determine the derivative of the product of two functions.

54. Without looking in the text, explain how to determine the derivative of the quotient of two functions.

2.4 DERIVATIVES OF COMPOSITE FUNCTIONS

None of the functions that we have differentiated have involved composite functions. In this section, we will learn how to find the derivative of a function that is the composite of two differentiable functions. The rule for finding these derivatives is called the chain rule.

We will begin by briefly reviewing composite functions. We will then look at a special case of the chain rule, called the power rule.

Suppose that f and g are two functions. The composite function of f and g is $(f \circ g)(x) = f(g(x))$, where the domain of f is the range of g.

EXAMPLE 2.21

(a) If $f(x) = x^2$ and $g(x) = x+1$, then $(f \circ g)(x) = (x+1)^2$

(b) If $f(x) = 2x^3 + 5x$ and $g(x) = x^4$, then

$$\begin{aligned}(f \circ g)(x) &= 2(x^4)^3 + 5(x^4)\\ &= 2x^{12} + 5x^4\end{aligned}$$

(c) If $g(w) = \sin 2w$ and $h(w) = 3w^2 + 1$, then

$$\begin{aligned}(g \circ h)(w) &= \sin[2(3w^2+1)]\\ &= \sin(6w^2+2)\end{aligned}$$

It is just as important to be able to recognize a composite function and to decompose it into its individual functions. In this case, you begin with $y = f \circ g$ and determine f and g.

EXAMPLE 2.22

(a) $y = (f \circ g)(x) = (x^2 - 2x + 1)^3$, where $f(x) = x^3$ and $g(x) = x^2 - 2x + 1$
(b) $y = (f \circ g)(x) = (x^5 - 4)^{-2}$, where $f(x) = x^{-2}$ and $g(x) = x^5 - 4$
(c) $y = (f \circ g)(x) = \sqrt{x^3 - 2x}$, where $f(x) = \sqrt{x} = x^{1/2}$ and $g(x) = x^3 - 2x$
(d) $y = (f \circ g)(x) = \ln x^5$, where $f(x) = \ln x$ and $g(x) = x^5$
(e) $y = (f \circ g)(x) = \tan 2x$, where $f(x) = \tan x$ and $g(x) = 2x$ ▪

The Power Rule

We will begin our exploration of the chain rule by looking at some special cases. In each of these cases, $f(x) = x^n$ and so $(f \circ g)(x) = [g(x)]^n$. We will use the product rule to look at the derivatives when $n = 2$, 3, and 4. Because it is easier, we will use the $\frac{d}{dx}$ notation to indicate what is being differentiated.

$n = 2$

$$\begin{aligned}\frac{d}{dx}[g(x)]^2 &= \frac{d}{dx}[g(x)g(x)] \\ &= g(x)g'(x) + g'(x)g(x) \\ &= 2g(x)g'(x)\end{aligned}$$

$n = 3$

$$\begin{aligned}\frac{d}{dx}[g(x)]^3 &= \frac{d}{dx}\{[g(x)]^2 g(x)\} \\ &= [g(x)]^2 g'(x) + \frac{d}{dx}[g(x)]^2 g(x)\end{aligned}$$

From the example when $n = 2$, we know that

$$\frac{d}{dx}[g(x)]^2 = 2g(x)g'(x)$$

and so we get

$$\begin{aligned}\frac{d}{dx}[g(x)]^3 &= [g(x)]^2 g'(x) + 2g(x)g'(x)g(x) \\ &= 3[g(x)]^2 g'(x)\end{aligned}$$

$n = 4$

$$\begin{aligned}\frac{d}{dx}[g(x)]^4 &= \frac{d}{dx}\{[g(x)]^3 g(x)\} = [g(x)]^3 g'(x) + \left\{\frac{d}{dx}[g(x)]^3\right\} g(x) \\ &= [g(x)]^3 g'(x) + 3[g(x)]^2 g'(x) g(x) \\ &= 4[g(x)]^3 g'(x)\end{aligned}$$

Look at these results:

$$\begin{aligned}\frac{d}{dx}[g(x)]^2 &= 2g(x)g'(x) \\ \frac{d}{dx}[g(x)]^3 &= 3[g(x)]^2 g'(x) \\ \frac{d}{dx}[g(x)]^4 &= 4[g(x)]^3 g'(x)\end{aligned}$$

These results seem to indicate a pattern. It is possible to show that the pattern results in the following rule, called the **general power rule**.

Rule 9: General Power Rule for Derivatives

If g is a function and its derivative g' exists, then

$$\frac{d}{dx}[g(x)]^n = n[g(x)]^{n-1}g'(x).$$

Notice that this is very similar to rule 3, which states that if $f(x) = x^n$, then $f'(x) = nx^{n-1}$. The main difference is the last factor, $g'(x)$. If $g(x) = x$, then $g'(x) = 1$, and rule 9 is the same as rule 3.

EXAMPLE 2.23

Find the derivatives of **(a)** $(x^2+1)^3$, **(b)** $(4-x^3)^5$, and **(c)** $(x^3-2x^2+1)^{-4}$.

Solutions

(a) If $y = (x^2+1)^3$, then this is $[g(x)]^3$, where $g(x) = x^2+1$. Since $g'(x) = 2x$ we have

$$y' = 3[g(x)]^2 g'(x) = 3(x^2+1)^2(2x)$$

(b) Here $y = (4-x^3)^5 = [g(x)]^5$, where $g(x) = 4-x^3$ and $g'(x) = -3x^2$.

$$\begin{aligned} y' &= 5[g(x)]^4 g'(x) \\ &= 5(4-x^3)^4(-3x^2) \\ &= -15x^2(4-x^3)^4 \end{aligned}$$

(c) $y = (x^3-2x^2+1)^{-4} = [g(x)]^{-4}$, where $g(x) = x^3-2x^2+1$ and $g'(x) = 3x^2-4x$.

$$\begin{aligned} y' &= -4[g(x)]^{-5} g'(x) \\ &= -4(x^3-2x^2+1)^{-5}(3x^2-4x) \end{aligned}$$

The general power rule is not always used alone. It can be used with other rules, such as the product rule or the quotient rule.

EXAMPLE 2.24

Find the derivatives of **(a)** $f(x) = 3x(x^3-1)^4$ and **(b)** $f(x) = \left(\dfrac{4x+1}{x^2-2}\right)^3$.

Solutions

(a) Here we have a product of $3x$ and $(x^3-1)^4$. We will first use the product rule.

$$f'(x) = 3x\left[\frac{d}{dx}(x^3-1)^4\right] + (x^3-1)^4(3)$$

EXAMPLE 2.24 (Cont.)

Now we use the general power rule to determine that

$$\frac{d}{dx}(x^3-1)^4 = 4(x^3-1)^3(3x^2) = 12x^2(x^3-1)^3$$

Putting this value into the formula for $f'(x)$, we get

$$\begin{aligned} f'(x) &= 3x[12x^2(x^3-1)^3] + (x^3-1)^4(3) \\ &= 36x^3(x^3-1)^3 + 3(x^3-1)^4 \\ &= [36x^3 + 3(x^3-1)](x^3-1)^3 \\ &= (39x^3-3)(x^3-1)^3 \end{aligned}$$

(b) Here $g(x) = \dfrac{4x+1}{x^2-2}$ and $f(x) = [g(x)]^3$

Using the general power rule, we see that

$$\begin{aligned} f'(x) &= 3[g(x)]^2 g'(x) \\ &= 3\left(\frac{4x+1}{x^2-2}\right)^2 g'(x) \end{aligned}$$

We will use the quotient rule to get

$$\begin{aligned} g'(x) &= \frac{(x^2-2)(4)-(4x+1)(2x)}{(x^2-2)^2} \\ &= \frac{(4x^2-8)-(8x^2+2x)}{(x^2-2)^2} \\ &= \frac{-4x^2-2x-8}{(x^2-2)^2} \end{aligned}$$

Putting this value for $g'(x)$ into the answer for $f'(x)$, we get

$$\begin{aligned} f'(x) &= 3\left(\frac{4x+1}{x^2-2}\right)^2 \left(\frac{-4x^2-2x-8}{(x^2-2)^2}\right) \\ &= -3\frac{(4x+1)^2(4x^2+2x+8)}{(x^2-2)^4} \end{aligned}$$

Note In Example 2.24(a), after we found the derivative as $f'(x) = 3x[12x^2(x^3-1)^3] + (x^3-1)^4(3)$, we continued until it was factored as $f'(x) = (39x^3-3)(x^3-1)^3$ because the factored form is more useful in applications.

The Chain Rule

The power rule is a special case of the chain rule. Another way of writing the composition function $y = (f \circ g)(x) = f(g(x))$ is

$$y = f(u) \quad \text{where } u = g(x)$$

The major change here is that y is now a function of u and u is a function of x. By substituting, we can get y as a function of x.

Rule 10: Chain Rule

If $y = f(u)$ and $u = g(x)$, and both $f'(u)$ and $g'(x)$ exist, then

$$y' = f'(u)g'(x)$$

or

$$y' = f'[g(x)] \cdot g'(x)$$

Note The chain rule is often written as

$$\frac{dy}{dx} = \frac{dy}{du} \cdot \frac{du}{dx}$$

EXAMPLE 2.25

Find y', if $y = (x^3 - 2x^2 + 1)^{-4}$.

Solution If $y = (x^3 - 2x^2 + 1)^{-4}$, then we will let $y = f(u) = u^{-4}$ and $u = g(x) = x^3 - 2x^2 + 1$. $f'(u) = -4u^{-5}$ and $g'(x) = 3x^2 - 4x$, so

$$\begin{aligned} y' &= f'(u)g'(u) \\ &= -4u^{-5}(3x^2 - 4x) \end{aligned}$$

Substituting for u, we get

$$y' = -4(x^3 - 2x^2 + 1)^{-5}(3x^2 - 4x)$$

This is the same answer we got when we worked this problem in Example 2.23(c).

EXAMPLE 2.26

Find y', if $y = \sqrt[3]{x^5 - 4x}$.

Solution $y = f(u) = \sqrt[3]{u} = u^{1/3}$ and $u = g(x) = x^5 - 4x$. $f'(u) = \frac{1}{3}u^{-2/3}$ and $g'(x) = 5x^4 - 4$, and so

$$\begin{aligned} y &= f'(u)g'(x) \\ &= \frac{1}{3}u^{-2/3}(5x^4 - 4) \\ &= \frac{1}{3}(x^5 - 4x)^{-2/3}(5x^4 - 4) \qquad \text{Substitute for } u. \end{aligned}$$

Hint It is possible to extend the chain rule. For example, suppose $y = f(u)$, $u = g(v)$, and $v = h(x)$. Then the composite function

$$y = (f \circ g \circ h)(x) = f(g(h(x)))$$

has a derivative

$$\begin{aligned} y' &= f'(u)g'(v)h'(x) \\ \text{or} \qquad &= \frac{dy}{du}\frac{du}{dv}\frac{dv}{dx} \end{aligned}$$

EXAMPLE 2.27

If $y = \left[6\left(\frac{5}{x^2}\right)^4 - 7\right]^{10}$, find y'.

Solution This looks more difficult than it is. First, if we let $v = h(x) = \frac{5}{x^2} = 5x^{-2}$, then we have $y = (6v^4 - 7)^{10}$. Now let $u = g(v) = 6v^4 - 7$ and we have $y = f(u) = u^{10}$. So, if $f(u) = u^{10}$, $g(v) = 6v^4 - 7$, and $h(x) = 5x^{-2}$, then

$$\begin{aligned} y' &= f'(u)g'(v)h'(x) \\ &= 10u^9(24v^3)(-10x^{-3}) \\ &= 10(6v^4 - 7)^9(24v^3)(-10x^{-3}) && \text{Substitute for } u. \\ &= 10[6(5x^{-2})^4 - 7]^9[24(5x^{-2})^3](-10x^{-3}) && \text{Substitute for } v. \\ &= 10\left[6\left(\frac{625}{x^8}\right) - 7\right]^9\left(24 \cdot \frac{125}{x^6}\right)\left(\frac{-10}{x^3}\right) \\ &= -\frac{300{,}000}{x^9}\left[6\left(\frac{625}{x^8}\right) - 7\right]^9 \end{aligned}$$

Application

EXAMPLE 2.28

After a sewage spill, the level of pollution in Herd Weyer Bay is estimated by $P(t) = \frac{100t + 25}{\sqrt{5t^2 + 10}}$ parts per million (ppm), where t is the time in days since the spill occurred.

(a) What is the rate of change in the level of pollution?

(b) What is the rate of change in the level of pollution after 1.5 days?

(c) What is the rate of change in the level of pollution after 2.5 days?

EXAMPLE 2.28 (Cont.)

Solution

(a) We find the rate of change in the level of pollution by finding the derivative of the level of pollution function $P(t) = \dfrac{100t+25}{\sqrt{5t^2+10}} = \dfrac{100t+25}{(5t^2+10)^{1/2}}$.

$$P'(t) = \frac{(5t^2+10)^{1/2}(100) - (100t+25)\frac{1}{2}(5t^2+10)^{-1/2}}{\left((5t^2+10)^{1/2}\right)^2}$$

$$= \frac{(5t^2+10)^{1/2}(100) - \left(50t+\frac{25}{2}\right)(5t^2+10)^{-1/2}}{5t^2+10}$$

$$= \frac{(5t^2+10)^{1/2}(100) - \left(50t+\frac{25}{2}\right)(5t^2+10)^{-1/2}}{5t^2+10} \cdot \frac{(5t^2+10)^{1/2}}{(5t^2+10)^{1/2}}$$

$$= \frac{(5t^2+10)(100) - \left(50t+\frac{25}{2}\right)}{(5t^2+10)^{3/2}}$$

$$= \frac{500t^2+1000-50t+\frac{25}{2}}{(5t^2+10)^{3/2}} = \frac{500t^2-50t+1012.5}{(5t^2+10)^{3/2}}$$

Thus, the pollution is changing at the rate of $P'(t) = \dfrac{500t^2-50t+1012.5}{(5t^2+10)^{3/2}}$ ppm/day.

(b) The rate of change in the level of pollution after 1.5 days is

$$P'(1.5) = \frac{500(1.5)^2-50(1.5)+1012.5}{(5(1.5)^2+10)^{3/2}} \approx 21.055 \text{ ppm/day.}$$

(c) The rate of change in the level of pollution after 2.5 days is

$$P'(2.5) = \frac{500(2.5)^2-50(2.5)+1012.5}{(5(2.5)^2+10)^{3/2}} \approx 15.145 \text{ ppm/day.}$$

So far, we have had ten rules for derivatives. These are summarized in the following box.

Ten Rules for Derivatives

1. If $f(x) = k$, k a constant, then $f'(x) = 0$.
2. If $f(x) = mx + b$, then $f'(x) = m$.
3. If $f(x) = x^n$, n a rational number, then $f'(x) = nx^{n-1}$.
4. If $F(x) = kf(x)$, k a constant, and $f'(x)$ exists, then $F'(x) = kf'(x)$.
5. If $F(x) = f(x) + g(x)$ and $f'(x)$ and $g'(x)$ exist, then $F'(x) = f'(x) + g'(x)$.
6. If $F(x) = f(x) - g(x)$ and both $f'(x)$ and $g'(x)$ exist, then $F'(x) = f'(x) - g'(x)$.
7. Product rule: If $F(x) = f(x)g(x)$, and both $f'(x)$ and $g'(x)$ exist, then
$$F'(x) = f(x)g'(x) + f'(x)g(x)$$
8. Quotient rule: If $F(x) = f(x)/g(x)$, $g(x) \neq 0$, and both $f'(x)$ and $g'(x)$ exist, then
$$F'(x) = \frac{g(x)f'(x) - f(x)g'(x)}{[g(x)]^2}$$
9. General power rule: If $g'(x)$ exists, then
$$\frac{d}{dx}[g(x)]^n = n[g(x)]^{n-1}g'(x)$$
10. Chain rule: If $y = f(u)$, $u = g(x)$, and both $f'(u)$ and $g'(x)$ exist, then $y' = f'(u)g'(x)$ or
$$\frac{dy}{dx} = \frac{dy}{du} \cdot \frac{du}{dx}$$

Exercise Set 2.4

In Exercises 1–22, find the derivative of the given function by using the general power rule.

1. $f(x) = (3x - 6)^4$
2. $g(x) = (7 - 2x)^3$
3. $h(x) = (5x - 7)^{-5}$
4. $k(x) = (9x + 5)^{-3}$
5. $f(x) = (x^2 + 3x)^4$
6. $h(x) = \sqrt{4x^2 + 7}$
7. $H(x) = \dfrac{1}{\sqrt{4x^2 + 7}}$
8. $g(x) = (x^7 - 9)^6$
9. $s(t) = (t^4 - t^3 + 2)^3$
10. $g(t) = (3t^5 + 2t^3 - t)^{10}$
11. $f(u) = (u^3 + 2u^{-4})^3$
12. $f(u) = (4u^2 - 3u^{-1})^2$
13. $g(x) = \left[(x - 2)(3x^2 - x)\right]^3$
14. $j(t) = 4t^3\sqrt{t^2 - 4}$
15. $h(v) = (v^2 + 1)^2(2v - 5)^3$
16. $g(x) = \left[\left(x^2 - 4x\right)\left(2x^3 - 7\right)\right]^3$
17. $h(x) = \left[\left(x^3 - 6x^2\right)\left(5x - 6x^2 + x^3\right)\right]^4$
18. $j(w) = (w^3 + 2w)^4(3w - 5)^{-2}$
19. $k(x) = (3x^2 + 2)^3\sqrt{x^3 - 7x}$
20. $h(x) = \left(\dfrac{3x + 4}{2x^2 - 1}\right)^4$

21. $g(v) = \dfrac{(v^2 - 4v)^3}{(v^3 - 9)^2}$

22. $f(u) = \left(\dfrac{4u^2 + 5}{6u^3 - 3u}\right)^5$

In Exercises 23–42, find the derivative of the given function by using the chain rule.

23. $y = f(u) = u^6, u = g(x) = x^5 + 4$
24. $y = f(u) = u^3, u = g(x) = 2x^3 - 5x$
25. $y = f(u) = \sqrt{u}, u = g(x) = 4x^2 - 5$
26. $y = f(u) = 4u^5, u = g(x) = \sqrt{3x^2 + 5x}$
27. $y = f(u) = \sqrt[3]{u^2}, u = g(x) = 7x^3 - 9x$
28. $y = f(u) = (u + 1)^2, u = g(x) = \dfrac{2}{x}$
29. $y = f(u) = (u^2 + 1)^3, u = g(x) = \dfrac{1}{x + 3}$
30. $y = f(u) = u^3 - 4u, u = x^4 + 5$
31. $y = f(u) = \sqrt[3]{u^2 - 2u}, u = x^3 + 4$
32. $y = f(u) = 2u^3 - 8u, u = 6x^2 - 5x - 1$
33. $y = g(x) = 4x^4 - 3x, x = 3t^2 - 4t$
34. $y = f(u) = 5u^3 - 7u^2 - 5u + 1, u = 6t^2 - 8t$
35. $y = (9x^2 + 4x)^6$
36. $y = 3(2x^2 - 5x + 1)^{5/3}$
37. $y = (11x^5 - 2x + 1)^{10}$
38. $y = \dfrac{8}{(x^3 - 5x + 2)^4}$
39. $y = \dfrac{7}{(9x^2 - 4)^8}$
40. $y = \left(\sqrt[3]{2x^5 - x^3}\right)^4$
41. $y = \sqrt[4]{(2x^2 - 5)^3}$
42. $y = \dfrac{1}{\sqrt[5]{(7x - 4x^3)^2}}$

Solve Exercises 43–56.

43. Find $\dfrac{dy}{dx}$, if $y = u^4$, $u = 2v^3 - 1$, and $v = \dfrac{4}{x^2}$.
44. Find $\dfrac{dy}{dt}$, if $y = 4u^2 - u$, $u = x^3 - 8$, and $x = 6t + 4$.
45. Find $\dfrac{dy}{dx}$, if $y = 3u^2$, $u = \dfrac{4}{v}$, and $v = x^5$.
46. Find an equation of the tangent line to y at $(4, 1)$, if $y = \sqrt[3]{(x - 5)^2}$.
47. Find an equation of the tangent line to y at $(2, -8)$, if $y = (4x^2 - 18)^3$.
48. Find an equation for the normal line to y at $(-3, 1)$, if $y = \dfrac{4}{\sqrt{25 - x^2}}$.
49. *Energy technology* The energy output of an electrical heater varies with time t according to the equation $E = 6\left(1 + 4t^2\right)^3$.
   (a) Find the power, P, in watts, generated by the heater if $P = E' = \dfrac{dE}{dt}$.
   (b) The answer in (a) is in watts. Rewrite your answer so that it is in kilowatts.
   (c) Find the power, in kilowatts, at $t = 2.5$ s
50. *Demography* It is estimated that t years from now the population of a certain city will be $P(t) = 10(45 + 3.5t)^2 - 1750t$.
   (a) What is the rate of change in the population?
   (b) How fast will the population be growing in 5 years?
51. *Environmental science* After a sewage spill, the level of pollution in San Juan Pedro Bay is estimated by $P(t) = \dfrac{250t^2}{\sqrt{t^2 + 15}}$ parts per million (ppm), where t is the time in days since the spill occurred.
   (a) What is the rate of change in the level of pollution?
   (b) What is the rate of change in the level of pollution after 5.0 days?
52. *Environmental science* It has been estimated that t years from now the level of pollution in the air will be $P(t) = \dfrac{0.6\sqrt{8t^2 + 11t + 60}}{(t + 1)^2}$ ppm.
   (a) What is the rate of change in the level of pollution?
   (b) Find the rate of change in the level of pollution in 10.0 years?
53. *Optics* When light passes from air into a transparent medium, the ratio of reflected light to incident light is given approximately by $R = \left(\dfrac{n - 1}{n + 1}\right)^2$, where n is the index of reflection for the medium. Find the derivative of R with respect to n.

54. *Electronics* The charge on a capacitor over a short interval of time, t, in seconds, is given by $q = 5t^2 + \sqrt{2-5t}$. The formula for the current is the derivative of the formula for the charge, that is, $I = q'$.
(a) What is the formula for the current?
(b) Find the value of I when $t = 0.25$ s?

55. *Electronics* In an RC circuit the current, I, is given by $I = \dfrac{V}{\sqrt{R^2 + {X_C}^2}}$, where V is the voltage in volts, R is the resistance in ohms, and X_C is the capacitative resistance in ohms. For this exercise, assume that $V = 120$ V and $R = 30\,\Omega$.
(a) What is the rate of change in the current with respect to the capacitative resistance?
(b) Find the value of I' when $X_C = 25\,\Omega$?

56. *Environmental technology* Some studies have shown that the average level of certain pollutants in the air is given by $L = 1 + 0.25x + 0.001x^2$ ppm when the population is x thousand people. It is estimated that t years from now the population will be $x = \dfrac{200}{\sqrt{9-0.4t}}$, in thousands of people.
(a) Find the rate of change in the level of pollutants with respect to the number of years from now.
(b) Find the rate of change in the level of pollutants when $t = 5$ years.
(c) Find the rate of change in the level of pollutants when $t = 10$ years.

In Your Words

57. Explain how to use the chain rule to differentiate the composition of two functions.

58. The general power rule for differentiation is a special case of the chain rule. Describe a function where you could use the chain rule but not the general power rule to find its derivative.

2.5 IMPLICIT DIFFERENTIATION

So far, we have considered only functions in the explicit form $y = f(x)$. If the functional relationship between the independent variable x and the dependent variable y is not of this form, they we say that x and y are related implicitly. Examples of x and y being implicitly related are

$$4x - 8y + 6 = 0 \qquad 4x^2 + 9y^2 - 36 = 0 \qquad y \le 0 \qquad xy + 9 = 0$$

Sometimes it is possible to form an explicit function from an expression where x and y are related implicitly. When this can be done, the relationship is solved for y, and a function $y = f(x)$ results. For example, the implicit equation $4x - 8y + 6 = 0$ results in the explicit function $y = \frac{1}{2}x + \frac{3}{4}$. In this section, we will learn how to take derivatives of implicit functions. The result of differentiating an implicit function is called an **implicit derivative** and the process is called **implicit differentiation.**

For many equations it is very difficult, or even impossible, to solve for y in terms of x. In these cases, implicit differentiation is a much easier method of finding the derivative. We will begin by taking the derivative in both the explicit and the implicit form.

We will begin with the implicit function $4x - 8y + 6 = 0$, which we write in the explicit form $y = \frac{1}{2}x + \frac{3}{4}$. The derivative of the explicit form is $y' = \frac{1}{2}$.

In taking an implicit derivative, it is not necessary to solve for y in terms of x. Instead, we will use the general power rule or the chain rule. One important realization is the different way to take the derivative of x^2 and of y^2. At present we know that $\frac{dx^2}{dx} = 2x$. But $\frac{dy^2}{dx} = 2y\frac{dy}{dx}$ because we are differentiating with respect to x. This is an application of the general power rule with $g(x)$ replaced by y. In general, if $y = f(x)$, then

$$\begin{aligned}\frac{d}{dx}[f(x)]^n &= n[f(x)]^{n-1}\frac{d}{dx}[f(x)]\\ &= ny^{n-1}\frac{dy}{dx}\\ &= ny^{n-1}y'\end{aligned}$$

Let's return to our example, $4x - 8y + 6 = 0$. We will differentiate each term with respect to x and get

$$\begin{aligned}\frac{d}{dx}(4x) - \frac{d}{dx}(8y) + \frac{d}{dx}(6) &= \frac{d}{dx}(0)\\ 4 - 8\frac{dy}{dx} + 0 &= 0\end{aligned}$$

Solving this for $\frac{dy}{dx}$ we get

$$\frac{dy}{dx} = y' = \frac{-4}{-8} = \frac{1}{2}$$

This is the same answer we got before. Notice that we used the chain rule to differentiate $8y$. Thus $\frac{d}{dx}(8y) = \frac{d}{dy}(8y)\frac{dy}{dx} = 8\frac{dy}{dx}$.

In general, use the following four steps in order to differentiate implicitly.

Four Steps for Implicit Differentiation

In the list below, we assume that x is the independent variable and that we are trying to determine dy/dx.

1. Differentiate both sides of the equation with respect to the independent variable, x.
2. Collect the terms with dy/dx on one side of the equation and place the remaining terms on the other side of the equals sign.
3. Factor out dy/dx.
4. Solve for dy/dx.

In the next example, we will again find the derivative using both the explicit and implicit methods.

EXAMPLE 2.29

Find $\dfrac{dy}{dx}$ of $4x^2+9y^2-36=0$.

Solution

Explicit form

This is the equation of an ellipse. If we solve this for y in terms of x, we get

$$y=\pm\sqrt{\frac{36-4x^2}{9}}=\pm\frac{1}{3}\sqrt{36-4x^2}$$

This can be written as two expressions:

$$y=\frac{1}{3}\sqrt{36-4x^2}\text{ and }y=-\frac{1}{3}\sqrt{36-4x^2}$$

Each separate expression is now a function. As you can see in Figure 2.6, $y=\frac{1}{3}\sqrt{36-4x^2}$ is the upper half of the ellipse and $y=-\frac{1}{3}\sqrt{36-4x^2}$ is the lower half.

If you want to find the slope of the tangent at a given point, you need to first determine which half of the ellipse the point is on and then differentiate the appropriate equation. We will differentiate the equation for the upper half:

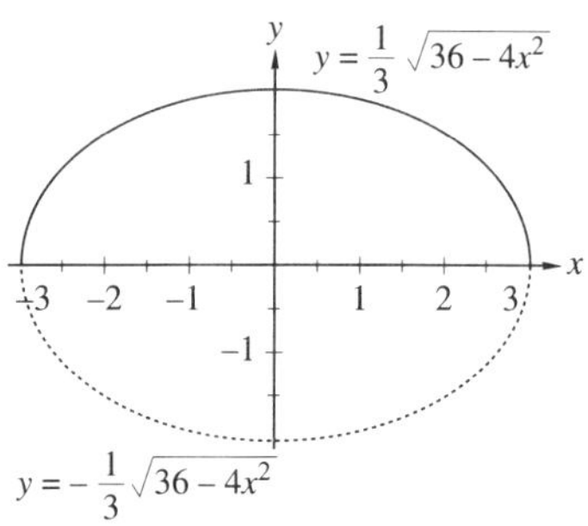

FIGURE 2.6

$$\begin{aligned}y&=\frac{1}{3}\sqrt{36-4x^2}=\frac{1}{3}(36-4x^2)^{1/2}\\ y'&=\frac{1}{6}(36-4x^2)^{-1/2}(-8x)\\ &=\frac{-4x}{3\sqrt{36-4x^2}}\end{aligned}$$

Let's compare this process to the implicit method.

Implicit form

The original equation is $4x^2+9y^2-36=0$. We will differentiate both sides with respect to x.

$$\begin{aligned}\frac{d}{dx}(4x^2+9y^2-36)&=\frac{d}{dx}(0)\\ 8x+18y\frac{dy}{dx}&=0\\ 8x+18yy'&=0\end{aligned}$$

Solving for y', we get

$$y'=\frac{-8x}{18y}=\frac{-4x}{9y}$$

This is certainly a much simpler equation. But, do we get the same results?

The easiest way to find out is to substitute one of the other equations for y. You will see that when $y=\frac{1}{3}\sqrt{36-4x^2}$ is substituted, you get

$$y=\frac{-4x}{3\sqrt{36-4x^2}}$$

It works! The results are not only the same, but by using implicit differentiation, we got an easier result to use.

EXAMPLE 2.30

Use implicit differentiation to find the derivative of $9x^2 - 4y^2 = 2x$.

Solution We differentiate both sides of the equation with respect to x.

$$\begin{aligned} \frac{d}{dx}(9x^2 - 4y^2) &= \frac{d}{dx}(2x) \\ 18x - 8y\frac{dy}{dx} &= 2 \\ \frac{dy}{dx} &= \frac{2 - 18x}{-8y} \\ \frac{dy}{dx} &= \frac{18x - 2}{8y} = \frac{9x - 1}{4y}, \text{ if } y \neq 0 \end{aligned}$$

EXAMPLE 2.31

Find $\frac{dy}{dx}$, if $xy^2 - x^2y + 4x - 5y = 16$.

Solution

$$\begin{aligned} \frac{d}{dx}(xy^2 - x^2y + 4x - 5y) &= \frac{d}{dx}(16) \\ \frac{d}{dx}(xy^2) - \frac{d}{dx}(x^2y) + \frac{d}{dx}(4x) - \frac{d}{dx}(5y) &= \frac{d}{dx}(16) \end{aligned}$$

We must use the product rule on the first two terms.

$$x\frac{d}{dx}(y^2) + y^2\frac{d}{dx}(x) - x^2\frac{d}{dx}y - y\frac{d}{dx}(x^2) + \frac{d}{dx}(4x) - \frac{d}{dx}(5y) = \frac{d}{dx}(16)$$

$$x\left(2y\frac{dy}{dx}\right) + y^2(1) - x^2\frac{dy}{dx} - y(2x) + 4 - 5\frac{dy}{dx} = 0$$

If $\frac{dy}{dx} = y'$, then we get

$$\begin{aligned} 2xyy' + y^2 - x^2y' - 2xy + 4 - 5y' &= 0 \\ (2xy - x^2 - 5)y' &= 2xy - y^2 - 4 \\ y' &= \frac{2xy - y^2 - 4}{2xy - x^2 - 5} \end{aligned}$$

if the denominator is not 0.

EXAMPLE 2.32

Find y', if $\sqrt[3]{x^2 - y} = x$.

Solution

$$\begin{aligned} \frac{d}{dx}(x^2 - y)^{1/3} &= \frac{d}{dx}(x) \\ \frac{1}{3}(x^2 - y)^{-2/3}\left[\frac{d}{dx}(x^2 - y)\right] &= \frac{d}{dx}(x) \end{aligned}$$

EXAMPLE 2.32 (Cont.)

$$\frac{1}{3}(x^2-y)^{-2/3}\left[2x-\frac{dy}{dx}\right]=1$$

And if $(x^2-y)^{-2/3}\neq 0$,

$$\begin{aligned} 2x-y' &= \frac{1}{\frac{1}{3}(x^2-y)^{-2/3}} \\ &= 3(x^2-y)^{2/3} \\ y' &= 2x-3(x^2-y)^{2/3} \end{aligned}$$

EXAMPLE 2.33

Find the slope of the tangent line to the graph of $x^3+3xy+12y^2=10$ at the point $(-2,-1)$.

Solution We first find the derivative, y'.

$$\begin{aligned} \frac{d}{dx}(x^3+3xy+12y^2) &= \frac{d}{dx}(10) \\ 3x^2+\frac{d}{dx}(3xy)+12\frac{d}{dx}y^2 &= 0 \\ 3x^2+(3xy'+3y)+24yy' &= 0 \\ (3x+24y)y' &= -3x^2-3y \\ y' &= \frac{-3x^2-3y}{3x+24y} \\ &= -\frac{x^2+y}{x+8y} \end{aligned}$$

Now, at the point $(-2,-1)$, we see that

$$y'=-\frac{(-2)^2+(-1)}{-2+8(-1)}=-\frac{4-1}{-2-8}=\frac{3}{10}$$

Thus, the slope of the tangent line to the graph of $x^3+3xy+12y^2=10$ at the point $(-2,-1)$ is $\frac{3}{10}$.

In Section 2.3, we learned how to use derivatives to calculate the equation of tangent and normal lines to a curve. When light enters a lens, the light forms an angle with the line perpendicular to the surface of the lens at the light's entry point.

The perpendicular line is **normal** to the surface of the lens. The next example shows how to use implicit differentiation to determine the equation of this normal line.

Application

EXAMPLE 2.34

A profile of a lens is shown in Figure 2.7a. This particular lens is described by the quadratic equation $x^2-xy+y^2=13$. Find the equation of the normal line to this lens at the point $(-1,3)$.

EXAMPLE 2.34 (Cont.)

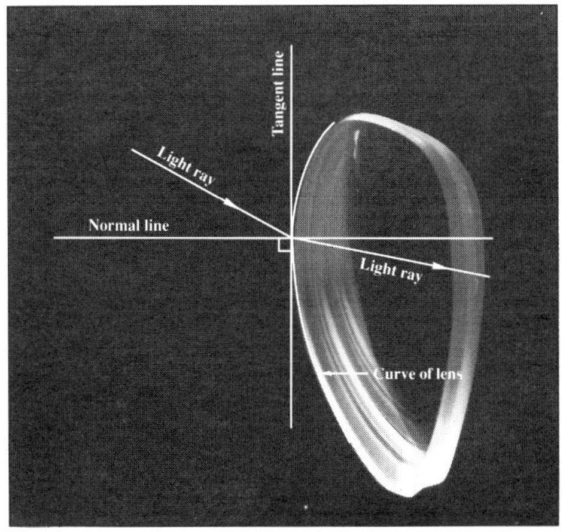

Courtesy of Ruby Gold

FIGURE 2.7a

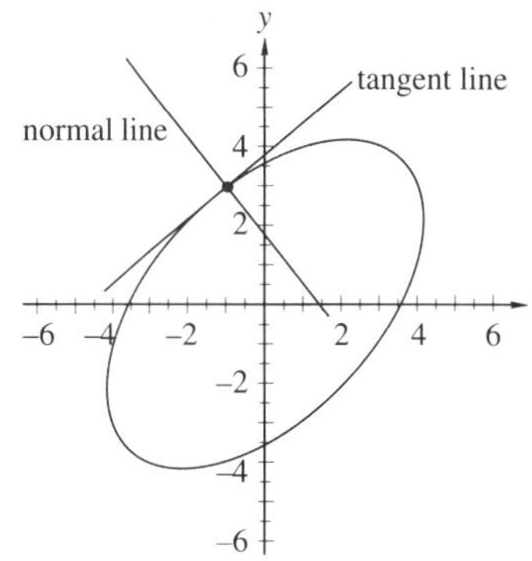

FIGURE 2.7b

Solution The sketch in Figure 2.7a shows a lens, the tangent and normal lines, where the normal line is the path of a "typical" light ray. In Figure 2.7b, we have sketched the lens on a coordinate system and shown the tangent and normal lines at the point $(-1, 3)$. We begin by implicitly differentiating the equation that describes the lens.

$$\frac{d}{dx}(x^2 - xy + y^2) = \frac{d}{dx}(13)$$
$$\frac{d}{dx}(x^2) - \frac{d}{dx}(xy) + \frac{d}{dx}y^2 = \frac{d}{dx}(13)$$

Using the product rule on the middle term, we obtain

$$\frac{d}{dx}(x^2) - \left[x\frac{d}{dx}(y) + y\frac{d}{dx}(x)\right] + \frac{d}{dx}y^2 = \frac{d}{dx}(13)$$
$$2x - x\frac{dy}{dx} - y + 2y\frac{dy}{dx} = 0$$

Letting $\frac{dy}{dx} = y'$, produces

$$2x - xy' - y + 2yy' = 0$$

or

$$(2y - x)y' = y - 2x$$

and so,

$$y' = \frac{y - 2x}{2y - x}$$

Now, at the point $(-1, 3)$, we see that

$$y' = \frac{3 - 2(-1)}{2(3) - (-1)} = \frac{3+2}{6+1} = \frac{5}{7}$$

EXAMPLE 2.34 (Cont.)

Thus, the slope of the tangent line to the lens at $(-1,3)$ is $\frac{5}{7}$, so the slope of the normal line at this point is $-\frac{7}{5}$. Using the point-slope form for the line, we see that the equation of the normal line is

$$y-3=-\frac{7}{5}(x+1)$$

$$\text{or} \quad 7x+5y=8$$

Exercise Set 2.5

In Exercises 1–34, use implicit differentiation to find $\frac{dy}{dx}$.

1. $4x+5y=0$
2. $6x-7y=3$
3. $x-y^2=4$
4. $x^2+y^2=9$
5. $x^2-y^2=16$
6. $x^2-2y^2=2x$
7. $9x^2+16y^2=144$
8. $16x^2-9y^2=144$
9. $4x^2-y^2=4y$
10. $x^2y+xy^2=x$
11. $xy+5xy^2=x^2$
12. $x^2y-3xy^2=y^3$
13. $x^2+4xy+y^2=y$
14. $x^3y-3xy^2=2x$
15. $x+5x^2-10y^2=3y$
16. $x^2-2xy+y^2=x$
17. $4x^2+y^3=9$
18. $y^4-9x^2=9$
19. $6y^3+x^3=xy$
20. $x^2y=7y$
21. $x^2y=x+1$
22. $x^2+\frac{1}{y}=2x$
23. $\frac{1}{x}+\frac{1}{y}=16$
24. $\frac{1}{x^2}-\frac{1}{y^2}=xy$
25. $\frac{3}{x+1}+x^2y=5$
26. $y^2=\frac{x}{y+1}$
27. $xy+\frac{y}{x}=x$
28. $x^3-8x^2y^2+y=9x$
29. $x^4-6x^2y^2+y^2=10x$
30. $(x+2y)^2=4x$
31. $\sqrt{x^2+y^2}=\frac{2y}{x}$
32. $\sqrt{x+xy}=x^2y^2$
33. $(x^2+y^2)^3=y$
34. $(x^2y+4)^2=3x$

Solve Exercises 35–44.

35. Find the equation of the line tangent to the circle $x^2+y^2=25$ at the point $(3,-4)$.
36. Find the equation of the line normal to the circle $x^2+y^2-2x-4y=20$ at the point $(-2,6)$.
37. Find the equation of the line tangent to the ellipse $9x^2+16y^2=100$ at the point $(-2,2)$.
38. Find the slopes of the lines tangent and normal to the parabola $y^2=-16x$ at the point $(-1,4)$.
39. Find the slopes of the lines tangent and normal to the hyperbola $12x^2-16y^2=192$ at the point $(-8,-6)$.
40. *Physics* The position of a particle at time t is described by the relation $s^3-4st+2t^3-5t=0$. Find the velocity, ds/dt, of the particle.
41. *Automotive technology* An automobile's position, s, in miles, on a track at time t, in minutes, is described by the relation $2s^2+\sqrt{st}-3t=0$.
    (a) Find $\frac{ds}{dt}$, the velocity of the car.
    (b) What is the position of the car when $t=2$?
    (c) What is the velocity of the car when $t=2$?
42. *Economics* The **Cobb-Douglas production formula**, $P=Cx^ay^{1-a}$, is frequently used by economists to relate the cost of a production process, P, to labor and capital, where C is a constant. Suppose that a firm's level of production is given by $P=20x^{1/5}y^{4/5}$, where x represents the units of labor and y the units of capital. Currently, the company is using 32 units of labor and 243 units of capital. If labor is increasing by 25% per month, what change in units of capital are needed to maintain the same level of production?

43. *Economics* The manager of an electronics store has determined that the number of digital satellites and the number of television sets sold weekly are related by the equation $0.7y^2 = 12x + xy$, where x is the number of digital satellites and y is the number of television sets.
(a) Find $\frac{dy}{dx}$ when $x = 10$ and $y = 25$.
(b) Interpret your answer.

44. *Business* The number of pairs of trousers, x, and the number of shirts, y, sold at a clothing store are related by the equation $48x = 12y + 0.01x^2y$.

(a) Find $\frac{dy}{dx}$ when $x = 10$ and $y = 40$.

(b) Interpret your answer.

In Your Words

45. Describe how to differentiate an implicit function.

46. Explain how an implicit function differs from an explicit function.

2.6 HIGHER ORDER DERIVATIVES

The derivative of a function, such as $f(x)$, is also a function, $f'(x)$, called the **derivative function** f'. At present, the only application of the derivative that we have examined is that of the slope of the curve or of a tangent to the curve. In Chapter 3, we will look at more applications of the derivative. As we work these applications there will be times when we need to take the derivative of a derivative.

When it exists, the derivative of the derivative f' is called the **second derivative** and is denoted by f''. It is possible to continue this process: The derivative of the second derivative is the **third derivative**, f''', and its derivative is the **fourth derivative**, $f^{(4)}$. The process can be continued indefinitely, as long as the resulting function has a derivative.

In general, if n is a positive integer, then $f^{(n)}$ represents the nth derivative of f and is found by starting with f and differentiating successively, n times. The integer n is called the **order** of the derivative $f^{(n)}$ and, as a group, these are known as **higher order derivatives**.

As with the derivative, there are many different notations. If $y = f(x)$, then

$$y'' = f''(x) = D_x^2 y = \frac{d^2y}{dx^2} = \frac{d^2}{dx^2}f(x)$$

all represent second derivatives,

$$y''' = f'''(x) = D_x^3 y = \frac{d^3y}{dx^3} = \frac{d^3}{dx^3}f(x)$$

all represent third derivatives,

$$y^{(4)} = f^{(4)}(x) = D_x^4 y = \frac{d^4y}{dx^4} = \frac{d^4}{dx^4}f(x)$$

all represent fourth derivatives, and in general,

$$y^{(n)} = f^{(n)}(x) = D_x^n y = \frac{d^ny}{dx^n} = \frac{d^n}{dx^n}f(x)$$

all represent the nth derivative.

EXAMPLE 2.35

Find the first four derivatives of $f(x) = 2x^5 - 3x$.

Solution

$$\begin{aligned} f'(x) &= 10x^4 - 3 \\ f''(x) &= 40x^3 \\ f'''(x) &= 120x^2 \\ f^{(4)}(x) &= 240x \end{aligned}$$

EXAMPLE 2.36

Find the first three derivatives of $f(x) = 4x^3 + \dfrac{2}{x}$.

Solution We will first rewrite $f(x)$ as $4x^3 + 2x^{-1}$.

$$\begin{aligned} f'(x) &= 12x^2 - 2x^{-2} \\ f''(x) &= 24x + 4x^{-3} \\ f'''(x) &= 24 - 12x^{-4} \end{aligned}$$

EXAMPLE 2.37

Use implicit differentiation to find y' and y'' of $x^2 + 2xy - y^2 = 25$.

Solution Using implicit differentiation, we get

$$\begin{aligned} \frac{d}{dx}(x^2 + 2xy - y^2) &= \frac{d}{dx}(25) \\ \frac{d}{dx}(x^2) + \frac{d}{dx}(2xy) - \frac{d}{dx}(y^2) &= \frac{d}{dx}(25) \\ 2x + 2y + 2xy' - 2yy' &= 0 \\ y' = -\frac{2x+2y}{2x-2y} &= \frac{x+y}{y-x}, \quad \text{if } y \neq x \end{aligned}$$

To find y'', we will use the quotient rule:

$$\begin{aligned} y'' &= \frac{d}{dx}\left(\frac{x+y}{y-x}\right) \\ &= \frac{(y-x)\dfrac{d}{dx}(x+y) - (x+y)\dfrac{d}{dx}(y-x)}{(y-x)^2} \\ &= \frac{(y-x)(1+y') - (x+y)(y'-1)}{(y-x)^2} \\ &= \frac{2y - 2xy'}{(y-x)^2} \end{aligned}$$

EXAMPLE 2.37 (Cont.)

Substituting $\frac{x+y}{y-x}$ for y', we get

$$\begin{aligned} y'' &= \frac{2y - 2x\left(\frac{x+y}{y-x}\right)}{(y-x)^2} \\ &= \frac{\frac{2y(y-x) - 2x(x+y)}{y-x}}{(y-x)^2} \\ &= \frac{2y^2 - 2yx - 2x^2 - 2xy}{(y-x)^3} \\ &= \frac{2y^2 - 4xy - 2x^2}{(y-x)^3} \\ &= \frac{-2(x^2 + 2xy - y^2)}{(y-x)^3} \\ &= \frac{-2(25)}{(y-x)^3} = \frac{-50}{(y-x)^3} \end{aligned}$$

This concludes our introduction to derivatives. In Chapter 3, we will look at many of the applications of derivatives.

Exercise Set 2.6

In Exercises 1–19, find the indicated higher derivatives of the given function.

1. $y = 4x^3 - 6x^2 + 3x - 10$; find y''.
2. $y = 9x^4 + x^3 - 10$; find y'''.
3. $y = 7x^5 - 3x^3 + x$; find $y^{(4)}$.
4. $f(x) = 10x - 1$; find $f''(x)$.
5. $f(x) = x + \frac{1}{x}$; find $f''(x)$.
6. $f(t) = t^3 - \frac{1}{t^2}$; find $f'''(t)$.
7. $f(t) = t^2 + \frac{1}{t+1}$; find $f''(t)$.
8. $h(x) = (x^2 + 2)(x - 1)$; find $h''(x)$.
9. $g(u) = \frac{u-1}{u^2}$; find $g'''(u)$.
10. $y = 3\sqrt[3]{x^2+1}$; find $\frac{d^2y}{dx^2}$.
11. $y = \sqrt{x+1}$; find $\frac{d^2y}{dx^2}$.
12. $y = (4x - x^2)^2$; find $\frac{d^2y}{dx^2}$.
13. $y = (2x + 1)^3$; find $\frac{d^3y}{dx^3}$.
14. $y = x^4 - 3x^2 + 2$; find $\frac{d^3y}{dx^3}$.
15. $y = \frac{1}{\sqrt[3]{x^2}}$; find $\frac{d^2y}{dx^2}$.
16. $y = \frac{x^3}{\sqrt[3]{x}}$; find $\frac{d^3y}{dx^3}$.
17. $f(x) = \frac{x}{x+1}$; find $D_x^4 f(x)$.
18. $f(t) = \frac{t+1}{t+2}$; find $D_t{}^3 f(t)$.
19. $f(w) = (3w^2 + 1)^{-2}$; find $D_w^2 f(w)$.

In Exercises 20–27, find the indicated derivatives.

20. $\frac{d^5}{dx^5}(x^4 - 3x^2 + 1)$

21. $\frac{d^4}{dx^4}(x^7 - 4x^5 - x^3)$

22. $\frac{d^8}{dx^8}(x^8 - 3x^2 + 2)$

23. $\frac{d^3}{dx^3}\left(x - \frac{1}{x^2}\right)$

24. $\frac{d^2}{dx^2}\left(\frac{x^2}{2x+3}\right)$

25. $\frac{d^2}{dx^2}\left(\frac{x^3}{3x-1}\right)$

26. $\frac{d^2}{dx^2}\left[(x^2+2x)(x-1)\right]$

27. $\frac{d^2}{dx^2}\left(\frac{x}{\sqrt{x+1}}\right)$

In Exercises 28–34, find y' and y'' in terms of x and y.

28. $x^2 + y^2 = 9$

29. $x^2 - 4y^2 = x$

30. $x^2 - y^2 = 16$

31. $xy^2 + 4x = y$

32. $xy^2 - 5x = 4y$

33. $x^2y = 4$

34. $xy + y^2 = 2x^2$

Solve Exercises 35–38.

35. *Police science* Due to the rapid increase in major crimes, the mayor of a large city plans to organize a major crime task force. It is estimated that for each 1,000 people in the city, the number of major crimes will be $N(t) = 62 + 5t^2 - 0.08t^{5/2}$, where t is the number of months after the task force has been organized and $0 \le t \le 12$. **(a)** Find $N'(t)$. **(b)** Find $N''(t)$.

36. *Meteorology* Meteorological results for a certain city indicate that for the month of April, the daily temperature in °F between midnight and 6:00 P.M. can be approximated by $T(t) = -0.04t^3 + 1.16t^2 - 9.3t + 56$, where t is the number of hours after midnight and $0 \le t \le 18$. **(a)** Find $T'(t)$. **(b)** Find $T''(t)$.

37. *Meteorology* A weather balloon is released from the ground and rises in the atmosphere. The balloon's height above the ground, in meters, from the moment it is released is given by $h(t) = 120t - 12t^2$, where the time, t, varies from 0 to 6 min. **(a)** Find $h'(t)$. **(b)** Find $h''(t)$.

38. *Automotive* A car accelerates from rest to a maximum velocity at $t = 45$ s and then decelerates to a stop at $t = 120$ s. The distance traveled in meters is given by $s(t) = 3t^2 - \frac{1}{60}t^3$. **(a)** Find $s'(t)$. **(b)** Find $s''(t)$. (Note: $s'(t)$ is the velocity and $s''(t)$ is the acceleration.)

In Your Words

39. If the derivative of a function describes the rate of change of the function, what information does the second derivative give?

40. Explain the difference in the meanings of the notations $f^{(4)}(x)$ and $f^4(x)$.

CHAPTER 2 REVIEW

Important Terms and Concepts

- Chain rule
- Derivative
  - At a point
  - Four-step method
  - Of a curve
  - Of polynomials
- Higher order derivatives
- Implicit differentiation
- Normal line
- Power rule
- Product rule
- Quotient rule
- Rules for derivatives
  - Chain rule
  - Constant function
  - Of a composite function

Linear function
Of a sum or difference
Power rule
Product rule
Quotient rule
Second derivative
Third derivative

Review Exercises

In Exercises 1–6, use the four-step method to differentiate the given function.

1. $f(x) = 2x^2 - x$
2. $g(x) = x^3 - 1$
3. $f(x) = x - 2x^3$
4. $k(x) = \dfrac{4}{x^2}$
5. $m(x) = \sqrt{x}$
6. $j(t) = \sqrt{t+4}$

In Exercises 7–32, find the first derivative of the given function.

7. $f(x) = 3x^2 + 2x - 1$
8. $g(x) = 4x^{-5} + 2x$
9. $H(x) = 3x^5 + \dfrac{4}{x^3} + 3\sqrt{x}$
10. $y = \dfrac{1}{x^2+1}$
11. $y = \dfrac{3x}{x+3}$
12. $y = (2x+1)(x^2+3)$
13. $y = (x-4)(x^3-3x)$
14. $F(x) = \dfrac{\sqrt{x+1}}{x}$
15. $g(x) = \dfrac{x+1}{\sqrt{x+1}}$
16. $G(x) = \dfrac{x^2}{\sqrt{x+1}}$
17. $h(x) = (x^2+4)(x-2)(x^3-2)$
18. $f(x) = (3x^2-4x)$
$(5x^3+2x-1)$
19. $F(x) = \sqrt{(x^3+7)(x+5)}$
20. $g(x) = (x+5)^3$
21. $y = (x^2+5x)^2(x^3+1)^{-2}$
22. $h(x) = (x^2+7x-1)^4$
23. $y = (x+1)^2(x-1)^3$
24. $y = (x+4)^3\sqrt{x-1}$
25. $y = \dfrac{5}{(x^2+1)^3}$
26. $y = \dfrac{4x}{(3x+4)^2}$
27. $y = \dfrac{x^2}{\sqrt{(x^2-2)^3}}$
28. $f(t) = \dfrac{4t^3-3}{t^2-2}$
29. $g(u) = (u^2 + \dfrac{1}{u} - \dfrac{4}{u^2})^{-4}$
30. $R(t) = \sqrt{t^2-2t+1}\sqrt[3]{t^2+1}$
31. If $y = u^3 - u$, $u = x^2+1$, find $\dfrac{dy}{dx}$.
32. If $y = (u+4)^3$, $u = 2t-5$, find $\dfrac{dy}{dt}$.

In Exercises 33–38, find the implicit derivatives.

33. $x^2 + 4xy + 4y^2 = 16$
34. $x^3 - 2xy = y^2$
35. $2xy - x^2 = y^2x$
36. $x^2y + yx^3 = 10$
37. $x^2y^3 + \dfrac{1}{y} = x$
38. $4x^2 + xy^2 + y^{-1} = 2x$

In Exercises 39–49, find the indicated derivative.

39. $f(x) = x^7 + 2x^4 + 3x$; find $f''(x)$.
40. $h(x) = \sqrt{x} - \dfrac{1}{\sqrt{x}}$; find $h'''(x)$.
41. $g(x) = 4x^3 - 5x + \dfrac{2}{x}$; find $g'''(x)$.
42. $y = 4x^5 - 2x^{-3}$; find $\dfrac{d^2y}{dx^2}$.
43. $y = 7x^3 + \dfrac{7}{x^3}$; find $\dfrac{d^3y}{dx^3}$.
44. $y = x^{-3} + 4x^{-1}$; find $\dfrac{d^3y}{dx^3}$.

45. $f(x) = 5x^3 + \sqrt{x}$; find $D_x^3 f(x)$.

46. $g(x) = (x^2+1)^3$; find $\frac{d^3}{dx^3} g(x)$.

47. $x^2 y = y + 2x$; find y''.

48. $y^2 = xy^2 + x^3$; find y'.

49. $xy = x^2 + y^2$; find y' and y''.

Solve Exercises 50–60.

50. Find the slopes of the tangent and the normal to the curve $y = 4x^2 - 9x$ at $(3, 9)$.

51. Find the slope of the tangent to the curve $f(x) = 7x^3 - 3x + 6$ at $(-1, 2)$

52. Find the slope of the tangent to the ellipse $4x^2 + 9y^2 = 40$ at $(-1, 2)$

53. Find the slopes of the tangent and the normal to the curve $4x^2 + 3y^2 - 5xy - 4x + 10y - 11 = 0$ at $(2, -1)$.

54. If $f(x) = x^3 - 3x - 1$, find all values of x for which $f'(x) = 0$.

55. If $g(x) = \frac{x^2}{x+1}$, find all values of x for which $g'(x) = 0$.

56. Find the equation of the tangent to the curve $y = x^2 + 4$ at $(-1, 5)$.

57. Find the equation of the normal to the ellipse $4x^2 + 5y^2 = 36$ at $(-2, 2)$.

58. If $y = \sqrt{x+4}$, find all values of x for which $y' = 0$.

59. Find all values of x where $f''(x) = 0$ if $f(x) = 3x^4 - 6x^2 + 2$.

60. If $y = x^3 - 9x^2 - 4$, find the value(s) of x for which $y'' = 0$.

CHAPTER 2 TEST

In Exercises 1–8, find the first derivative of the given functions.

1. $f(x) = 5x^3 - 4$

2. $g(x) = 3x^{-4} + 5x^2$

3. $h(x) = \sqrt[3]{x^6 - 2}$

4. $j(x) = (2x+1)(3x^2 - x)$

5. $k(x) = \frac{2x+1}{\sqrt{4x^2+1}}$

Solve Exercises 6–10.

6. Find the second and third derivatives of $f(x) = 5x^4 - 4x^{-3}$.

7. Find y', if $2xy + y^3 x = x^3$.

8. Find the equation of the tangent to the curve $g(x) = 4x^3 - 2x^2 + 4$ at $(1, 6)$.

9. A particle moving in a straight line is at a distance of $s(t) = 4.5t^2 + 27t$ ft from its starting point after t s. If the velocity $v(t)$ of the particle at time, t, is $v(t) = s'(t)$, find the velocity of this particle.

10. The population of a certain bacterial culture after t hours is approximated by $P(t) = \frac{t^2 - 5t}{2\sqrt{t} + 7}$, where $P(t)$ is in thousands of bacteria. **(a)** Find an expression for the rate of growth of these bacteria. **(b)** Determine the rate of growth after 5 hours.

CHAPTER

3

Applications of Derivatives

A telephone company is asked to provide service to the house shown here. In Section 3.4, you will learn where to connect this line so that it will cost the least.

Courtesy of Michael A. Gallitelli, Metroland Photo Inc.

In Chapter 2, we learned how to take derivatives of many functions. In this chapter, we will look at some of the applications of derivatives. We will begin by looking at motion applications. We will then look at ways to determine maxima and minima of functions. Finally, we will learn two tests that use derivatives.

3.1 RATES OF CHANGE

One idea that is important to the study of mathematics is **change**. As a technician, one of your responsibilities may be to observe something and measure it at certain time intervals. When these observations are completed, it will be necessary to see if any changes resulted and how those changes occurred. One of the powerful aspects of calculus is that it can be used to determine the rate at which something changes.

Velocity, Speed, and Acceleration

Suppose the distance x of an object p from the origin at any time t is given by $s(t)$. It is often convenient to consider **rectilinear motion** where the object or particle moves along a straight line. Usually we use a horizontal axis to represent rectilinear motion and select the origin as the initial position of the particle. Thus, $x = s(t)$ represents the distance of the object p from the origin 0 at t seconds. (See Figure 3.1.) Because $x = s(t)$ gives the position of p at time t, it is often referred to as a **position function**.

FIGURE 3.1

In order to measure the velocity of p at some given time c, we look at the average velocity of p from c to $c+h$. Now, the average velocity is the distance traveled divided by the time it took to go that distance. In this case, the average velocity.

$$\bar{v} = \frac{s(c+h) - s(c)}{h}$$

As you can see, if we take the limit as h approaches 0, we will get the **instantaneous velocity** of p at $t = c$ or $v(c)$. Thus,

$$v(c) = \lim_{h \to 0} \frac{s(c+h) - s(c)}{h}$$

But, the expression on the right-hand side is the derivative of s and so

$$v(c) = s'(c)$$

Thus, the **velocity function** for p is the derivative of the position function s. In other words,

$$v(t) = s'(t) = D_t s = \frac{ds}{dt}$$

The units of s and t determine the units for the velocity. If s is in miles and t in hours, the velocity is in miles per hour (mph). If s is in meters or feet and t is in seconds, the velocity is in meters per second (m/s) or feet per second (ft/s).

We should probably point out that the velocity of an object is not the same as the speed of an object.

Note

Speed indicates the magnitude of the velocity and is defined as the absolute value of velocity, $|v(t)|$. Thus, speed is always nonnegative. Velocity can be positive, negative, or zero.

If the velocity is positive, the object is moving in a positive direction. If it is negative, the object is moving in the opposite direction. Thus, speed tells us how fast the object is moving while velocity tells us both the speed and the direction.

Now, if we differentiate the velocity, we get $v'(t) = s''(t)$. This measures the instantaneous rate of change of the velocity and is known as the **acceleration**. If we let $a(t)$ denote the acceleration, then

$$a(t) = v'(t) = s''(t) = \frac{d^2s}{dt^2}$$

The units for the acceleration are cm/s^2 (centimeters per second per second), m/s^2 (meters per second per second), mi/h^2 (miles per hour per hour), and so on.

EXAMPLE 3.1

The position function s of a particle p on a coordinate line is given by

$$s(t) = t^3 - 9t^2 + 24t - 10$$

where t is in seconds and $s(t)$ in millimeters. Where is the particle when $t = 3$ and what is its velocity and acceleration at that time?

Solution The position is given by

$$\begin{aligned} s(3) &= 3^3 - 9(3^2) + 24(3) - 10 \\ &= 27 - 81 + 72 - 10 \\ &= 8 \text{ mm or } 8 \text{ mm to the right of the origin} \end{aligned}$$

The first derivative of s, $s'(t)$, is the velocity function

$$v(t) = s'(t) = 3t^2 - 18t + 24$$

and when $t = 3$,

$$\begin{aligned} v(t) &= 3(3^2) - 18(3) + 24 \\ &= 27 - 54 + 24 = -3 \text{ mm/s} \end{aligned}$$

The object is moving with a speed of 3 mm/s in a negative direction. The acceleration function is the second derivative of the position function, and so

$$a(t) = s''(t) = 6t - 18$$

and when $t = 3$,

$$a(3) = 6(3) - 18 = 0 \text{ mm/s}^2$$

The following table shows the values for the position, velocity, and acceleration of the particle in Example 3.1, each second from $t = 0$ through $t = 6$.

t	0	1	2	3	4	5	6
s(t)	−10	6	10	8	6	10	26
v(t)	24	9	0	−3	0	9	24
a(t)	−18	−12	−6	0	6	12	18

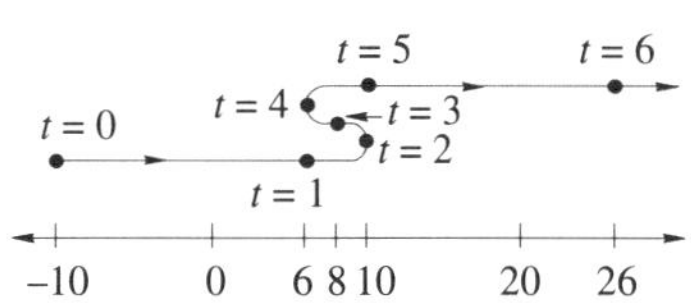

FIGURE 3.2

The motion of an object such as the particle in Example 3.1 can often be represented by a drawing like that shown in Figure 3.2. This curve does not show the path of the particle, but only shows how the particle moves. The direction of the motion is indicated by the arrows.

Not all motion is horizontal. The motion of a projectile fired straight upward is another example of rectilinear motion. In this instance, the axis is vertical rather than horizontal, the origin is at ground level, and the positive direction is upward.

Application

EXAMPLE 3.2

A projectile is fired straight upward with a velocity of 192 ft/s. Its distance above the ground at t seconds is given by $s(t) = -16t^2 + 192t$. Find the time when the projectile hits the ground. What is its velocity and acceleration at the time of impact?

Solution We are given the distance function as $s(t) = -16t^2 + 192t$. When $s(t) = 0$, the projectile will be at ground level, so we need to solve $s(t) = -16t^2 + 192t = 0$. We can factor t out of the expression to obtain

$$t(-16t + 192) = 0$$

and so $s(t) = 0$ when $t = 0$ or when $t = 12$. Since $t = 0$ when the projectile is fired, it must strike the ground when $t = 12$.

The velocity is represented by $v(t) = s'(t) = -32t + 192$ and the acceleration by $a(t) = s''(t) = v'(t) = -32$. When $t = 12$, $v(12) = -192$ and $a(t) = -32$. So, the velocity is 192 ft/s downward and the acceleration is 32 ft/s^2 downward.

Electricity

Time is not the only variable that can be used to study rates of change. In the next example, we will look at the rate at which the current changes with respect to the resistance.

Application

EXAMPLE 3.3

The current I in a certain electrical circuit is given by the formula $I = \dfrac{120}{R}$, where R represents the resistance. What is the rate of change of I with respect to R when the resistance is $60\,\Omega$?

Solution We are given $I = \dfrac{120}{R}$ and want to find $\dfrac{dI}{dR}$ when $R = 60$.

$$\begin{aligned} I &= \frac{120}{R} = 120R^{-1} \\ \frac{dI}{dR} &= -120R^{-2} \\ &= \frac{-120}{R^2} \end{aligned}$$

So, when $R = 60$, $\dfrac{dI}{dR} = \dfrac{-120}{60^2} = \frac{-1}{30}$. Thus, when $R = 60\,\Omega$, the current is decreasing at a rate of $\frac{1}{30}$ A/Ω (amperes per ohm).

Exercise Set 3.1

In Exercises 1–8, position functions of points moving rectilinearly are defined. Find the functions for the velocity and acceleration at time t. Make a table of values for each of the three functions at each second. (You might want to write a computer program to help you.) Illustrate the motion by means of a diagram like the one shown in Figure 3.2.

1. $s(t) = 3t^2 - 12t + 5$ from $t = 0$ to $t = 5$
2. $s(t) = 3t^2 - 18t + 10$ from $t = 0$ to $t = 5$
3. $s(t) = t^3 - 12t + 2$ from $t = -4$ to $t = 4$
4. $s(t) = t^3 - 27t + 20$ from $t = -4$ to $t = 4$
5. $s(t) = t + 4/t$ from $t = 1$ to $t = 4$
6. $s(t) = t + 8/t$ from $t = 1$ to $t = 4$
7. $s(t) = 2\sqrt{t} + \dfrac{1}{\sqrt{t}}$ from $t = 1$ to $t = 4$
8. $s(t) = t^3 - 8$ from $t = 0$ to $t = 4$

In Exercises 9–12, an object is fired straight upward. Its height is represented by the given formula. (a) When will it hit the ground? (b) What are its velocity and acceleration when it hits the ground? (c) What was its height when it was fired?

9. $s(t) = 144t - 16t^2$ ft
10. $s(t) = 250 + 256t - 16t^2$ ft
11. $s(t) = 29.4t - 4.9t^2$ m
12. $s(t) = 180 + 98t - 4.9t^2$ m

Solve Exercises 13–28.

13. *Electronics* The repulsion F, in dynes, between a certain pair of electrical charges varies with the distance apart, s, in centimeters.
    (a) If $F(s) = \dfrac{50}{s^2}$, what is the rate of change of F with respect to s?
    (b) What is the rate of change when $s = 3$ cm?
14. *Electronics* The current at a given point in a circuit i, in amperes (A), is defined as the instantaneous rate of change of the charge q, in coulombs (C), at any given time t, in seconds (s). The expression for the charge is given by
    $$q = 25 + 10t - 2.0t^2$$
    (a) What is the general expression for the current?
    (b) What is the initial charge; that is, what is the charge when $t = 0.0$ s?
    (c) What is the current at $t = 0.0$ s and $t = 4.0$ s?
15. *Physics* If the relationship between F, the temperature in degrees Fahrenheit (°F), and C, the temperature in degrees Celsius (°C), is given by $F = \frac{9}{5}C + 32$, find **(a)** $\dfrac{dF}{dC}$, **(b)** $\dfrac{dC}{dF}$, and **(c)** $\dfrac{dC}{dF}$ at $C = 20$°C.
16. *Thermodynamics* When metal is heated it expands. The area of a circle is related to its radius.
    (a) If a circular metal plate is heated, what is the rate of change of the area with respect to the radius?
    (b) What is the rate when $r = 1.5$ cm?
    (c) What is the rate when $r = 3$ cm?
17. *Hydrology* As water leaks out of a tank, the quantity Q of water in gallons remaining in the tank after t min is given by the formula $Q(t) = 900 - 50t + 0.5t^2$.
    (a) What is the rate of change in the quantity of water?
    (b) What is the rate of change when $t = 4$ min?
    (c) What is the rate of change when $t = 8$ min?
18. *Physics* An object rolls down a ramp. The distance in millimeters it rolls in t seconds is given by the formula $s(t) = 5t^2 + 2t$.
    (a) What is the formula for its velocity?
    (b) What is the formula for its acceleration?
    (c) What distance has it rolled after 3 s, and what are its velocity and acceleration at that time?
    (d) At what time is its velocity 46 mm/s?
19. *Physics* The position of a ball that is projected upward on a frictionless inclined plane is given by $s(t) = 8.0t - 2.0t^2$, where $s(t)$ is in meters (m) and t is time in seconds (s).
    (a) What is the initial velocity?

(b) How far will the ball move up the inclined plane before it stops and starts to roll back down the plane?
(c) How long will it take for the ball to roll up the plane and return to the initial position?

20. Find the rate of change of the area A of a circle with respect to its circumference C.

21. *Electronics* The induced voltage in a coil v, in volts (V), is given by $v = -N\dfrac{d\phi}{dt}$, where N is the number of turns in the coil, ϕ is the magnetic flux in webers (Wb), and t is time in seconds (s). If a coil of 50 turns is connected by a magnetic flux of $\phi = 2.4t^{1/2} - 0.4t^2$, find
(a) the general expression for the induced voltage and
(b) the induced voltage at $t = 1.0$ s and $t = 9.0$ s.

22. The height of a cone h is twice the radius r of the base of the cone.
(a) Find the instantaneous rate of change of the volume V with respect to the radius.
(b) What is $\dfrac{dV}{dr}$ when $r = 24.3$ mm?

23. *Hydrology* A stone dropped in a pond makes a circular ripple that travels out from the point of impact at 1.7 m/s. At what rate is the area of the circle increasing when $t = 8$ s?

24. *Electricity* The current i, in amperes (A), in a resistance of $r\,\Omega$ when the voltage V, in volts (V), is given by $i = \dfrac{V}{r}$. The resistance changes with time t in seconds (s) according to $r = 3.0t^{1/2}$ and the voltage changes according to $V = t^2 - 1.2$.
(a) Give an expression for i in terms of t.
(b) Find $\dfrac{di}{dt}$.
(c) Find $\dfrac{di}{dt}$ at $t = 2.4$ s.

25. *Physics* The pressure of a gas P in kilopascals (kPa) varies inversely with the volume V in cubic meters. If the pressure of the gas is $P = 5.21$ kPa when the volume is $1.23\ \text{m}^3$,
(a) state the equation for P, and
(b) find the instantaneous rate of change of the pressure with respect to the volume.

26. *Physics* The angular displacement of a rotating wheel θ in radian measure (rad) is given by $\theta(t) = 3.4t - 1.2t^2$, where t is time in seconds (s).
(a) Find the angular velocity, $\omega(t)$.
(b) Find the angular acceleration, $\alpha(t)$.
(c) What is the initial angular velocity and acceleration?

27. *Automotive* A car accelerates from rest to a maximum velocity at $t = 45$ s and then decelerates to a stop at $t = 120$ s. The distance traveled in meters is given by $s(t) = 3t^2 - \frac{1}{60}t^3$.
(a) What is the velocity as a function of time t of the car during this period?
(b) What is the acceleration as a function of time t of the car during this period?
(c) What is the speed of the car during this period?

28. *Meteorology* A weather balloon is released from the ground and rises in the atmosphere. The balloon's height above the ground, in meters, for the first six minutes after it is released is given by $h(t) = 120t - 12t^2$.
(a) What is the velocity of the balloon during this period?
(b) What is the acceleration of the balloon during this period?

In Your Words

29. Describe how the position, velocity, and acceleration of an object are related.

30. What is the relationship between the speed and the velocity of an object?

3.2 EXTREMA AND THE FIRST DERIVATIVE TEST

In Section 3.1, we looked at several applications of derivatives, particularly as they apply to the distance function. We were unable to determine when the distance, velocity, or acceleration is the greatest or least, or when the distance or speed stops increasing and begins decreasing. In this section, we will look at methods for using calculus to determine this type of information.

We will begin by examining the graph of a function f shown in Figure 3.3. This graph might have resulted when a recording instrument was used to measure something over a period of time. Several times have been marked because they are of interest. For example, the graph decreased in the time interval $[a, t_1]$ and increased in the interval $[t_1, t_2]$, decreased in $[t_2, t_3]$, and so on.

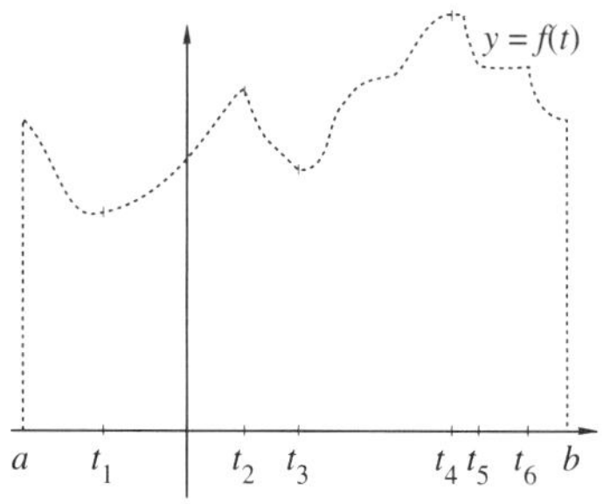

FIGURE 3.3

You can also see that, in the time period immediately surrounding t_1, the function is at a low point at t_1. Similarly, you can see that the function reaches a high point at t_2, another low point at t_3, and another high point at t_4. Because t_1 represents a place where the graph is at its lowest in an interval around the point, it is referred to as a **local minimum** or **relative minimum**. Another local minimum is at t_3. Similarly, t_2 and t_4 each represent a **local maximum**, because each one is a place where the graph is at its highest over some interval. Now let's look at all of these ideas more closely.

Increasing and Decreasing Functions

The first thing we noticed in Figure 3.3 was that over some intervals the function increased, over some it decreased, and on one interval it neither increased nor decreased.

Look at the graph of $y = f(x)$ in Figure 3.4. As the values of x increase (going from left to right) on the interval $[a, b]$, the corresponding values of $f(x)$ also increase and the curve rises. More formally we would state the following.

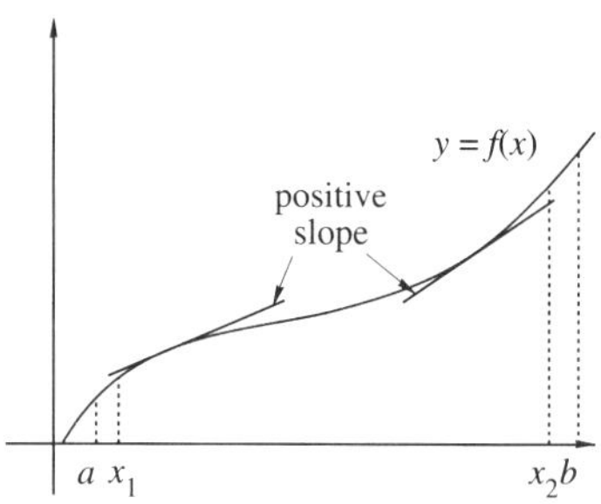

FIGURE 3.4

> **Increasing Function**
>
> If x_1 and x_2 are any two points in the interval $[a, b]$ with $x_1 < x_2$ and if $f(x_1) < f(x_2)$, then f is an **increasing function** on the interval.

As you can see, the curve has a positive slope at every point in this interval and so $f'(x) > 0$ for all values of x in $[a, b]$.

In a similar manner, we should look at the function in Figure 3.5. The graph shows a decreasing function. In particular, as the values of x increase on the interval $[c, d]$, the corresponding values of f decrease and the curve falls.

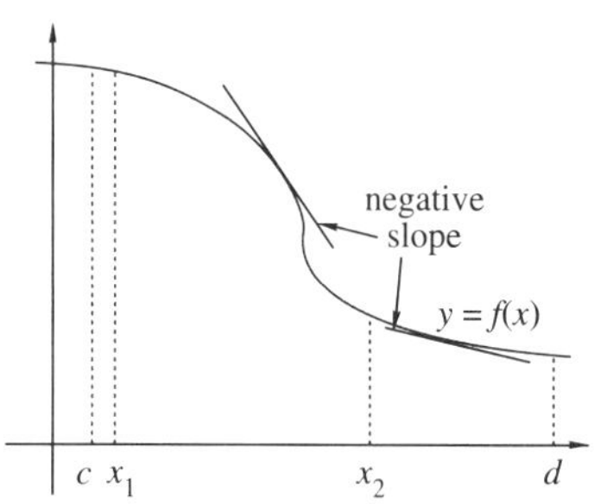

FIGURE 3.5

> **Decreasing Function**
>
> If x_1 and x_2 are any two points in the interval $[a, b]$ with $x_1 < x_2$ and if $f(x_1) > f(x_2)$, then f is a **decreasing function** on the interval.

The tangent lines of a decreasing function all have a negative slope and so $f'(x) < 0$ for all points in the interval $[c, d]$.

This leads us to an important rule dealing with increasing and decreasing functions.

Rule 1: Test to See if a Function is Increasing or Decreasing

If $f'(x) < 0$ on an interval, then f is a decreasing function on that interval.

If $f'(x) > 0$ on an interval, then f is an increasing function on that interval.

If $f'(x) = 0$ on an interval, then f is a constant function on that interval.

Determining the intervals when a derivative is positive or negative will often require your ability to solve inequalities. The next two examples will show how to determine when a function is increasing and when it is decreasing.

EXAMPLE 3.4

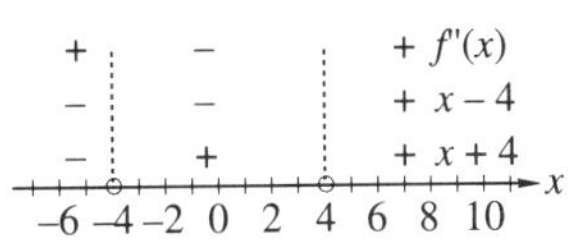

FIGURE 3.6

Find the intervals on which $f(x) = \frac{4}{3}x^3 - 64x$ is increasing or decreasing.

Solution Rule 1 says that we can determine when f is decreasing or increasing by looking at f'. The derivative of f is

$$\begin{aligned} f'(x) &= 4x^2 - 64 \\ &= 4(x^2 - 16) \\ &= 4(x+4)(x-4) \end{aligned}$$

The sign chart in Figure 3.6 shows where $f'(x)$ is positive and where it is negative. Since $f'(x)$ is negative on the interval $(-4, 4)$, we can say the f decreases on $(-4, 4)$. Similarly, f increases over the intervals $(-\infty, -4)$ and $(4, \infty)$.

EXAMPLE 3.5

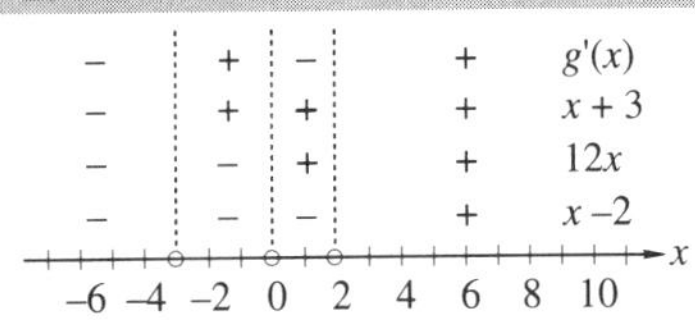

FIGURE 3.7

Find the intervals on which $g(x) = 3x^4 + 4x^3 - 36x^2$ is increasing or decreasing.

Solution The derivative of $g(x)$ is

$$\begin{aligned} g'(x) &= 12x^3 + 12x^2 - 72x \\ &= 12(x^3 + x^2 - 6x) \\ &= 12x(x^2 + x - 6) \\ &= 12x(x+3)(x-2) \end{aligned}$$

From Figure 3.7 we can see that $g'(x) < 0$ on the intervals $(-\infty, -3)$ and $(0, 2)$, so $g(x)$ is decreasing on those intervals. Similarly, since $g'(x) > 0$ on $(-3, 0)$ and $(2, \infty)$, $g(x)$ is increasing on these two intervals.

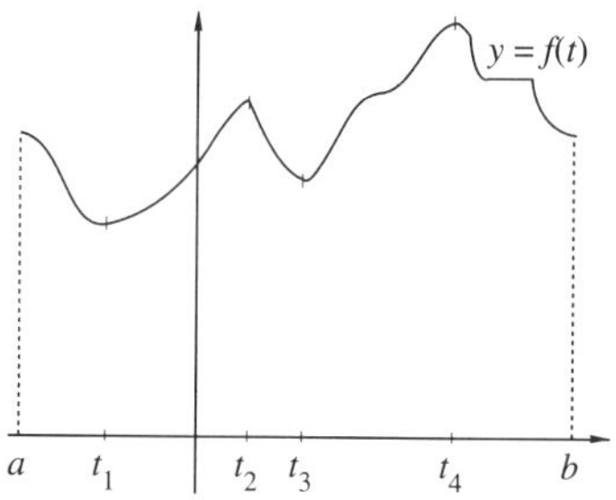

FIGURE 3.8

Extrema and Critical Points

Let's look again at the graph that was used to begin this section. The graph is repeated in Figure 3.8. At t_2 and t_4 we see that the function is higher than any other "nearby" points on the curve. The function is lower than any other nearby points at t_1 and t_3. Since these may not be the highest (or lowest) points on the entire curve, we say that f has a local maximum or a relative maximum at t_2 and t_4. If we can show that the curve does have a highest point then it is called the **absolute maximum**. Just as we described a function as increasing or decreasing over an interval, we say that a function has a local maximum at a point if the value of the function at that point is the highest value in the interval. When finding the maxima or minima over an interval you must always check the endpoints of the interval.

As you look at the local maxima and minima in Figure 3.8, there is something else you should notice. To the left of each relative minimum the function decreases; to the right, the function increases. Just the opposite is true for relative maxima.

EXAMPLE 3.6

Determine the relative maxima and minima of

$$g(x) = 3x^4 + 4x^3 - 36x^2$$

Solution This is the same function we examined in Example 3.5. We found that on

$$(-\infty, -3),\ g \text{ is decreasing.}$$
$$(-3, 0),\ g \text{ is increasing.}$$
$$(0, 2),\ g \text{ is decreasing.}$$
$$(2, \infty),\ g \text{ is increasing.}$$

Thus, we have a minimum when $x = -3$, a maximum when $x = 0$, and a minimum when $x = 2$.

The graph of g, as shown in Figure 3.9, supports the conclusions we reached using calculus.

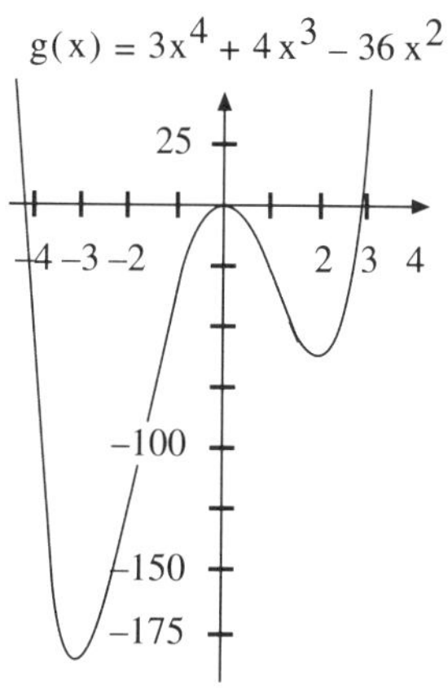

FIGURE 3.9

In the previous example, we began a practice we will follow from now on. We will let maximum refer to either an absolute or a relative maximum. We will also use minimum for either an absolute or relative minimum. Collectively, maximum and minimum values are referred to as **extreme values** or **extrema**. (A single maximum or minimum is called an **extremum**.)

Consider the function $h(x) = x^{2/3}$. Its graph is shown in Figure 3.10. Looking at the graph, we can see that $h(x)$ has a minimum when $x = 0$. But,

$$h'(x) = \frac{2}{3}x^{-1/3} = \frac{2}{3x^{1/3}}$$

When $x = 0$, $h'(x)$ is not defined.

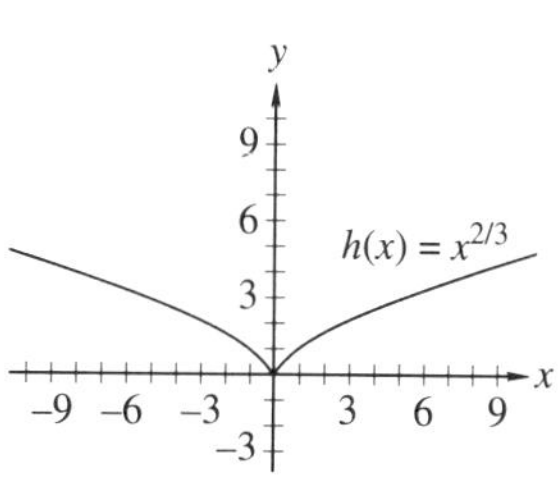

FIGURE 3.10

The example above leads to a definition of critical values (and critical points) for a function.

Critical Values and Critical Points

Extrema may occur at values of x in the domain of a function $y = f(x)$ for which $f'(x) = 0$ or $f'(x)$ is not defined. These values of x are referred to as **critical values**. If x is a critical value, then the point $(x, f(x))$ is called a **critical point**.

First Derivative Test

We are now ready to talk about a method for testing critical values to see if they are extrema. This method is called the **first derivative test** for $f(x)$ and it has four steps.

First Derivative Test

1. Find $f'(x)$.
2. Determine the critical values.
3. Use the critical values to determine intervals when f is decreasing ($f'(x) < 0$) or increasing ($f'(x) > 0$).
4. At each critical value c, determine if $f'(x)$ changes sign as x increases through c. To do this, select a value of x less than c and one greater than c. If, as x increases,
   (a) $f'(x)$ changes from $+$ to $-$, then $f(c)$ is a maximum;
   (b) $f'(x)$ changes from $-$ to $+$, then $f(c)$ is a minimum;
   (c) the sign of $f'(x)$ does not change, then $f(c)$ is neither a maximum nor a minimum.

EXAMPLE 3.7

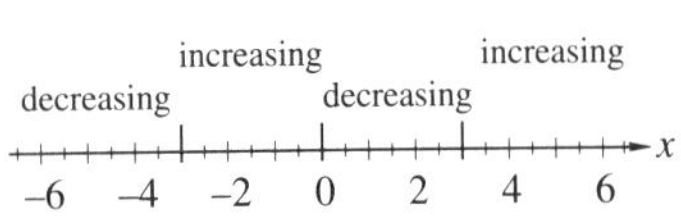

FIGURE 3.11

If $f(x) = x^4 - 18x^2$, use the first derivative test to find the intervals where f is increasing or decreasing and to locate all extrema.

Solution

1. $f'(x) = 4x^3 - 36x = 4x(x+3)(x-3)$
2. Letting $f'(x) = 0$ gives the critical values when $x = 0, -3$, and 3.
3. There are four intervals to consider. Using the same procedures we used in Examples 3.4 and 3.5, we find that on

$$\begin{aligned}&(-\infty, -3),\ f'(x) < 0, \text{ so } f \text{ is decreasing}\\&(-3, 0),\ f'(x) > 0, \text{ so } f \text{ is increasing}\\&(0, 3),\ f'(x) < 0, \text{ so } f \text{ is decreasing}\\&(3, \infty),\ f'(x) > 0, \text{ so } f \text{ is increasing}\end{aligned}$$

   This is summarized in Figure 3.11.

EXAMPLE 3.7 (Cont.)

4. At $x = -3$, there is a minimum since $f'(x)$ changes from $-$ to $+$. When $x = 0$, there is a maximum since $f'(x)$ changes from $+$ to $-$. At $x = 3$, there is a minimum since $f'(x)$ changes from $-$ to $+$.

   The graph of this function, $f(x) = x^4 - 18x^2$, is shown in Figure 3.12.

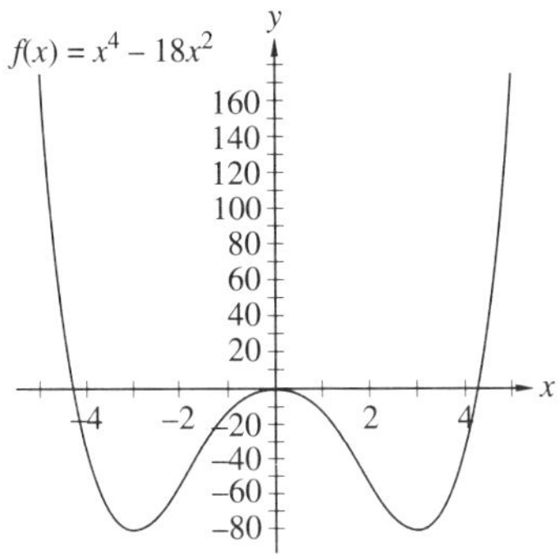

FIGURE 3.12

EXAMPLE 3.8

If $g(x) = x + \dfrac{9}{x+2}$, use the first derivative test to find all extrema.

Solution Notice that -2 is not in the domain of g and that the graph of g has a vertical asymptote at $x = -2$.

1. $$g'(x) = 1 - \frac{9}{(x+2)^2}$$
$$= \frac{(x+2)^2}{(x+2)^2} - \frac{9}{(x+2)^2}$$
$$= \frac{(x+2)^2 - 9}{(x+2)^2}$$
$$= \frac{x^2 + 4x + 4 - 9}{(x+2)^2}$$
$$= \frac{x^2 + 4x - 5}{(x+2)^2}$$
$$= \frac{(x+5)(x-1)}{(x+2)^2}$$

2. $g'(x) = 0$ when $(x+5)(x-1) = 0$ or when $x = -5$ or $x = 1$. $g'(x)$ is undefined when $(x+2)^2 = 0$ or when $x = -2$. However, since -2 is not in the domain of g, it is not a critical value. The only critical values are $x = -5$ and $x = 1$.

3. There are four intervals to consider. Remember that $(x+2)^2$ is never negative. Solving we get the following results.

$(-\infty, -5)$, $g'(x) > 0$, so g is increasing
$(-5, -2)$, $g'(x) < 0$, so g is decreasing
$(-2, 1)$, $g'(x) < 0$, so g is decreasing
$(1, \infty)$, $g'(x) > 0$, so g is increasing

This is summarized in Figure 3.13.

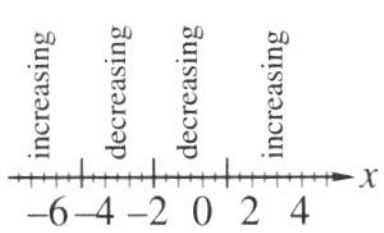

FIGURE 3.13

EXAMPLE 3.8 (Cont.)

4. At $x = -5$, there is a maximum since $g'(x)$ changes from + to −. At $x = 1$, there is a minimum since $g'(x)$ changes from − to +. The value $x = -2$ is not in the domain of g. There is no extrema at that point. The graph of g is shown in Figure 3.14.

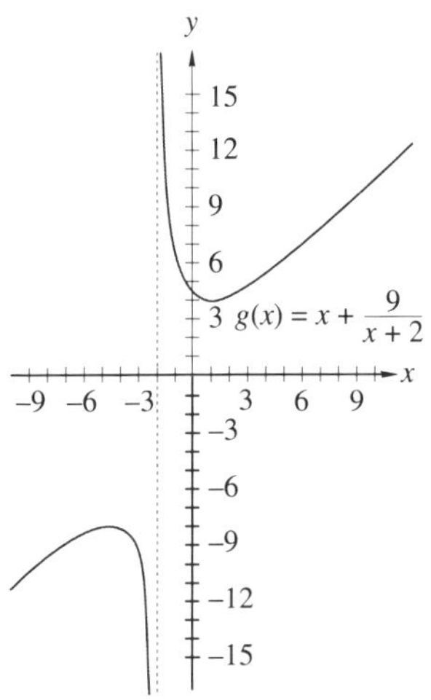

FIGURE 3.14

Exercise Set 3.2

In Exercises 1–26, determine when the function is increasing or decreasing, locate all critical values, and determine all extrema. Do not sketch the graph of the function.

1. $f(x) = x^2 - 8x$
2. $g(x) = 10x - x^2$
3. $m(x) = 16 - x^2$
4. $h(x) = 3x^2 - 12x + 10$
5. $j(x) = 15 + 16x - 4x^2$
6. $k(x) = x^2 - 4x + 11$
7. $f(x) = 4x^2 - 16x + 1$
8. $s(t) = -t^2 + 7t + 4$
9. $q(t) = t^3 - 3t$
10. $v(t) = 21t + 9t^2 - t^3$
11. $y = \frac{1}{4}x^4 - 18x^2 + 16$
12. $t = 3.2 - 12.5w^2 + 0.25w^4$
13. $f(z) = (z+4)^3$
14. $g(x) = \left(2x^2 - 8\right)^2$
15. $g(v) = 3v^4 - 4v^3 - 12v^2 + 12$
16. $y = \dfrac{5}{x}$
17. $s(t) = \dfrac{6}{t+1}$
18. $h(x) = x + \dfrac{1}{x}$
19. $g(x) = x^2 - \dfrac{2}{x}$
20. $j(x) = x + \dfrac{1}{x+3}$
21. $k(x) = \dfrac{x^2}{1-x}$
22. $m(b) = 4\sqrt{b}$
23. $f(t) = t\sqrt{1-t^2}$
24. $g(s) = (s+2)^2(s-1)^{2/3}$
25. $y = x(x-1)^{2/3}$
26. $f(x) = 9\sqrt[3]{x} - 4x$

Solve Exercises 27–36.

27. *Physics* An object is fired straight upward with its height, given in feet, represented by $s(t) = 250 + 256t - 16t^2$. **(a)** At what time is the altitude a maximum? **(b)** What is the maximum altitude?

28. *Physics* An object is fired straight upward with its height, given in meters, represented by $s(t) = 180 + 98t - 4.9t^2$. **(a)** At what time is its height a maximum? **(b)** What is the maximum height?

29. *Electricity* The power P in watts (W) in a resistor is given by $P = 4.7(t - 2.0)^2$, where t is time in seconds (s).
(a) What is the lowest power in the resistor?
(b) When does this occur?

30. *Electricity* The current i in amperes (A) at a given point in a circuit is given by $i = 4.8t - 1.2t^2$ where t is time in seconds (s).
(a) At what time is the current maximum?
(b) What is the maximum current?

31. *Electricity* When two resistors R_1 and R_2 are connected in parallel, their total resistance R is

$$R = \frac{R_1 R_2}{R_1 + R_2}$$

If $R_1 = 9\,\Omega$ and R_2 is a variable resistor, then

$$R = \frac{9R_2}{9 + R_2}$$

(a) What values of R_2 provide the minimum and maximum values of R?
(b) What are the minimum and maximum values of R?

32. *Physics* A particle moves along the x-axis according to the position function $x(t) = 8t^3 - t^4$. Find the velocity at the instant the acceleration is a maximum.

33. *Business* The profit from the sale of n items of a certain product is given by $P(n) = (4 - 0.2n)(2.5n + 7)$, where $P(n)$ is in hundreds of dollars and $1 \leq n \leq 20$.
(a) If the marginal profit is $P'(n)$, find the marginal profit function.
(b) How many items must be sold to make the most profit?
(c) What is the maximum profit?

34. *Energy technology* The energy output in joules of an electrical heater varies with time, t, according to the equation $E = 0.01\left(6t - t^2\right)^3$, where $0 \leq t \leq 6$.
(a) When is the energy output a maximum?
(b) What is the maximum energy output?
(c) When are the energy outputs at their lowest levels?

35. *Police science* Due to the rapid increase in major crimes, the mayor of a large city plans to organize a major crime task force. It is estimated that for each 1,000 people in the city, the number of major crimes will be $N(t) = 132 + 4\sqrt{t} + 2t^{3/2} - 0.8t^{5/2}$, where t is the number of months after the task force has been organized and $0 \leq t \leq 6$. **(a)** Find $N'(t)$. **(b)** When is the number of crimes at its peak?

36. *Meteorology* Meteorological results for a certain city indicate that for the month of April, the daily temperature in °F between midnight and 6:00 P.M. can be approximated by $T(t) = -0.04t^3 + 1.26t^2 - 9.3t + 56$, where t is the number of hours after midnight and $0 \leq t \leq 18$.
(a) Find $T'(t)$.
(b) When is the temperature the highest?
(c) When is the temperature the lowest?

In Your Words

37. Explain the first derivative test and what information it gives about a function.

38. **(a)** What determines whether a function is increasing or decreasing?
(b) How can you use calculus to determine whether a function is increasing or decreasing?

3.3 CONCAVITY AND THE SECOND DERIVATIVE TEST

In Section 3.2, we used the first derivative to determine when a function is increasing or decreasing and to locate its extrema. In this section, we will learn some uses for the second derivative. In order to be able to use the second derivative, we need to take a closer look at a curve and its shape.

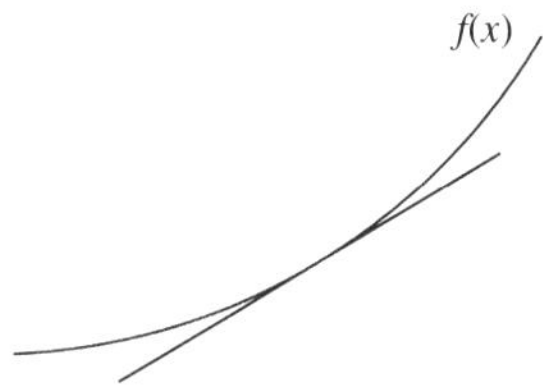

FIGURE 3.15a

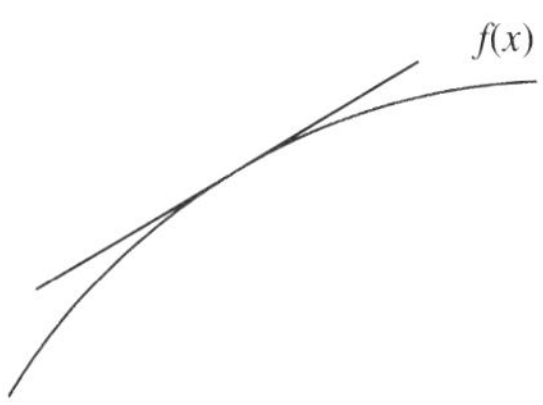

FIGURE 3.15b

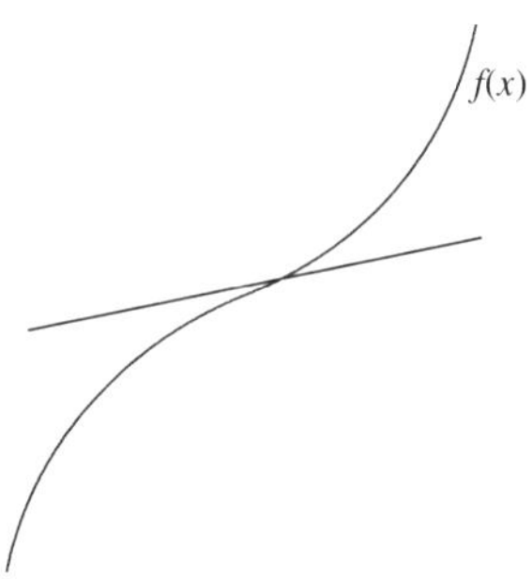

FIGURE 3.15c

Concavity

We will begin by looking at the three curves $f(x)$ in Figures 3.15a–c. In all three cases a tangent line has been drawn. Looking at the tangents, we can see that they all have positive slopes and so $f'(x) > 0$. These figures illustrate the three possible situations. In Figure 3.15a, the curve lies above the tangent line; in Figure 3.15b, the curve lies below the tangent line; and in Figure 3.15c, the curve crosses the tangent line. This example is not unique. We could have the same three situations if the first derivative were negative or if it were zero.

The three situations demonstrated in Figures 3.15a–c give us the chance to introduce some new terminology. If there is an open interval around a point where the graph of a function f is above the tangent lines for all points in that interval, then the graph of f is **concave up** at that point. This is demonstrated in Figure 3.15a and in Figure 3.16. There is an open interval (a, b) that contains the point c. The graph of $y = f(x)$ passes through $(c, f(c))$ and is above the tangent line through that point. Pick any other point in (a, b) and draw the tangent line through that point. If the curve is above that tangent line, then the curve is concave up. In Figure 3.16, the points x_1 and x_2 represent any two points in (a, b), except c. The tangent lines to the curve at each of the points $(x_1, f(x_1))$ and $(x_2, f(x_2))$ are drawn. In both cases the curve is above the tangent line and so the curve is concave up.

We have a similar situation if the curve is below the tangent lines as in Figure 3.15b. In this case, if there is an open interval around a point where the graph of a function is below the tangent lines for all points in that interval, then the graph of the function is **concave down** at that point.

Test for Concavity

It would be rather cumbersome to have to examine the tangent lines to a graph every time we wanted to determine if the graph was concave up or down. Fortunately, a test has been developed that allows us to determine the concavity of a curve. The test of concavity uses the second derivative and states the following.

Test for Concavity

The graph of a function f is

(a) concave up on any interval where $f''(x) > 0$

(b) concave down on any interval where $f''(x) < 0$

Hint

You might want to use these happy and sad faces to help remember how to use the second derivative to check the concavity of a function.

If the second derivative is positive, the graph is concave upward.

If the second derivative is negative, the graph is concave downward.

EXAMPLE 3.9

Determine when $f(x) = x^3 - 6x^2 + 3x + 5$ is concave up and when it is concave down.

Solution We find the second derivative of f is $f''(x) = 6x - 12$. Solving this, we see that $f''(x) < 0$ when $x < 2$ and $f''(x) > 0$ when $x > 2$. Thus, $f(x)$ is concave down on the interval $(-\infty, 2)$ and concave up on $(2, \infty)$. ▪

We have two items that have not been examined. What happens when a curve crosses a tangent line as in Figure 3.15c? What happens when $f''(x) = 0$?

It turns out that we can use Figure 3.15c to help answer both of these questions. To the left of the tangent line the curve is concave down and so the second derivative is negative. To the right of the tangent line the curve is concave up and so the second derivative is positive. At the tangent line in Figure 3.15c the curve changes from concave down to concave up. A point on the graph at which the concavity changes is called an **inflection point**.

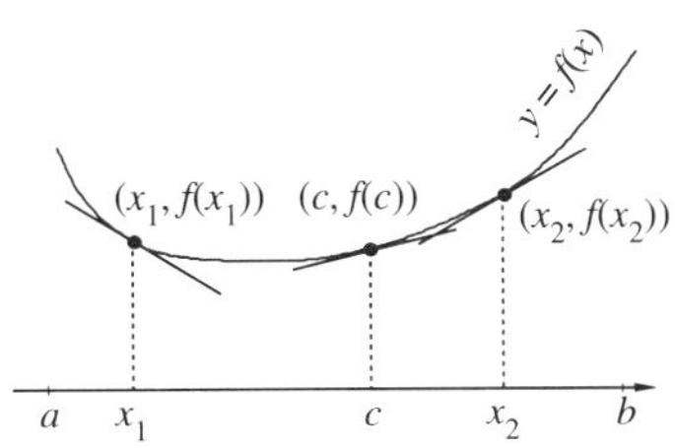

FIGURE 3.16

Locating Inflection Points

There are two ways to locate possible inflection points.

- Determine when the second derivative is zero.
- Determine when the second derivative is undefined.

Once you have found all of these points, look at the sign of the second derivative on both sides of each point. If the signs are different, then the concavity changes and it is a point of inflection.

EXAMPLE 3.10

Locate all inflection points of $f(x) = x^3 - 6x^2 + 3x + 5$.

Solution The second derivative is $f''(x) = 6x - 12$. Solving this, we see that $f''(x) = 0$ when $x = 2$. In Example 3.9, we showed that $f''(x) < 0$ when $x < 2$ and $f''(x) > 0$ when $x > 2$. Since the sign of f'' is negative to the left of 2 and positive to the right of 2, the point $(2, f(2)) = (2, -5)$ is an inflection point. The graph of f is shown in Figure 3.17. ▪

The fact that the second derivative is zero at some point does not mean that it is an inflection point. Consider the graph of $f(x) = x^4$. The second derivative is $f''(x) = 12x^2$ and $f''(x) = 0$ when $x = 0$. But $f''(x) > 0$ for all $x \neq 0$. The sign of the second derivative is the same on both sides of 0. Thus, there is no inflection point when $x = 0$, as can be seen in the graph of $f(x) = x^4$ in Figure 3.18.

Second Derivative Test

The second derivative can sometimes tell us more than the concavity of a curve and the location of possible inflection points. Suppose a curve is concave up on some interval. If there is a point c in this interval with a first derivative of zero, then it must

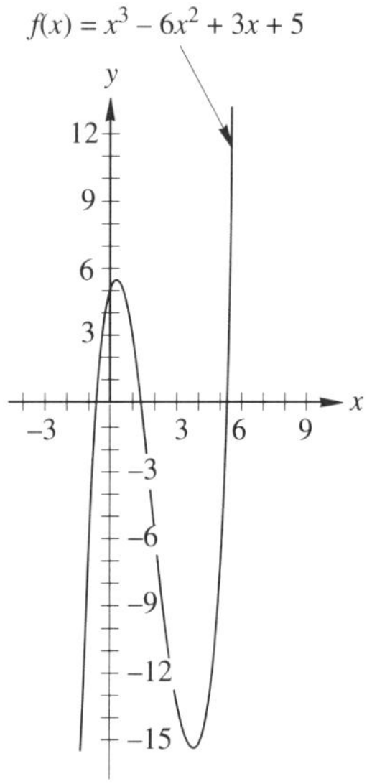

FIGURE 3.17

have a horizontal tangent line. Since the curve is concave up, the tangent line is below the curve. Thus the point is minimum. This forms the basis for what is known as the **second derivative test**.

> **Second Derivative Test**
>
> If f is a function whose second derivative exits, and if there is a point c where $f'(c) = 0$:
>
> **(a)** if $f''(c) < 0$, then $f(c)$ is a maximum,
> **(b)** if $f''(c) > 0$, then $f(c)$ is a minimum,
> **(c)** if $f''(c) = 0$, then the second derivative will not tell if $f(c)$ is a maximum or minimum.

EXAMPLE 3.11

Use the second derivative test to find all extrema of $g(x) = 2x^3 - 15x^2 + 36x - 25$.

Solution Using the first derivative $g'(x) = 6x^2 - 30x + 36$, we find that $x = 2$ and $x = 3$ are critical values. The second derivative is $g''(x) = 12x - 30$. Evaluating the second derivative at each critical value, we get $g''(2) = -6$ and $g''(3) = 6$. Since $g''(2) < 0$, there is a maximum at 2 and since $g''(3) > 0$, there is a minimum at 3. Remember that the minimum occurs when $x = 3$ so the location of the minimum is at $(3, g(3))$ or $(3, 2)$. Similarly, the maximum is at $(2, g(2))$ or $(2, 3)$.

In tabular form, we get the following results from the first derivative:

Interval	$g'(x)$	$g(x)$
$(-\infty, 2)$	$+$	increasing
$(2, 3)$	$-$	decreasing
$(3, \infty)$	$+$	increasing

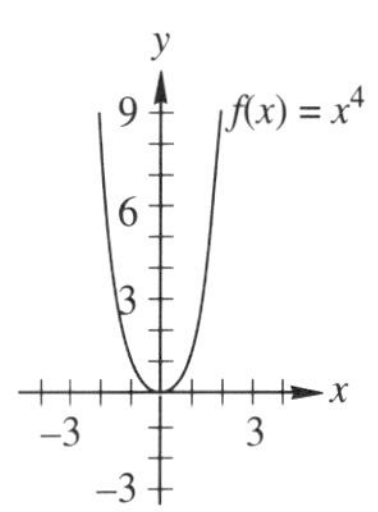

FIGURE 3.18

Setting $g''(x) = 0$, we see that the only possible inflection point is when $x = \frac{30}{12} = \frac{5}{2} = 2.5$. Using the concavity test, we get the following results:

Interval	$g''(x)$	$g(x)$
$(-\infty, 2.5)$	$-$	concave down
$(2.5, \infty)$	$+$	concave up

The graph of g is shown in Figure 3.19.

EXAMPLE 3.12

Sketch the graph of $h(x) = 3 + x + 3x^{2/3}$.

Solution $h'(x) = 1 + \dfrac{2}{x^{1/3}} = \dfrac{x^{1/3} + 2}{x^{1/3}}$. The critical values are when $h'(x) = 0$ and when $h'(x)$ is not defined. Now, since a fractional expression is 0 only when the

EXAMPLE 3.12 (Cont.)

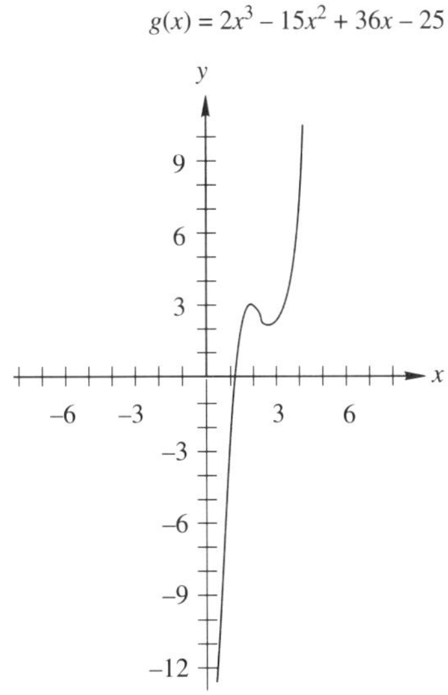

FIGURE 3.19

numerator is 0, then $h'(x) = \dfrac{x^{1/3}+2}{x^{1/3}} = 0$ is equivalent to $x^{1/3}+2=0$ or $x^{1/3} = -2$ and $x = -8$. Now, $h'(x)$ is not defined whenever the denominator is 0, or when $x^{1/3} = 0$, which is when $x = 0$. Since $x = 0$ is in the domain of h, the critical values are $x = -8$ where $h'(x) = 0$ and $x = 0$ where $h'(x)$ is not defined.

The second derivative is $h''(x) = -\dfrac{2}{3x^{4/3}}$. When $x = -8$, then $h''(-8) = -\dfrac{2}{3\,(-8^{4/3})} = -\frac{2}{48} < 0$, and so $(-8, 7)$ is a maximum. When $x = 0$, then $h''(0) = -\dfrac{2}{3\,(0^{4/3})}$, which is not defined and the second derivative test fails. Using the first derivative test, we see that when $x > 0$, $h'(x) > 0$ and so h is increasing. When x is in the interval $(-8, 0)$, $h'(x) < 0$ and so h is decreasing. Thus, $x = 0$ produces the minimum $(0, h(0))$ or $(0, 3)$. Summarizing our findings and adding those of the test for concavity, we get the following results.

Interval	$h'(x)$	h	Interval	$h''(x)$	h
$(-\infty, -8)$	$+$	increasing	$(-\infty, 0)$	$-$	concave down
$(-8, 0)$	$-$	decreasing	$(0, \infty)$	$-$	concave down
$(0, \infty)$	$+$	increasing			

The graph of this function is shown in Figure 3.20. ▪

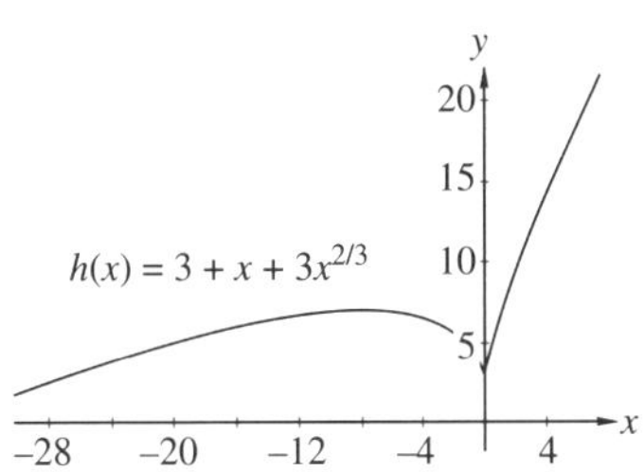

FIGURE 3.20

In Example 3.12, the domain of $h(x) = 3 + x + 3x^{2/3}$ was all real numbers. It is important to notice the domain. For example, graphing `h(x) = 3 + x + 3x^(2/3)` on a graphing calculator or a computer graphing program may result in something like the graph in Figure 3.21a. However, graphing h as `h(x) = 3 + x + 3(x^2)^(1/3)` produces the correct graph shown in Figure 3.21b.

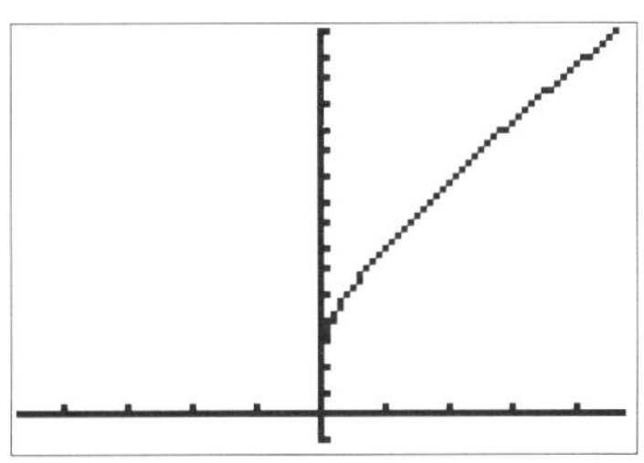

[–3.1, 3.1, 1] x [-1, 16, 1]

FIGURE 3.21a

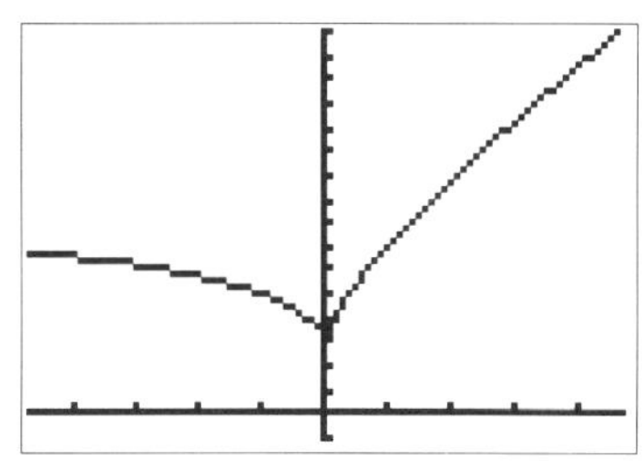

[–3.1, 3.1, 1] x [-1, 16, 1]

FIGURE 3.21b

In algebra you learned how to find some horizontal and vertical asymptotes. Basically, in order to test for horizontal asymptotes of $f(x)$, we evaluate $\lim_{x\to\infty} f(x)$ and $\lim_{x\to-\infty} f(x)$. Vertical asymptotes occur when the function $f(x)$ is not defined. Consider the next example.

EXAMPLE 3.13

Sketch the graph of $k(x) = \dfrac{x}{x^2+4}$.

Solution We will first check for asymptotes.

$$\lim_{x\to\infty} \frac{x}{x^2+4} = \lim_{x\to\infty} \frac{1/x}{1+4/x^2} = \frac{0}{1} = 0$$

and $\displaystyle\lim_{x\to-\infty} \frac{x}{x^2+4} = 0$. So, the x-axis, or $y = 0$, is a horizontal asymptote.

Finding the derivatives of $k(x)$, we get

$$k'(x) = \frac{4-x^2}{(x^2+4)^2} \text{ and } k''(x) = \frac{2x^3-24x}{(x^2+4)^3} = \frac{2x(x^2-12)}{(x^2+4)^3}$$

From the first derivative we see that the critical values are ± 2. In tabular form, we get the following results from the first derivative.

Interval	$k'(x)$	$k(x)$
$(-\infty, -2)$	$-$	decreasing
$(-2, 2)$	$+$	increasing
$(2, \infty)$	$-$	decreasing

Thus, when $x = -2$, we have a minimum at $(-2, -0.25)$, and at $x = 2$, we have a maximum at $(2, 0.25)$.

Setting $k''(x) = 0$, we see that the possible inflection points are when $x = 0$ and $x = \pm\sqrt{12} = \pm 2\sqrt{3}$. Using the concavity test, we get the following results.

Interval	$k''(x)$	$k(x)$
$(-\infty, -2\sqrt{3})$	$-$	concave down
$(-2\sqrt{3}, 0)$	$+$	concave up
$(0, 2\sqrt{3})$	$-$	concave down
$(2\sqrt{3}, \infty)$	$+$	concave up

Thus, we have points of inflection at all three points. The sketch of the graph is shown in Figure 3.22. ∎

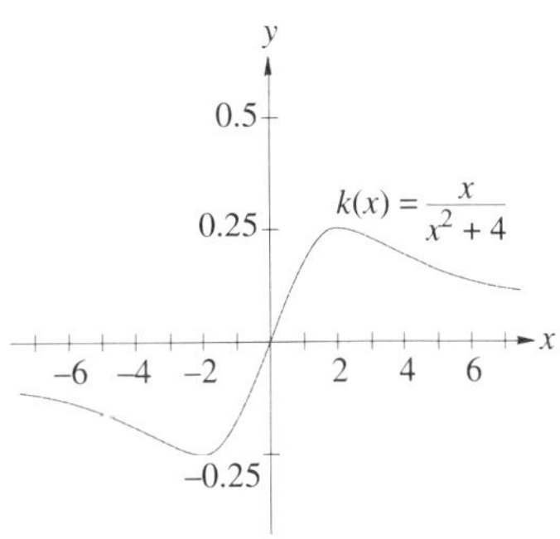

FIGURE 3.22

Exercise Set 3.3

In Exercises 1–28, find the extrema, determine the concavity, locate the inflection points, and sketch the graph.

1. $f(x) = -3x^2 + 12x$
2. $g(x) = 3x^2 - 6x + 5$
3. $g(x) = x^3 - 12x + 12$
4. $f(x) = x^3 + 3x^2 - 5$
5. $j(x) = x^3 - 6x^2 + 12x - 4$
6. $k(x) = x^3 - 6x^2 - 20$
7. $h(x) = -2x^3 + 3x^2 - 12x + 5$
8. $m(x) = x^4 - 4x^3 + 15$
9. $j(x) = 4x^2 - \frac{4}{3}x^3 - x^4$
10. $f(x) = 3x^4 - 4x^3 + 8$
11. $h(x) = (1-x)^3$
12. $g(x) = 3x^5 - 5x^3$
13. $h(x) = \sqrt[3]{x} + 2$
14. $j(x) = x - x^{2/3}$
15. $k(x) = x + \dfrac{4}{x}$
16. $m(x) = x - \dfrac{9}{x}$
17. $n(x) = x^2 - \dfrac{2}{x}$
18. $f(x) = \dfrac{1}{x-2}$
19. $g(x) = \dfrac{x^3}{3} - \dfrac{x^2}{2} - 6x$
20. $h(x) = x^4 - 32x + 48$

21. $j(x) = 2\sqrt{x} - x$
22. $k(x) = -\sqrt{x^2 - 4}$
23. $g(x) = 2 + x^{2/3}$
24. $h(x) = 2x - 3x^{2/3}$
25. $f(x) = x^{2/3}(x - 10)$
26. $g(x) = \dfrac{2}{x^2 - 4}$
27. $h(x) = \dfrac{x}{x^2 + 1}$
28. $j(x) = \dfrac{x^{2/3}}{x - 4}$

Solve Exercises 29–32.

29. *Agriculture* The catfish population of a commercial catfish farm can be approximated by $F(t) = 30 + 0.621t^2 - 0.009t^3$, where t is the number of weeks since this pond was started, $0 \le t \le 52$. To produce the maximum yield of catfish, the best time to harvest the catfish is at the point of the fastest growth in the fish population. This point is an inflection point. When is the best time to harvest this fish population?

30. *Economics* When a company starts a successful advertising campaign, sales and the rate at which sales grow both increase. At the **point of diminishing returns**, even though more money is spent on advertising, sales continue to grow but at a slower rate. The point of diminishing returns occurs at an inflection point where the sales curve changes from concave upward to concave downward. Find the point of diminishing returns for the sales function $S(x) = 97 + 2.1x^2 - 0.1x^3$, where x represents thousands of dollars on advertising, $0 \le x \le 10$, and S is sales in thousands of dollars.

31. *Police science* Due to the rapid increase in major crimes, the mayor of a large city plans to organize a major crime task force. It is estimated that for each 1,000 people in the city, the number of major crimes will be $N(t) = 67.4 + 3t^2 - 0.8t^{5/2}$, where t is the number of months after the task force has been organized and $0 \le t \le 12$.
(a) Find the maximum of $N(t)$.
(b) When is the maximum rate of increase in $N(t)$?

32. *Meteorology* Meteorological results for a certain city indicate that for the month of April, the daily temperature in °F between midnight and 6:00 P.M. can be approximated by $T(t) = -0.04t^3 + 1.26t^2 - 9.3t + 56$, where t is the number of hours after midnight and $0 \le t \le 18$.
(a) What are the maximum and minimum temperatures?
(b) What is the maximum rate of increase in the temperature?

In Your Words

33. Explain the second derivative test and what information it gives about a function.

34. What is an inflection point?

≡ 3.4 APPLIED EXTREMA PROBLEMS

Up to this point, most of the applications for extrema have been made to curve sketching. We have used derivatives to help determine the high and low points on the curve, when it is increasing or decreasing, and other useful information. A curve is a picture of a function and functions are often used to represent real situations. In this section, we will look at some ways to solve problems that involve using derivatives.

As each example is worked, you should make sure that you understand what was done. The most important part will be in understanding how to set up the function; then, being aware of how derivatives are used to find possible solutions and to determine the correct answer should be understood.

Application

EXAMPLE 3.14

A company needs to erect a fence around a rectangular storage yard next to a warehouse. They have 400 m of available fencing. If they do not fence in the side next to the warehouse, what is the largest area that can be enclosed?

Solution The clue to the desired answer is "largest area." We need to develop a function for the area and then determine when it is a maximum.

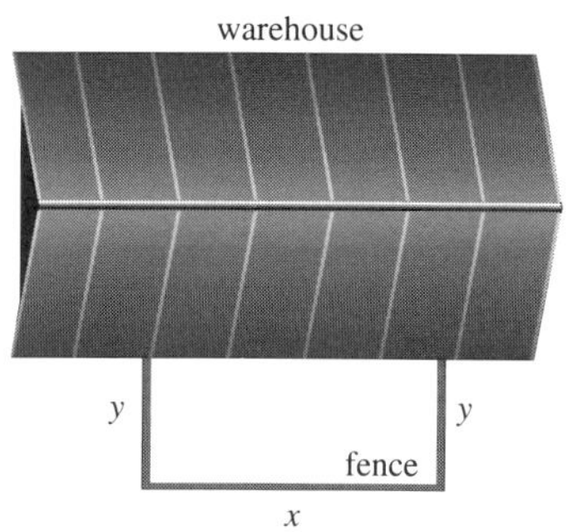

FIGURE 3.23

Look at Figure 3.23. The fenced-in rectangular yard has dimensions that have been labeled x and y. The area of the fenced-in region is $A = xy$. We need to write A in terms of a single variable. The total amount of fencing is $400\text{ m} = y + x + y$. Thus, we have

$$2y + x = 400$$

or

$$x = 400 - 2y$$

If we substitute this in the area formula, we obtain

$$\begin{aligned} A &= xy \\ &= (400 - 2y)y \\ &= 400y - 2y^2 \end{aligned}$$

The area is now expressed in terms of one variable, y, and so we have

$$A(y) = 400y - 2y^2$$

We want to know when $A(y)$ is a maximum. Taking the first derivative, we get

$$A'(y) = 400 - 4y$$

The only critical value occurs when $A'(y) = 0$, which is when $y = 100$.

Is the area a maximum when $y = 100$? Using the second derivative test, we see that $A''(y) = -4$. Since $A''(100) < 0$, the area is a maximum when $y = 100$.

We have one other item to check—is this an acceptable solution? We know that $y > 0$. We can also determine that the largest possible value for y is 200 m. Thus, the answer must be in the interval $[0, 200]$.

Now, when $y = 100$, $x = 200$, and so the maximum area is $A = (200)(100) = 20\,000\text{ m}^2$.

Application

EXAMPLE 3.15

When a battery or generator of voltage V and an internal resistance r is connected to an external resistance R, the total resistance in the circuit is $R + r$ and the current flow is $I = \dfrac{V}{R+r}$. The power P dissipated in the load R is $P = I^2R$. If the voltage is 100 V and $r = 5\,\Omega$, what value of R will produce the maximum load power? (See Figure 3.24.)

EXAMPLE 3.15 (Cont.)

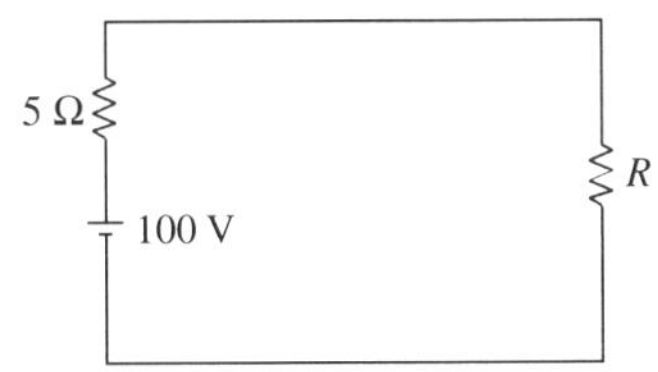

FIGURE 3.24

Solution We need to develop a formula in terms of the external resistance R. Since $I = \frac{V}{R+r}$, we have $I = \frac{100}{R+5}$. Substituting this into the power formula, we get

$$P = I^2R = \left(\frac{100}{R+5}\right)^2 R = \frac{10\,000R}{(R+5)^2}$$

Differentiating produces

$$\begin{aligned} P' &= \frac{(R+5)^2(10\,000) - 10\,000R(2)(R+5)}{(R+5)^4} \\ &= \frac{(R+5)10\,000 - 20\,000R}{(R+5)^3} \\ &= \frac{50\,000 - 10\,000R}{(R+5)^3} \end{aligned}$$

Since $R > 0$, the only critical value occurs when $P' = 0$ or when $R = 5$. Using the first derivative test, we see that this is a maximum and so the maximum load power occurs when $R = 5\,\Omega$.

We could generalize this last example to show that the maximum load power occurs when the external load is equal to the internal resistance.

Application

EXAMPLE 3.16

A container is to be made in the form of a right circular cylinder and is to contain 1 L. What dimensions of the container will require the least amount of material?

Solution We know that the volume of a right circular cylinder is given by the formula $V = \pi r^2 h$ and the total surface area is $A = 2\pi rh + 2\pi r^2$. We will assume that the thickness of the material is negligible. We want to minimize the area. The formula for the area is given in terms of two variables, h and r. We know that $V = 1\text{ L} = 1\,000\text{ cm}^3$. We can solve the equation $1\,000 = \pi r^2 h$ for either r or h and substitute the result into the area formula. If we solve for h we get

$$h = \frac{1\,000}{\pi r^2}$$

Substituting, we get the area as a function of r:

$$\begin{aligned} A(r) &= 2\pi r\left(\frac{1\,000}{\pi r^2}\right) + 2\pi r^2 \\ &= \frac{2\,000}{r} + 2\pi r^2 \end{aligned}$$

Differentiating with respect to r, we have

$$A'(r) = -2\,000r^{-2} + 4\pi r$$

EXAMPLE 3.16 (Cont.)

Critical values occur when $r = 0$, because $A'(r)$ is not defined when $r = 0$, and when $A'(r) = 0$. If we set the derivative equal to zero, we obtain

$$\frac{2\,000}{r^2} = 4\pi r$$

$$\text{or} \qquad r^3 = \frac{2\,000}{4\pi} = \frac{500}{\pi}$$

$$\text{and so} \qquad r = \sqrt[3]{\frac{500}{\pi}}$$

Using the second derivative test, we see that

$$A''(r) = 4\,000r^{-3} + 4\pi$$

When $r = 0$, $A''(r)$ is not defined, but we know that $r > 0$. (Why? What happens if $r = 0$?)

When $r = \sqrt[3]{\frac{500}{\pi}}$, $A''(r) > 0$, so the area is a minimum.

When $r = \sqrt[3]{\frac{500}{\pi}} \approx 5.4193\,\text{cm}$, $h = \frac{1\,000}{\pi r^2} \approx 10.8385\,\text{cm}$ and the area is $A \approx 553.5810\,\text{cm}^2$.

Application

EXAMPLE 3.17

A telephone company is asked to provide telephone service to a customer whose house is 0.5 km away from the road along which the telephone lines run. The nearest telephone box is 5 km down the road. The cost to connect the telephone line is \$80 per kilometer along the road and \$100 per kilometer away from the road. At which point along the road should the company connect the line in order to minimize the cost?

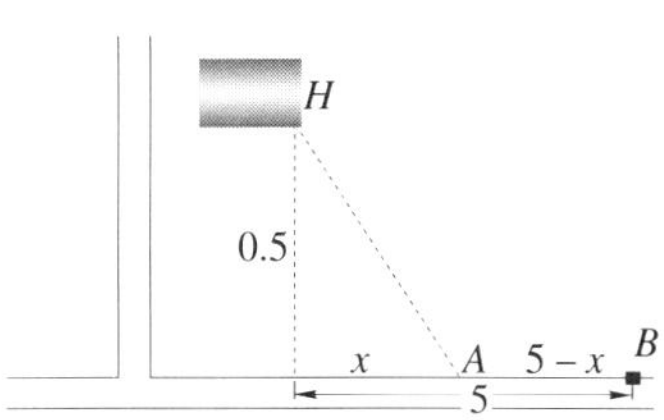

FIGURE 3.25

Solution Examine the sketch of this problem in Figure 3.25. We let $5 - x$ represent the distance from the box B to the place where the connection is made, A. The distance x is from the connection A to the point on the road closest to the house, H. The telephone line will go from B to A and then from A to H. The length AH is $\sqrt{x^2 + 0.5^2} = \sqrt{x^2 + 0.25}$. The total cost C is $C(x) = 80(5 - x) + 100\sqrt{x^2 + 0.25}$. We want $C(x)$ to be a minimum.

$$C'(x) = -80 + \frac{100x}{\sqrt{x^2 + 0.25}}$$

$$= \frac{-80\sqrt{x^2 + 0.25} + 100x}{\sqrt{x^2 + 0.25}}$$

EXAMPLE 3.17 (Cont.)

The critical value occurs when the numerator is 0; that is, when

$$80\sqrt{x^2+0.25} = 100x$$
$$6\,400(x^2+0.25) = 10\,000x^2$$
$$1\,600 = 3\,600x^2$$
$$\frac{16}{36} = x^2$$

or when

$$x = \pm\frac{2}{3}$$

When $x = \frac{2}{3}$ km, the cost will be a minimum. The minimum cost is \$430. ▪

When solving extrema problems, there are some helpful procedures to use. If you go back and look at the examples in this section, you will see that we used each of them.

Hint

Hints for Solving Extrema Problems

1. When possible, draw a figure to illustrate the problem. Label all the parts that are important to the problem.
2. Write an equation for the quantity that is going to be a maximum or a minimum.
3. If more than one variable is involved, eliminate all but one of them.
4. Take the derivative and locate the critical values.
5. Use the first and/or second derivative test on each critical value to see if it provides a maximum or a minimum.
6. If the function is defined for a limited range of values, examine the endpoints for possible extrema values.

Exercise Set 3.4

Solve Exercises 1–36.

1. *Agriculture* A company has 10,000 ft of available fencing. It wants to use the fencing to enclose a rectangular field. One side of the field is bordered by a river. If no fencing is placed on the side next to the river, what is the largest area that can be enclosed?

2. *Agriculture* If the company in Exercise 1 decides to fence the side bordering the river, what is the largest area that can be enclosed?

3. *Agriculture* A rectangular field contains an area of 6,000,000 ft^2. One side of the field borders a river. Because of difficulties caused by the river, it costs \$2.00/ft to fence along the river and \$1.00/ft to fence the other three sides. **(a)** What dimensions will give the minimum cost? **(b)** What is the cost?

4. *Agriculture* A 200-m piece of link fencing is to be cut so that one piece encloses a square section of garden and the other piece encloses a circular section. Where should the fencing be cut if the enclosed areas are to be a maximum?

5. *Industrial design* An open box is to be formed by cutting a square from each corner of a square piece of cardboard 24 cm on a side. After the squares are removed from each corner, the sides are folded up as shown in the following figure. How large a square must be removed from each corner in order for the box to have the largest volume?

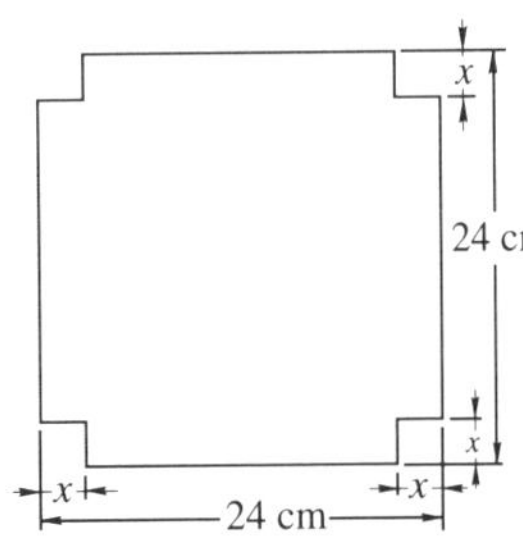

6. *Business* A rectangular sign is to be painted on the end of a semicircular-shaped building that has a diameter of 20.00 m. What dimensions will produce a sign with the largest area?

7. What is the largest rectangle that can be inscribed in the first quadrant of the ellipse $9x^2 + 16y^2 = 144$?

8. *Industrial design* An open rectangular box is to be made from a piece of cardboard 8 in. wide and 15 in. long by cutting a square from each corner and bending up the sides. How large a square should be removed from each corner in order for the box to have the largest volume?

9. *Industrial design* A box with a square base and an open top is to hold 32 in.^{3} What dimensions will require the least amount of material?

10. *Industrial design* If the box in Exercise 9 is to be closed on top, what dimensions will require the least amount of material?

11. *Industrial design* The total surface area of a right-circular cylindrical tin is $480\,\text{cm}^2$. What are the dimensions of the cylinder that has the largest volume?

12. *Mechanical engineering* The cutting speed s (in meters per minute) of a saw used to cut a certain type of material is given by $s(t) = \sqrt{t - 9t^2}$, where t is the time in seconds. What is the maximum speed of this product?

13. *Printing* A page of a book is to have $580\,\text{cm}^2$ with a 3-cm margin at the bottom and sides and a 2-cm margin at the tope. Find the dimensions of the page that will allow the largest printed area.

14. Find the point on the curve $y = 2x^2$ that is closest to the point $(5, 1)$.

15. *Construction* The strength of a rectangular beam is directly proportional to the width of the beam and the square of its depth. What are the dimensions of the strongest beam that can be cut from a log whose cross-section is a circle of radius 15 in.? What are the dimensions of the strongest beam cut from a circular log whose cross-section has a radius of r?

16. *Sheet metal technology* A metal trough is to be made by bending a rectangular sheet of metal in the middle to form a "V." If the piece of metal is 300 mm wide, what angle between the sides will give the maximum capacity?

17. *Sheet metal technology* A gutter is to be made from a long sheet of metal 320 mm wide by bending up equal widths along the edges into vertical positions. What dimensions will give the largest capacity?

18. *Electronics* The current I in amperes (A) through a resistor of $R\,\Omega$ with a constant voltage E in volts (V), is given by

$$I = \frac{E}{\sqrt{R^2 + X^2}}$$

(a) Assuming that R is constant, what value of X will make the current in the resistor a maximum?

(b) What is this maximum current?

19. *Electricity* The output of a battery is given by the formula $P = VI - RI^2$, where V is the voltage, I the current, and R the resistance. Find the current for which the output is a maximum if $V = 10\,\text{V}$ and $R = 4.0\,\Omega$.

20. *Chemistry* The rate of a certain autocatalytic reaction is given by $v(x) = 5x(a - x)$ for x in $[0, a]$, where a is the original amount of the substance produced by the reaction. Determine when the reaction is a maximum?

21. *Construction* A power company wishes to install a temporary power line. The purchase price of the wire is proportional to the cross-sectional area A of the wire. The total cost C of the wire is given by the formula $C(A) = bA + \frac{c}{A}$, where b and c are constants. What cross-sectional area will produce the least cost?

22. *Architecture* A window above the entrance to a room that has a cathedral ceiling is to be designed in the shape of a rectangle with a semicircle on top. If the area of the rectangular part is to be $2.00\,\text{m}^2$, what is the diameter of the semicircle that will result in a window with the smallest perimeter?

23. *Architecture* What would be the diameter of the semicircular part of the window in Exercise 22 if the perimeter is 8.00 m and the total area of the window is to be a maximum?

24. *Business* The cost of operating a truck on an interstate highway is $0.35 + \frac{s}{300}$ dollars per mile, where s is the speed of the truck in miles per hour. The driver's salary of \$12 per hour is extra. **(a)** At what speed should the truck driver operate the truck to make an 800-mi trip for the least cost? **(b)** What is the cost of this 800-mi trip?

25. *Electricity* The power P in watts (W) in a circuit is given by $P = I^2R$, where I is current in amperes (A) and R is resistance in ohms (Ω). If $R = 25.4\,\Omega$ and $I = 4.5 - t$, where t in time is seconds (s), find
(a) the time when there is minimum power, and
(b) the minimum power.

26. *Electricity* The current in a circuit is given by $\frac{dq}{dt}$, where q is the charge in coulombs (C) and t is the time in seconds (s). If $q = -4.36t^2 + 2.14t^3$, find the minimum current in the circuit and the time at which this occurs.

27. *Lighting technology* The intensity of illumination at a point varies inversely as the square of the distance between the point and the light source. Two lights are 8 m apart. One of the lights has an intensity four times that of the other. At which point between the lights is the illumination the least?

28. *Aerodynamics* The drag D on an airplane traveling at velocity v is $D = av^2 + \frac{b}{v^2}$, where a and b are positive constants. At what speeds does the airplane have the least drag?

29. *Lighting technology* The deflection Y of a beam of length L at a horizontal distance x from one end is given by $Y = k(2x^4 - 5Lx^3 + 3L^2x^2)$, where k is a positive constant. For what value of x does the maximum deflection occur?

30. *Mechanical engineering* The efficiency E of a screw is given by $E = \frac{T(1 - T\mu)}{T + \mu}$ where μ is the coefficient of friction and T is the tangent of the pitch angle of the screw. For what value of T is the efficiency the greatest?

31. *Civil engineering* A proposed tunnel has a fixed cross-sectional area. The tunnel will have a horizontal floor, vertical walls of equal height, and the ceiling will be a semicircle. The ceiling cost four times as much per square meter as the floor and walls. What ratio of the diameter of the semicircular ceiling to the height of the walls will produce the least cost?

32. A person in a rowboat is 4 km offshore, where the nearest point P is on a straight shoreline of the lake. A town is 20 km down the shore road from P. This person can row 2.5 km/h and can walk 4.5 km/h. At which point should the boat be landed in order to arrive in town in the shortest time?

33. *Petroleum engineering* An oil company wants to lay a pipeline from its offshore drilling rig to a storage tank on shore. The rig is 5 mi offshore, and the storage tank is 12 mi down the straight shoreline. The cost of laying pipe underwater is \$1,200 per mi and along the shoreline the cost is \$600 per mi. How far from the storage tank should they locate the connection of the underwater pipe to the shoreline pipe?

34. *Package design* A manufacturer is designing a closed rectangular shipping crate with a square base. The volume of the crate is to be $36\,\text{ft}^3$. Material for the top costs \$1 per square foot, material for the sides costs $\$0.75/\text{ft}^2$, and material for the bottom costs $\$1.50/\text{ft}^2$.
(a) Find the dimensions of the box that will minimize the total cost of material.
(b) Determine the minimum cost.

35. *Construction* A rectangular warehouse is being constructed with a total floor area of 4,200 ft^2. A single interior wall will partition the area into storage space and office space. It costs \$175/ft to construct an exterior wall and \$115/ft to erect the interior wall.
 (a) What are the dimensions of the warehouse that will minimize the cost?
 (b) Determine the minimum cost.

36. *Package design* The U.S. Postal Service has a limit of 108 in. on the combined length and girth of a rectangular package that can be mailed. Find the dimensions of the package of maximum volume that has a square cross-section. (The girth is defined as the smallest perimeter of a rectangular cross-section of the box.)

In Your Words

37. Without looking in the text, describe how you solve a problem that asks you to find a maximum or a minimum.

38. Write an application in your technology area of interest that requires you to find a maximum or a minimum. Give your problem to a classmate and see if he or she understands and can solve your problem using the techniques of this and earlier chapters. Rewrite the problem as necessary to remove any difficulties encountered by your classmate.

3.5 RELATED RATES

It is not unusual to have two or more quantities that vary with time and an equation that expresses some relationship between them. Normally, the values of these quantities at some time are given. We are also given all but one of the rates of change. The problem is to find the rate of change that is not given. We do this by taking the derivative of the expression (with respect to time) that relates the variables. The following examples show the basic method for solving these problems.

EXAMPLE 3.18

A ladder 30 ft long leans against a vertical wall. The foot of the ladder is slipping away from the wall at the rate of 4 ft/s. How fast is the top of the ladder sliding down the wall at the instant when the foot of the ladder is 18 ft from the wall?

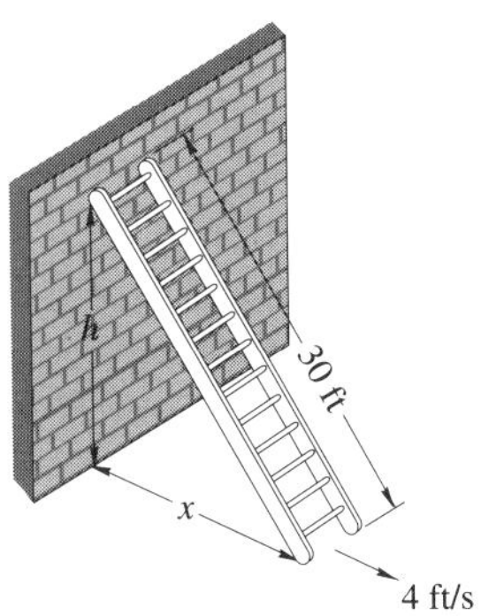

FIGURE 3.26

Solution Figure 3.26 shows the ladder against the wall. The constants are the length of the ladder, 30 ft, and the rate at which the foot of the ladder moves, 4 ft/s. The quantities that change are the height of the top of the ladder, h, and the distance of the foot of the ladder from the wall, x.

We want to find a rate. If h denotes the height of the top of the ladder above the ground in feet and t denotes the time in seconds, then we want to find dh/dt: the rate at which the height of the ladder is changing (dh) with respect to time (dt). In particular, we want to find dh/dt when $x = 18$ ft.

We have a right triangle with the right angle where the wall meets the ground. The ladder is the hypotenuse of the triangle. Thus, $x^2 + h^2 = 30^2$. Using implicit differentiation with respect to time t, we get

$$\frac{d(x^2)}{dt} + \frac{d(h^2)}{dt} = 0$$

or

$$2x\frac{dx}{dt} + 2h\frac{dh}{dt} = 0$$

EXAMPLE 3.18 (Cont.)

Thus, the rate we want to find is

$$\frac{dh}{dt} = \left(\frac{-x}{h}\right)\left(\frac{dx}{dt}\right)$$

When $x = 18$, $h = 24$, and $\frac{dx}{dt} = 4$ ft/s, we obtain

$$\frac{dh}{dt} = -\frac{18}{24}(4) = -3 \text{ ft/s}$$

Thus, the top of the ladder is falling at the rate of 3 ft/s.

We solved Example 3.18 for the general case when the foot of the ladder was x-feet from the bottom of the wall. Once we had this answer, we solved it for the specific case when $x = 18$. Thus, by first solving for x we are then able to determine the rate for any value of x.

As with the extrema problems, there are some guidelines to help us solve related rate problems. Some of the guidelines are virtually the same as those for extrema, but are restated here to keep the guidelines as a group.

Hint

Hints for Solving Related Rates Problems

1. Make a diagram of the problem. Which quantities change; which remain constant?
2. Write down what you are to find. (In Example 3.18, it was a rate.) Express it in terms of a variable (in Example 3.18, dh/dt). Show the variable in the sketch (h in Figure 3.26).
3. Identify the other variables in the problem. Show them in the sketch. Write down any numerical information that is given about them.
4. Write equations that relate the variables.
5. Use substitution and differentiation to establish additional relationships among the variables and their derivatives.
6. Substitute numerical values for the variables and the derivatives. Solve for the unknown rate.

Caution

A common error is to introduce specific values for the rates and variable quantities too early in the solution. Remember to first get a general formula that involves the rates of change at any time t. After you have the general formula, specific values can be substituted to get the desired solution.

Application

EXAMPLE 3.19

A weather balloon in the shape of a sphere is being inflated at the rate of 12 m^3/min. Find the rate at which the surface area of the balloon is changing when the radius is 6 m.

Solution The variables in this problem are:

$$\begin{aligned} t &= \text{time in minutes} \\ r &= \text{radius in meters of the balloon at time } t \\ V &= \text{volume in cubic meters at time } t \\ S &= \text{surface area in square meters at time } t \end{aligned}$$

We have the following rates of changes:

$$\begin{aligned} \frac{dr}{dt} &= \text{rate of change of radius with respect to time} \\ \frac{dV}{dt} &= \text{rate of change of volume with respect to time} \\ \frac{dS}{dt} &= \text{rate of change of surface area with respect to time} \end{aligned}$$

We are told that $\frac{dV}{dt} = 12\,\text{m}^3/\text{min}$. We want to find $\frac{dS}{dt}$ when $r = 6$ m.

The volume of a sphere is $V = \frac{4}{3}\pi r^3$ and the surface area is $S = 4\pi r^2$. Differentiating each of these with respect to t, we get

$$\frac{dV}{dt} = 4\pi r^2 \frac{dr}{dt}$$

$$\text{and} \qquad \frac{dS}{dt} = 8\pi r \frac{dr}{dt}$$

Since we want to find $\frac{dS}{dt}$ and we know the value of $\frac{dV}{dt}$, we will solve the $\frac{dV}{dt}$ equation for $\frac{dr}{dt}$ and substitute this into the $\frac{dS}{dt}$ equation. Solving for $\frac{dr}{dt}$ we get

$$\frac{dr}{dt} = \frac{1}{4\pi r^2} \cdot \frac{dV}{dt}$$

Substituting, this gives

$$\frac{dS}{dt} = 8\pi r \left(\frac{1}{4\pi r^2} \right) \frac{dV}{dt} = \frac{2}{r} \frac{dV}{dt}$$

When $r = 6$ m and $\frac{dV}{dt} = 12\,\text{m}^3/\text{min}$, we get

$$\frac{dS}{dt} = \frac{2}{6} \cdot 12 = 4\,\text{m}^2/\text{min}$$

The surface area is increasing at a rate of 4 m^2/min when the radius is 6 m. ▪

Application

EXAMPLE 3.20

Water is running out of a conical funnel at a rate of $5\,\text{cm}^3/\text{s}$. If the radius of the base of the funnel is 10 cm and the altitude is 20 cm, find the rate at which the water level is dropping when it is 5 cm from the top.

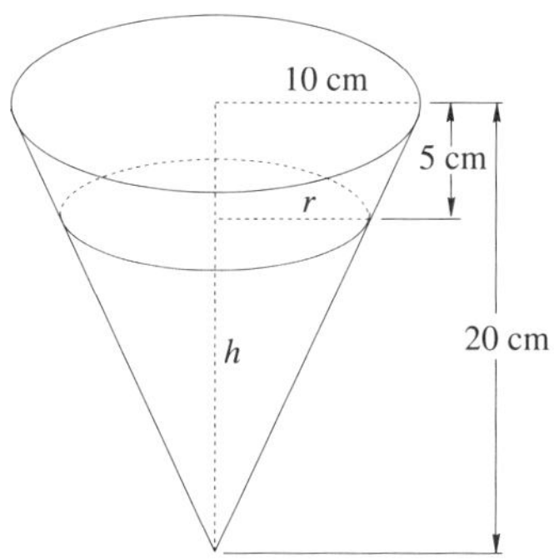

FIGURE 3.27

Solution A sketch of this problem is given in Figure 3.27. The variables are:

t = time in seconds
r = radius in cm of water level at time t
h = height in cm of surface of water at time t
V = volume in cm^3 of water in cone at time t

We have the following rates of change:

$$\frac{dh}{dt} = \text{rate of change of height of water with respect to time}$$
$$\frac{dV}{dt} = \text{rate of change of volume of water with respect to time}$$

By similar triangles, we have $\frac{r}{10} = \frac{h}{20}$ or $r = \frac{1}{2}h$. The volume of a cone is $V = \frac{1}{3}\pi r^3$, and substituting for r, we get

$$V = \frac{1}{3}\pi\left(\frac{1}{2}h\right)^3 = \frac{1}{24}\pi h^3$$

Differentiating with respect to t we get

$$\frac{dV}{dt} = \frac{dV}{dh}\cdot\frac{dh}{dt} = \frac{1}{8}\pi h^2\frac{dh}{dt}$$

We are given $\frac{dV}{dt} = -5$. This is negative because the volume is decreasing. When the water level is 5 cm from the top, $h = 20 - 5 = 15$ cm and so we have

$$-5 = \frac{1}{8}\pi(15)^2\frac{dh}{dt}$$
$$\text{or}\quad \frac{dh}{dt} = -\frac{40}{225\pi} = \frac{-8}{45\pi}\ \text{cm/s}$$

Application

EXAMPLE 3.21

When air changes volume adiabatically (without the addition of heat), the pressure P and volume V satisfy the formula $PV^{1.4} = k$, where k is a constant. At a certain instant, the pressure is $75\,\text{lb/in.}^2$ and the volume is $25\,\text{in.}^3$ and is decreasing at the rate of $3\,\text{in.}^3/\text{s}$. How rapidly is the pressure changing at that time?

EXAMPLE 3.21 (Cont.)

Solution Since the equation is given, a drawing is not needed for this problem. The variables are:

$$t = \text{time in seconds}$$
$$P = \text{pressure in lb/in.}^2$$
$$V = \text{volume in in.}^3$$

We have the following rates of change:

$$\frac{dV}{dt} = \text{rate of change of volume with respect to time}$$
$$\frac{dP}{dt} = \text{rate of change of pressure with respect to time}$$

From the formula $PV^{1.4} = k$ and the fact that when $P = 75$, $V = 25$, we can determine that $k \approx 6{,}794.81$.

We are given $\frac{dV}{dt} = -3$ when $V = 25$ and need to find $\frac{dP}{dt}$. We know that $P = kV^{-1.4}$ or

$$\begin{aligned}\frac{dP}{dt} = \frac{dP}{dV} \cdot \frac{dV}{dt} &= -1.4kV^{-2.4}\frac{dV}{dt} \\ &= \frac{-1.4(6{,}794.81)(-3)}{(25)^{2.4}} \\ &= 12.6\end{aligned}$$

The pressure is increasing at the rate of 12.6 lb/in.^{2} ▪

Exercise Set 3.5

Solve Exercises 1–44.

1. Gas is escaping from a balloon at the rate of 20 L/min. How fast is the surface area changing when the radius is 300 cm? (1 liter = 1000 cm^3)
2. *Construction* Sand is emptied down a chute at the rate of 10 ft^3/s forms a conical pile whose height is always twice the radius. At what rate is the radius changing when the height is 4 ft?
3. Water is emptied from a spherical tank of radius 24 ft. The water in the tank is 8 ft deep and decreasing at the rate of 2 ft/min. At this time, what is the rate of change of the radius of the top of the water?
4. *Environmental technology* A circular oil slick of uniform thickness is caused by a spill of 10 m^3 of oil. The thickness of the oil slick is decreasing at the rate of 1 mm/h. At what rate is the radius of the oil slick changing when the radius is 12 m?
5. *Transportation* A helicopter flies parallel to the ground at an altitude of 0.75 km and at a speed of 3 km/min. If the helicopter flies in a straight line that will pass directly over you, at what rate is the distance between you and the helicopter changing 1 min after it passes over you?
6. *Recreation* A rectangular swimming pool 15 m long and 10 m wide is 4 m deep at one end and 1 m deep at the other, with a constant drop from the shallow end to the deep end. Water is pumped into the pool at the rate of 50 L/min. At what rate is the water rising when it is 2 m deep at the deep end? (1 000 L = 1 m^3)

7. *Landscaping* A fish pond is in the shape of the bottom half of a sphere with a radius of 5 ft. Water is pumped into the pond at the rate of $10\,\text{ft}^3/\text{min}$. **(a)** At what rate is the water rising when it is 2 ft deep? **(b)** What is the rate when the water is 3 ft deep? **(c)** What is the rate when the water is 4 ft deep? [Note: $V_{\text{liquid}} = \frac{1}{3}\pi h^2(3R - h) = \frac{1}{6}\pi h(3r^2 + h^2)$.]

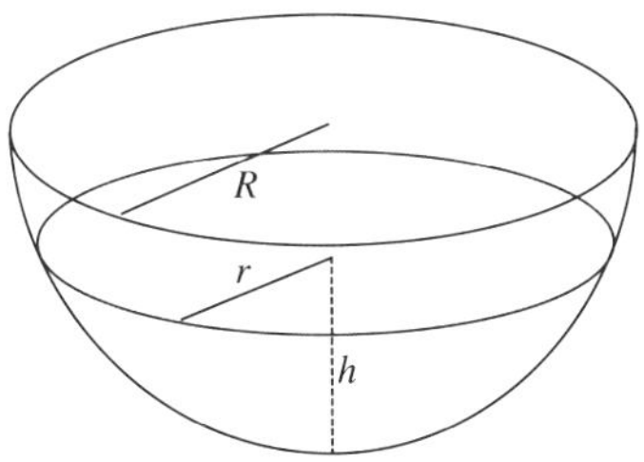

8. *Landscaping* When the pond in Exercise 7 is drained, the water drains at the rate of $45\,\text{ft}^3/\text{min}$. **(a)** At what rate does the water drop when it is 4 ft deep? **(b)** What is the rate when it is 2 ft deep?

9. A horizontal trough 10.00 m long has a cross-section in the shape of an equilateral triangle with a top width of 0.500 m. If the trough is filled with water at the rate of 25.0 L/min, how fast is the surface of the water rising when the depth is 0.175 m?

10. The base of a triangle is increasing at the rate of 5.00 cm/s and the height is decreasing at the rate of 4.00 cm/s. At what rate is the area of the triangle changing when the base is 2.625 m and the height is 1.375 m?

11. *Environmental science* Air is pumped into a spherical weather balloon at the rate of $60\,\text{cm}^3/\text{s}$. How fast is the surface area changing when the radius is 8 cm?

12. The height of a right-circular cone with radius of 2.43 m is increasing at the rate of 0.425 m/h. How fast is the lateral area of the cone changing when the height is 4.27 m?

13. A balloon is in the shape of a cylinder with a hemisphere on each end. The height of the cylinder is four times its radius. Air is pumped into the balloon at the rate of $60\,\text{cm}^3/\text{s}$. How fast is the surface area changing when the radius is 8 cm?

14. *Thermodynamics* When a metal plate is heated it expands. If the shape of the plate is circular and the radius is increasing at the rate of 0.2 mm/s, at what rate is the area of the top surface changing when the radius is 20 mm?

15. *Thermodynamics* If a metal plate in the shape of a square is heated and the length of one side is increasing at the rate of 0.3 mm/s, at what rate is the area of the top surface changing when the length of one side is 40 mm?

16. At a certain instant when air is changing adiabatically, the pressure is $60\,\text{dynes/cm}^2$ and is increasing at a rate of $4\,\text{dynes/cm}^2/\text{s}$. If the volume at that time is $100\,\text{cm}^3$, what is the rate at which the volume is changing?

17. *Physics* The pressure P in kilopascals (kPa) and volume V in cubic meters of a gas is given by $PV^{1.2} = 400$. If the pressure is increasing at a rate of 10% of itself per hour, find the rate at which the volume is changing expressed as a percent of itself.

18. *Meteorology* A weather balloon is rising vertically at the rate of 5 ft/s. An observer is 500 ft from a point on the ground directly below the balloon. At what rate is the distance between the balloon and the observer changing when the altitude of the balloon is 2,000 ft?

19. *Space technology* A radar station is following the path of a rocket which has been launched vertically from a site 2.430 km away. How fast is the rocket rising when it is 3.750 km high, if the distance between the rocket and the radar site is increasing at the rate of 325.0 m/s?

20. *Physics* **Boyle's law** for gases states that $PV = k$, where P is the pressure and V the volume, and k is a constant. At a certain instant the volume is $50\,\text{in.}^3$, the pressure $20\,\text{lb/in.}^2$, and the pressure is increasing at the rate of $1.5\,\text{lb/in.}^2$ every minute. At what rate is the volume changing at this instant?

21. *Physics* The displacement s in meters and the velocity v of an object in meters per second are related according to $v^2 = 1\,200 - 36.0s$. **(a)** Find the acceleration of the object. **(b)** What is the acceleration when $s = 2.45$ m?

22. A 10-m ladder is leaning against a vertical wall. If the base of the ladder slips away from the wall at the rate of 0.5 m/s, how fast is the top of the ladder going down the wall when the base of the ladder is **(a)** 2 m from the wall? **(b)** 3 m from the wall? **(c)** 4 m from the wall? **(d)** 6 m from the wall?

23. *Navigation* Two radar stations are at points A and B. Point B is 8 mi east of A. The two stations are tracking a ship. At a certain instant, the ship is 8 mi from A and this distance is increasing at the rate of 14 mph. At the same instant, the ship is also 6 mi from B and this distance is increasing at the rate of 2 mph. **(a)** Where is the ship? **(b)** How fast is it moving? **(c)** In what direction is it moving?

24. *Transportation* A car leaves an intersection and travels due north with a speed of 80.0 km/h. Another car leaves the same intersection 0.5 h later going due west at a speed of 100.0 km/h. How fast are the two cars separating

(a) 1.00 h after the second car started; and

(b) 2.00 h after the second car started?

25. A person 6 ft tall walks at the rate of 5 ft/s away from a street light that is 20 ft above the ground. **(a)** At what rate is the top of the person's shadow moving? **(b)** At what rate is the length of the shadow changing when the person is 10 ft from the base of the light? **(c)** What is the rate of change of the shadow when the person is 15 ft from the base of the light?

26. *Physics* A 75-lb weight is attached to a rope 50 ft long, which passes over a pulley at P, 20 ft above the ground. The other end of the rope is attached to the bumper of a truck at a point 2 ft above the ground. If the truck moves at the rate of 8 ft/s, how fast is the weight rising when it is 10 ft above the ground?

27. *Electricity* The electric resistance R of a certain resistor is a function of the temperature T, as shown by $R = 6 + 0.008T^2$, where R is in ohms and T in degrees Celsius. If the temperature is increasing at the rate of 0.01°C/s, how fast is the resistance changing when $T = 40°\text{C}$?

28. *Transportation* Two cars leave the same point, traveling on straight roads that are perpendicular. One car is traveling east at the rate of 30 mph and the other car is traveling north at the rate of 40 mph. How fast is the distance between them changing **(a)** after 3 min (0.05 h) and **(b)** after 6 min (0.10 h)?

29. *Electronics* The impedance Z in ohms (Ω) of a circuit is given by

$$Z = \frac{RX}{R+X}$$

If R is constant at $3.00\,\Omega$ and X is increasing at a rate of $1.45\,\Omega$/min, what is the rate at which Z is changing when $X = 1.05\,\Omega$?

30. *Electronics* The impedance Z in a series circuit is given by $Z^2 = R^2 + X^2$, where X is the reactance. If $X = 12\,\Omega$ and R increases at the rate of $3.0\,\Omega$/s, what is the rate at which Z is changing when $R = 6.0\,\Omega$?

31. *Electricity* If two variable resistances R_1 and R_2 are linked in parallel, then the effective resistance R is given by $\frac{1}{R} = \frac{1}{R_1} + \frac{1}{R_2}$. At a certain instant, R_1 is $4.00\,\Omega$ and increasing at the rate of $0.50\,\Omega$/s. At the same time, R_2 is $5.00\,\Omega$ and increasing at the rate of $0.40\,\Omega$/s. What is the rate of change of R?

32. *Electricity* The power P, in watts, in a circuit varies according to $P = RI^2$. If $R = 80.00\,\Omega$ and I varies at the rate of 0.24 A/s, what is the rate of change of P when $I = 2.5$ A?

33. *Electronics* The resistance in a resistor is given by $R = 35.0 + 0.0174T^2$, where R is in ohms and T is in degrees Celcius. How is the resistance changing when the temperature of the resistor is 47.0°C and is decreasing at the rate of 1.25°C/min?

34. *Aeronautical engineering* In the air, the speed of sound v, in m/s, is given by the formula $v = 331\sqrt{\frac{T}{273}}$, where T is the temperature in degrees Kelvin. (n degrees Celsius $= n + 273$ degrees Kelvin.) As an airplane increases altitude, the outside temperature decreases at the rate of 5.8°C/km. What is the rate of change in the speed of sound when the temperature outside the airplane is −30°C?

35. *Acoustical engineering* The **Doppler effect** is the apparent change in frequency of a sound source when there is relative motion between the source and the observer. When a person stands still, the apparent change in frequency f in hertz (Hz) of an object moving toward the person is given by the formula

$$f = f_s \frac{v_L}{v_L - v_s}$$

where f_s is the frequency of sound, in Hz, from the source; v_s is the velocity of the source in m/s; and v_L is the speed of sound in m/s. At 21°C, the speed of sound is about 343 m/s. A train is moving at a speed of 29 m/s toward an observer at the instant it sounds it horn at a frequency of 200 Hz. **(a)** What is the frequency of sound to the observer? **(b)** What is the apparent rate of change of sound to the observer?

36. *Medical technology* Blood flows faster the closer it is to the center of a blood vessel. According to **Poiseuille's laws**, the velocity, V, of blood is given by $V = k\left(R^2 - r^2\right)$, where R is the radius of the blood vessel, r is the distance of a layer of blood flow from the center of the vessel, and k is a constant.
(a) Find $\frac{dV}{dt}$ when $k = 475$. Treat r as a constant.
(b) Suppose that a cross-country skier's blood vessel has radius $R = 0.015$ mm and that the cold weather is causing the vessel to contract at a rate of $\frac{dR}{dt} = -0.001$ mm/min. How fast is the velocity of the blood changing?

37. *Medical technology* One of the treatments for high blood pressure is to give a patient some medication to dilate the blood vessels. Use **Poiseuille's law** with $k = 637.5$ to find the rate of change of the blood velocity when $R = 0.0250$ mm and R is changing at 0.002 mm/min.

38. *Police science* Crime rates are influenced by temperature. In a certain town with a population of 100,000, the crime rate has been approximated at $C = 0.1\,(T - 47)^2 + 95$, where C is the number of crimes per month and T is the average monthly temperature. The average temperature was 51°F, and by the end of the month the average monthly temperature was rising at the rate of 7°F/mo. How fast is the crime rate rising at the end of April?

39. *Electronics* The impedance of a circuit containing a resistance and an inductance in parallel is given by $Z = \frac{RX}{R+X}$. If $R = 7.5\,\Omega$ and X is decreasing at the rate of 2.5 Ω/s, at what rate is Z changing when $X = 4.0\,\Omega$?

40. *Petroleum engineering* An oil spill on the open sea in calm weather spreads in a circular pattern. If the radius of the oil slick increases at the rate of 0.9 m/s, how fast is the area of the spill increasing when its radius is 225 m?

41. *Electronics* The impedance of a circuit containing a resistance and an inductance in parallel is given by $Z = \frac{RX}{R+X}$. If $R = 7.5\,\Omega$ and Z is increasing at the rate of 2.0 Ω/s, at what rate is X changing when $X = 3.5\,\Omega$?

42. *Petroleum engineering* Oil is leaking from a tanker at the rate of 5 000 L/min. The leak results in an oil spill that is circular on the ocean's surface. The depth of the spill varies linearly from a maximum of 5.0 cm at the origin of the spill to a minimum of 0.5 cm at the outside edge of the oil slick. How fast is the radius of the slick increasing 4 h after the tanker started leaking?

43. *Heating technology* A boiler pipe has a circular cross-section. The inner radius of the pipe was originally 1 in. but is decreasing at the rate of 0.1 in./yr due to mineralization. At what rate is the cross-sectional area of its opening decreasing at the instant when the radius is 0.5 in.?

44. *Construction technology* A concrete pier is being constructed to support a bridge. The pier is in the shape of the frustum of a square pyramid. The bottom base of the pier measures 3 m on each side, the top base measures 2 m on a side, and the height is 2 m. If the concrete is poured at a constant rate of 0.01 m^3/s, at what rate is the height h of the poured concrete rising when $h = 1$ m?

In Your Words

45. Without looking in the book, describe the steps used to solve a related rate problem.

46. Explain how you can tell a problem involves related rates.

3.6 NEWTON'S METHOD

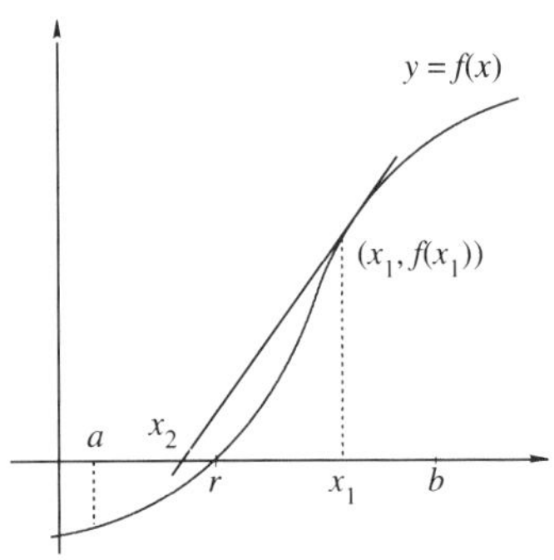

FIGURE 3.28

One method for finding the roots of a polynomial isby using linear interpolation. In this section, we will learn a faster method that uses derivatives. This new technique is known as **Newton's method** or the **Newton-Raphson method** for finding roots.

In using the linear interpolation method, we used the locator theorem to determine that a root existed between the real numbers a and b. We will begin Newton's method the same way. For a continuous function f over an interval, we have found two real numbers, a and b, in the interval, such that one of the values of $f(a)$ and $f(b)$ is positive and the other is negative.

Assume that $a < b$. We will select some point x_1 in the interval $[a,b]$. We know that the slope of the tangent to the graph of f at x_1 is given by $f'(x_1)$. We also know that the equation of this tangent line is $y - f(x_1) = f'(x_1)(x - x_1)$ or $y = f(x_1) + f'(x_1)(x - x_1)$. Suppose that x_2 is the x-intercept of this tangent line as shown in Figure 3.28. Then we know that $(x_2, 0)$ is a point on the tangent line. Thus, the tangent line satisfies the equation

$$0 = f(x_1) + f'(x_1)(x_2 - x_1)$$

If $f'(x_1) \neq 0$, we can rewrite this equation as

$$x_2 = x_1 - \frac{f(x_1)}{f'(x_1)}$$

The process can be continued. We can use x_2 to determine a new point x_3, and so on until we get a point x_n, where $f(x_n)$ is within some desired distance of zero. In general, if x_n is the nth approximation, then

$$x_{n+1} = x_n - \frac{f(x_n)}{f'(x_n)}$$

provided $f'(x_n) \neq 0$.

EXAMPLE 3.22

Use Newton's method to approximate the root between -4 and -3, if $P(x) = x^3 + x^2 - 7x + 3$.

Solution We will let our first guess, x_1, be -3. The results of the successive approximations are given in the following table. $P'(x) = 3x^2 + 2x - 7$

i	x_i	$P(x_i)$	$P'(x_i)$	$x_{i+1} = x_i - \frac{P(x_i)}{P'(x_i)}$	$P(x_{i+1})$
1	−3.0000	6.0000	14.0000	−3.4286	−1.5487
2	−3.4286	−1.5487	21.4087	−3.3563	−0.0490
3	−3.3563	−0.0490	20.0816	−3.3539	−0.0009

■

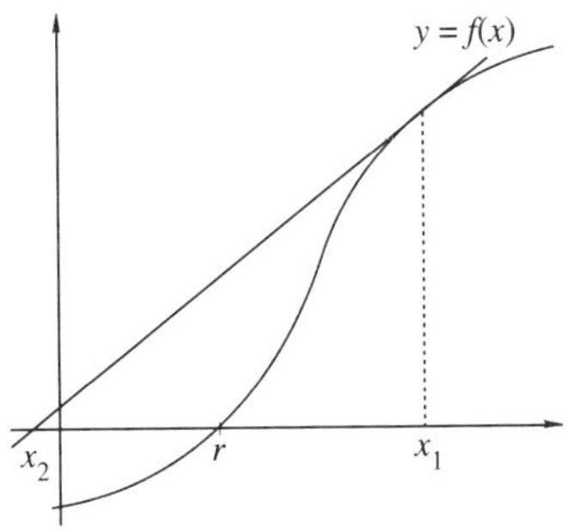

FIGURE 3.29

Linear interpolation takes five steps to get $x_6 = -3.3535$ and $P(x_6) = 0.0062$. Newton's method took three steps to get $x_4 = -3.3539$ and $P(x_4) = -0.0009$. It appears that Newton's method is almost twice as fast as the linear interpolation method.

Newton's method does not always work. In particular, if the first choice of x_1 is not sufficiently close to the root r, then the second approximation, x_2, will be farther away from r instead of closer. This is indicated in Figure 3.29. In general, you do not want to select a point where the derivative is close to zero. As a rule of thumb, whenever $|f'(x_1)| \leq 1$, then select a different choice for x_1.

EXAMPLE 3.23

Use Newton's method to approximate $\sqrt{11}$.

Solution This is equivalent to approximating the positive real root of $f(x) = x^2 - 11$. Since $f(3) = -2$ and $f(4) = 5$, we know there is a root between 3 and 4. Using $f'(x) = 2x$ and selecting $x_1 = 3$, we get the following results.

i	x_i	$P(x_i)$	$P'(x_i)$	$x_{i+1} = x_i - \dfrac{P(x_i)}{P'(x_i)}$	$P(x_{i+1})$
1	3.0000	−2.0000	6.0000	3.3333	0.1109
2	3.3333	0.1109	6.6667	3.3167	0.0005
3	3.3167	0.0005	6.6334	3.3166	−0.0002

Thus, $\sqrt{11} \approx 3.3166$. ■

Exercise Set 3.6

In Exercises 1–7, use Newton's method to approximate the real roots to four decimal places.

1. $x^3 + 5x - 8 = 0$
2. $x^3 - 4x + 2 = 0$
3. $x^4 - x - 4 = 0$
4. $x^4 - 2x^3 - 3x + 2 = 0$
5. $x^4 = 125$
6. $x^3 = 91$
7. $x^5 = 111$

Solve Exercises 8–9.

8. *Sheet metal technology* An open-topped barrel in the shape of a right circular cylinder is to be fabricated from sheet metal. The manufacturer has specified that it holds a volume of $V = 10\,000\,\text{cm}^3$ and has a total surface area of $A = 2\,400\,\text{cm}^2$.
   (a) Show that the radius r of the barrel must satisfy the equation $2V + \pi r^3 - Ar = 0$, where V and A are the area and volume given above.
   (b) Use Newton's method to find the radius of the barrel accurate to 4 decimal places.
9. *Air traffic control* One method for clearing fog from an airport runway is to ignite long troughs filled with fuel that have been placed along the sides of the runway. A trough is to be formed from a rectangular sheet of metal that measures $20\,\text{m} \times 0.7\,\text{m}$. The cross-section of the trough is an arc of a circle with radius r and subtended angle θ. If each trough is to hold $0.85\,\text{m}^3$ of fuel, use Newton's method to determine the size of θ.

In Your Words

10. What is Newton's method and how is it used?

11. What are the advantages and disadvantages of Newton's method compared to using a graphing calculator?

3.7 DIFFERENTIALS

In studying the derivative of a function $y = f(x)$, we used the notations $f'(x)$ and dy/dx to represent the derivative. The symbols dy and dx are called **differentials**. They may be given their own meaning as we'll see in this section.

Since $f'(x) = \dfrac{dy}{dx}$, we can solve this equation for dy and get $dy = f'(x)dx$, which is called the **differential of** f. In this equation, $dx \neq 0$.

EXAMPLE 3.24

Find the differential of the function $f(x) = 3 + x + 14x^{2/7}$.

Solution Here $f'(x) = 1 + 4x^{-5/7}$. Thus, the differential is

$$dy = f'(x)\,dx = \left(1 + 4x^{-5/7}\right) dx$$

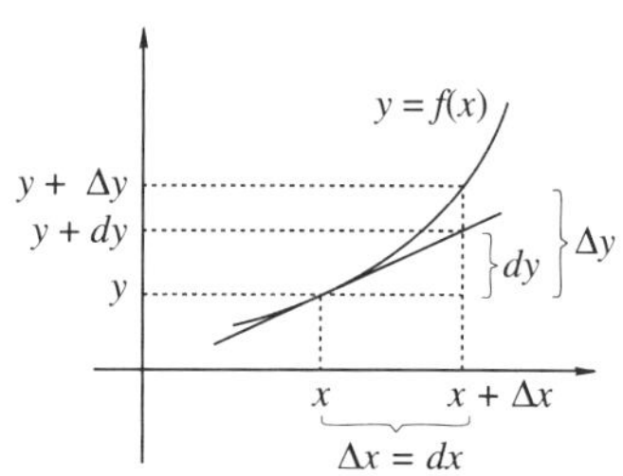

FIGURE 3.30

Figure 3.30 shows the change in height of a point that moves along the tangent line to the function $y = f(x)$ at the point $(x, f(x))$, as x changes from x to $x + \Delta x$. This change in height is called dy. The actual change in the value of y is the increment Δy. If we think of x as a fixed point, then dy is a linear function of the increment Δx. For this reason, dy is called the **linear approximation** to the actual increment Δy. We can approximate $f(x + \Delta x)$ by writing dy in place of Δy. Thus, we see that

$$f(x + \Delta x) = y + \Delta y \approx y + dy$$

Since $y = f(x)$ and $dy = f'(x)\Delta x$, we get the **linear approximation formula**:

$$f(x + \Delta x) \approx f(x) + f'(x)\Delta x$$

EXAMPLE 3.25

If $y = x^3$, determine Δy and dy, if $x = 2$ and $\Delta x = 0.01$.

Solution Since $f(x + \Delta x) = y + \Delta y$ and $y = f(x)$, we can see that $\Delta y = f(x + \Delta x) - f(x)$. Thus, when $x = 2$ and $\Delta x = 0.01$, we have

$$\begin{aligned}\Delta y &= (x + \Delta x)^3 - x^3 \\ &= (2.01)^3 - 2^3 \\ &= 8.1206 - 8 = 0.1206\end{aligned}$$

We also know that $dy = f'(x)dx$. If $f(x) = x^3$, then $f'(x) = 3x^2$. Since $dx = \Delta x$, we have

$$dy = 3(2)^2(.01) = 0.12$$

As you can see, dy is a good approximation for Δy.

Application

EXAMPLE 3.26

A spherical ball-bearing has a radius of 8 mm when it is new. What is the approximate volume of metal lost after it wears down to a radius of 7.981 mm?

Solution The exact volume of the metal lost is the change in volume ΔV of the sphere, where $V = \frac{4}{3}\pi r^3$. The change in radius is $\Delta r = 7.981 - 8.000 = -0.019$ mm. The change in volume is approximated by dV, where

$$\begin{aligned} dV &= 4\pi r^2 dr \\ &= 4\pi(8^2)(-0.019) \approx -15.281 \end{aligned}$$

The approximate loss in volume is 15.281 mm^3. ■

Application

EXAMPLE 3.27

A hemispherical dome has a radius of 25 ft. It is going to be given a coat of paint 0.01 in. thick. What is the approximate volume of paint needed for this job?

Solution What we are looking at is the change in volume of this hemisphere when the radius changes from 25.000 ft to 25.00083 ft (0.01 in. $\approx$ 0.00083 ft) or $\Delta r = 0.00083$ ft. The volume of a hemisphere is one-half that of a sphere, or $V = \frac{2}{3}\pi r^3$. Again, using differentials we see that $dV = 2\pi r^2 dr$. When $r = 25$ and $dr = 0.00083$, we have

$$\begin{aligned} dv &= 2\pi(25^2)(0.00083) \\ &\approx 3.2594 \end{aligned}$$

About 3.26 ft^3 of paint are needed. ■

Differentials can be used to estimate errors in results.

Application

EXAMPLE 3.28

A cylindrical can with an open top has an inside height of exactly 100 mm and a radius of 25 mm with a maximum error of 0.1 mm. Estimate the maximum error in the calculated volume of this can.

Solution If r denotes the radius and h the height, then $V = \pi r^2 h$. If the maximum error in the radius is denoted as dr, then we can see that $dV = 2\pi rh\,dr$. If $r = 25$, $h = 100$, and $dr = 0.1$, then

$$\begin{aligned} dV &= 2\pi(25)(100)(0.1) \\ &= 500\pi \approx 1\,570.8\text{ mm}^3 \end{aligned}$$

Thus, the maximum error in volume due to the error on the radius is 1 570.8 mm^3. ■

If there is an error in measurement, we define the **relative** or **average error** as the ratio of the actual error to the actual size of the quantity. This is often written in differential form as

$$\text{relative error} = \frac{dy}{y}$$

The relative error is usually expressed as a percentage.

In Example 3.28, the actual error in the radius was 0.1 mm when the radius was 25 mm. Thus the relative error in measuring the radius is $\frac{dr}{r} = \frac{0.1}{25} = 0.004 = 0.4\%$. The actual volume of the can is $62\,500\pi$ with a possible error of 500π. The relative error is $\frac{500\pi}{62\,500\pi} = 0.008 = 0.8\%$.

Exercise Set 3.7

In Exercises 1–6, find the differential of the given functions.

1. $f(x) = x^4 - x^2$
2. $g(x) = 3x^2 + 2x$
3. $h(x) = 5x^3 - x^2 + x$
4. $j(x) = (x^2 - 4)^3$
5. $k(x) = \sqrt[3]{4 - 2x}$
6. $m(x) = \frac{x}{x - 4}$

In Exercises 7–12, find Δy and dy for the given values of x and Δx.

7. $y = x^2 - x,\ x = 3,\ \Delta x = 0.2$
8. $y = x^3 + x,\ x = 2,\ \Delta x = 0.1$
9. $f(x) = 3x^2 - 4x + 1,\ x = 5,\ \Delta x = 0.15$
10. $g(x) = x^4,\ x = 4.1,\ \Delta x = 0.1$
11. $h(x) = \frac{1}{x^3},\ x = 3,\ \Delta x = 0.2$
12. $j(x) = \frac{x}{x + 1},\ x = 0.5,\ \Delta x = 0.1$

Solve Exercises 13–36.

13. *Physics* A tank filled with water is in the shape of a right-circular cone. The radius of the cone is measured to be 2 m with a maximum error of 0.01 m. The height is exactly 4 m. **(a)** What is the approximate volume of the water in the tank? **(b)** What is the maximum error in this volume? **(c)** What is the relative error in the volume?
14. *Physics* The radius of a circular disk is estimated to be 200 mm with a maximum error of 1.5 mm. **(a)** What is the maximum error in calculating the area of one side of the disc? **(b)** What is the relative error in the radius? **(c)** What is the relative error in the area?
15. The side of a square is measured to be 24 in. with a possible error of 0.02 in. **(a)** What is the maximum error in the area of the square? **(b)** What is the relative error?
16. A cube has an edge measuring 1.452 m. If the error in measurement is ± 0.5 mm, what is the approximate error in the volume?
17. The radius of a hemispherical dome is measured as 100 m with a maximum error of 10 mm. What is the maximum error in calculating its surface area?
18. With what accuracy must the radius of the dome in Exercise 17 be measured in order to have a percentage error of at most 0.01% in its calculated surface area?

19. The diameter of a sphere is measured to be 12.4 m. If measurement error is plus or minus half the unit of the most precise digit, that is ± 0.05 m, **(a)** what is the approximate error in the volume of the sphere due to measurement error? **(b)** What is the approximate relative error?

20. If the diameter in Exercise 19 is given as 12.40 m, what is the approximate error in the volume and the relative error in volume due to measurement error?

21. *Energy technology* The **Stefan-Bolzmann law** for the emission of radiant energy from the surface of a body is given by $R = \sigma T^4$, where R is the rate of emission per unit area, T is the temperature in degrees Kelvin, and σ is the Stefan-Bolzmann constant. If a relative error of 0.02 is made in T, what is the maximum error in R?

22. *Electricity* A wire has a resistance of R ohms given by $R = \frac{250}{r^2}$, where r is the radius of the wire. There is a possible error in the radius of $\pm 0.4\%$.
    (a) What is the approximate error in the resistance? (Express your answer in terms of r.)
    (b) What is the relative error?

23. *Electricity* In a dc circuit, $V = IR$. If a circuit has a constant voltage of 30 V, what is the approximate change in the current if the resistance changes from 10.0 to 10.1 Ω?

24. *Electricity* The equivalent resistance R when a resistance of 20 Ω and a variable resistance R_v are connected in parallel is

$$R = \frac{20R_v}{20 + R_v}$$

If the variable resistance is changed from 10.0 to 10.1 Ω, what is the approximate change in R?

25. *Electricity* The resistance in a resistor is given by $R = 35.0 + 0.0174T^2$, where R is in ohms (Ω) and T is the temperature in degrees Celsius (°C). If the error in reading the temperature at $T = 125$°C is $\pm - 0.5$°C, what is the approximate error in the resistance?

26. *Construction* A coat of paint 0.5 mm thick is applied to the surface of a spherical storage tank with a radius of 20 m. How many liters of paint are needed?

27. *Mechanical engineering* A company manufactures stainless-steel, spherical ball-bearings with a radius of 8 mm. The percentage error in the radius must be no more than 0.5%. What is the approximate percentage error for the surface area of the ball bearing?

28. *Material science* Stainless steel has a mass of 0.00794 g/mm^3.
    (a) What is the mass of one of the ball-bearings in Exercise 27?
    (b) What is the maximum error in the mass?

29. *Industrial technology* In a precision manufacturing process, ball bearings must be made with a radius of 0.6 mm, with a maximum error in the radius of ± 0.015 mm. Estimate the maximum error in the volume of the ball bearing.

30. *Pharmacology* The concentration of a certain drug in the bloodstream x hours after being administered is approximately $C(x) = \frac{5x}{9 + x^2}$. Use differentials to approximate the changes in concentration as x changes from **(a)** 1 h to 1.5 h, **(b)** 2 h to 2.25 h, and **(c)** 3 h to 3.25 h.

31. *Medical technology* A tumor is approximately spherical in shape. If the radius of the tumor changes from 15 mm to 15.5 mm, use differentials to find the approximate change in volume of the tumor.

32. *Petroleum engineering* The top of an oil slick is in the shape of a circle. Find the approximate increase in the area of the slick's top if its radius changes from 2.1 km to 2.3 km.

33. *Business* A company estimates that the revenue in dollars from the sale of n computers is given by

$$R(n) = 2345 + 0.02n^2 + 0.00015n^3$$

Use differentials to approximate the change in revenue from the sale of one more computer when 500 have been sold.

34. *Meteorology* A spherical weather balloon is being inflated. Use differentials to approximate the change in the volume of the balloon as the radius changes from 20.0 in. to 21.5 in.

35. *Machine technology* A plate in the shape of a right circular cylinder 3.2 cm thick was originally made 30.0 cm in diameter. If 0.2 cm is trimmed all around its edge, use differentials to approximate the volume of the metal that was trimmed.

36. *Acoustical engineering* The frequency, f, in Hertz, of the fundamental note on a guitar string is related to the tension, T, in the string; the length, L, of the string; and the linear density, μ, of the string according to the formula

$$f = \frac{1}{2L}\sqrt{\frac{T}{\mu}}$$

Estimate the amount by which the frequency will change if the tension is adjusted 1% during tuning. Let $L = 1.15$ m, $\mu = 0.0036$ kg/m, and $T = 620$ N.

In Your Words

37. What is meant by the differential terms dy and dx?

38. Explain the difference between dy and Δy.

3.8 ANTIDERIVATIVES

Until now, we have always differentiated a function to arrive at a solution. For example, if we were given a position function, we took its first derivative to get its velocity function and its second derivative to gets its acceleration function. Is it possible to reverse this process? For example, if we know the velocity function for an object, could we determine its position function? As we will see in this section, it is sometimes possible to reverse the differentiation process.

Suppose we have a function F and the derivative of F is f. Thus, $F' = f$. In this case, F is called the **antiderivative** of f.

EXAMPLE 3.29

Find an antiderivative of $f(x) = 5x^3$.

Solution We want to find $F(x)$, where $F'(x) = f(x) = 5x^3$. When we differentiate a polynomial, we reduce the power of x by 1. We also multiply the coefficient by the power of x. Thus, the derivative of x^n is nx^{n-1}. Since the power of f is 3, then this must be one less than the power of F. Thus, F is of the form $F(x) = kx^4$. Now $F'(x) = 4kx^3 = 5x^3$ and so $k = \frac{5}{4}$. Thus $F(x) = \frac{5}{4}x^4$.

But, we have not found a unique solution. For example, the derivative of $\frac{5}{4}x^4 + 9$ is also $5x^3$, and the derivative of $\frac{5}{4}x^4 - 7$ is $5x^3$. In fact, since the derivative of a constant C is 0, the antiderivative of $5x^3$ is $\frac{5}{4}x^4 + C$.

What we have discovered is that, in general, if $f(x) = kx^n$, then

$$F(x) = \frac{k}{n+1}x^{n+1} + C, \text{ if } n \neq -1$$

Notice that

$$F'(x) = \frac{k(n+1)}{n+1}x^{n+1-1} = kx^n = f(x).$$

So F is the antiderivative of f. A special case occurs when $n = 0$. Then $f(x) = k$ and $F(x) = kx + C$.

> **Antiderivatives Differ by a Constant**
>
> If G and F are both antiderivatives of the same function, f, then
>
> $$G(x) = F(x) + C \qquad \text{where } C \text{ is a constant.}$$

EXAMPLE 3.30

Find the antiderivative of $g(x) = 3x^2 - 2x + 5$.

Solution Using this new method to take the antiderivative of each term, we get

$$G(x) = x^3 - x^2 + 5x + C$$

You should check to see that $G'(x) = g(x)$.

EXAMPLE 3.31

Find the antiderivative of $h(x) = 4x^2 + 2x^{-2}$.

Solution The antiderivative of $4x^2$ is $\frac{4}{3}x^3$ and of $2x^{-2}$ is $-2x^{-1}$, so $H(x) = \frac{4}{3}x^3 - 2x^{-1} + C$.

We can apply antiderivatives to many problems. For example, we know that the acceleration due to the Earth's gravity is about 32 ft/s^2 or 9.8 m/s^2 downward. We also know that the acceleration is the derivative of the velocity or that the velocity function is an antiderivative of the acceleration. Similarly, the position function is an antiderivative of the velocity function. With enough information we can determine the value of the constant of the antiderivative. This is demonstrated in the next example.

Application

EXAMPLE 3.32

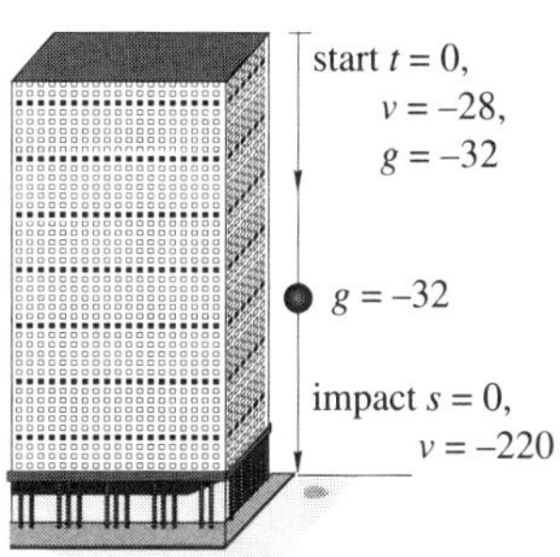

FIGURE 3.31

An object is thrown straight downward from the top of a building with an initial speed of 28 ft/s. The object strikes the ground with a speed of 220 ft/s. How tall is the building?

Solution First, let's set up a coordinate system with the ground level at $y = 0$, as shown in Figure 3.31. We know that acceleration is given by $a(t) = -32$. The velocity is the antiderivative of the acceleration and so $v(t) = -32t + C$. When the object is thrown, $t = 0$ and we are told that $v(0) = -28$. (It is negative because the velocity is downward.) Thus, we can see that $C = -28$ and $v(t) = -32t - 28$.

The object strikes the ground with a velocity of 220 ft/s downward or -220. Thus, when it strikes the ground, $v(t) = -32t - 28 = -220$. Solving for t, we see that it strikes the ground when $t = 6$ s.

The height of the object is given by the position function $s(t)$, and is the antiderivative of the velocity. Thus, $s(t) = -16t^2 - 28t + k$. When the object

EXAMPLE 3.32 (Cont.) hits the ground, $t = 6$ and $s(t) = 0$, so $-16(6^2) - 28(6) + k = 0$. Solving, we get $k = 744$ and $s(t) = -16t^2 - 28t + 744$. At $t = 0$ when the object is released, $s(0) = 744$ and so the height of the building is 744 ft. ■

Exercise Set 3.8

In Exercises 1–16, find the antiderivative of the given function.

1. $f(x) = 7$
2. $g(x) = 4x$
3. $h(x) = 2 - 3x^2$
4. $j(x) = x^2 - 3x + 5$
5. $k(x) = 4x^3 - 3x^2 + 2x + 9$
6. $m(x) = \frac{4}{5}x^3 - 6x^2$
7. $f(x) = x^{-4} + 2x^{-3} - x^{-2} + 5$
8. $g(x) = \dfrac{1}{x^3} + 2$
9. $h(x) = \sqrt{x} + x - \dfrac{1}{x^2}$
10. $j(x) = 3x^2 - \dfrac{1}{\sqrt[3]{x}} - \dfrac{4}{x^2}$
11. $k(x) = 2x^{-1.4} + 3.5x^{-6} + \dfrac{1.2}{x^{1.3}}$
12. $m(x) = \dfrac{2}{\sqrt{x}} + 3x^2\sqrt{x}$
13. $s(t) = t^2 + 2t$
14. $a(t) = \frac{2}{3}t - \frac{3}{4}t^2 + \sqrt{t^3}$
15. $v(t) = 42t - 5$
16. $p(v) = \frac{1}{3}v^3 - \dfrac{3}{v^4}$

Solve Exercises 17–36.

17. *Physics* A ball is dropped from a building that is 600 ft high. **(a)** How long does it take it to reach the ground? **(b)** What is its velocity when it hits the ground?

18. *Physics* A ball is dropped from a building that is 140 m high. **(a)** How long does it take to reach the ground? **(b)** What is its velocity when it hits the ground?

19. *Physics* A ball is thrown straight down from a building with an initial speed of 25 ft/s. It hits the ground with a speed of 185 ft/s. **(a)** How long was the ball in the air? **(b)** How high is the building?

20. *Physics* A ball is thrown straight down from a building with an initial speed of 12.1 m/s. It hits the ground with a speed of 66 m/s. **(a)** How long was the ball in the air? **(b)** How high is the building?

21. *Physics* A ball is thrown straight upward from the top of a building with an initial speed of 160 ft/s. It strikes the ground with a speed of 384 ft/s. **(a)** How long is it in the air? **(b)** How high is the building?
22. *Physics* A ball is thrown straight upward from the top of a building with an initial speed of 45 m/s. It strikes the ground with a speed of 120 m/s. **(a)** How long is it in the air? **(b)** How high is the building?
23. *Transportation* A car starting from rest has a constant acceleration of 3.2 m/s^2. How far has the car traveled when the speed is 32 m/s?
24. *Automotive engineering* A car's brakes are applied when it is going 45 mph and provides a constant deceleration of 20 ft/s^2. **(a)** How long does it take the car to stop? **(b)** How far does it travel before it stops? (Change 45 mph to ft/s.)
25. *Aeronautical engineering* A runway is 500 m long. If a plane lands at an end of the runway with a speed of 50 m/s, what is the constant acceleration that would bring the plane to a stop at the other end of the runway?

26. *Automotive engineering* The angular acceleration of a wheel is given by $\alpha(t) = 4 - t$ rad/s^2. Angular velocity is given by $\omega(t)$ and angular displacement by $\theta(t)$. If $\theta(0) = 0$ rad, and $\omega(0) = 10$ rad/s, find
(a) an expression for angular displacement,
(b) the angular displacement when the wheel turns in the opposite direction, and
(c) the time when the angular velocity is -14 rad/s.

27. *Machine technology* When the motor turning a flywheel is shut off, the angular velocity of the wheel is given by $\omega(t) = 12.4 - 3.4t + 0.30t^2$ rad/s.
(a) During the first 3 s after the motor has been shut off, how many revolutions does the wheel complete?
(b) When the angular acceleration is $\alpha(t) = -1$, how many revolutions has the wheel completed?

28. *Physics* The acceleration of a ball starting from rest and moving in a straight line is proportional to the time t in seconds (s). If the ball moves 24 m in the first 4 s, find an expression for the displacement $s(t)$.

29. *Electronics* An expression for the current i in a circuit in amperes (A) is given by $i = 4.4t - 2.1t^2$, where t is in seconds. If $i = \frac{dq}{dt}$, where q is the charge in coulombs (C), find **(a)** an expression for q at $t = 0.0$ s, $q = 5.0$ C, and **(b)** the charge at $t = 3.2$ s.

30. *Electronics* The current $i(t)$, in amperes (A), flowing to a capacitor is given by $i(t) = 6\sqrt{t}$, where t is time in seconds (s). If $i(t) = q'(t)$, where q is the charge in coulombs (C), find the charge on the capacitor at $t = 0.25$ s, if $q(0.16) = 0.347$ C.

31. *Electronics* The voltage in volts (V) across a coil is given by $v(t) = -N\phi'(t)$, where N is the number of turns in the coil, $\phi(t)$ is the magnetic flux in webers (Wb), and t is the time in seconds (s). If $v(t) = 2t - 4t^{1/3}$ and $\phi(0) = 0.020$ Wb when N is 200 turns, find **(a)** a general expression for $\phi(t)$ and **(b)** $\phi(0.729)$.

32. A general expression for the slope of a tangent at any point on a certain curve is given by $m = 4x + 2$. If the curve goes through the point $(-1, 5)$, state the equation of the curve.

33. The slope of a normal line to a curve $y = f(x)$ is given by $m = \sqrt{x}$. Find $f(x)$, if $f(4) = -3$.

34. *Industrial engineering* A stamping machine produces computer cases at a rate that varies over the working day according to the equation $\frac{dN}{dt} = 175 + 2t - 0.2t^2$, where t is the time in hours. **(a)** Find an expression for N and **(b)** calculate N for an 8-hour day if $N = 0$ at $t = 0$.

35. *Thermodynamics* The temperature, T, in °C in an industrial furnace varies from its center to a point outside. The rate of temperature change is given by the equation

$$\frac{dT}{dx} = -\frac{5750}{(x+1)^3}$$

Find an expression for T at a distance of x meters from the center if $T = 2900$°C at $x = 0$.

36. *Ecology* Biologists are treating a stream contaminated with bacteria. The level of contamination is changing at a rate of $\frac{dN}{dt} = -\frac{870}{t^2} - 210$ bacteria/cm^3/day, where t is the number of days since the treatment began. Find a function $N(t)$ to estimate the level of contamination if the level after 1 day was about 7320 bacteria/cm^3.

In Your Words

37. Explain how an antiderivative differs from a derivative.

38. A function has an infinite number of antiderivatives. How can you determine which is the "correct" antiderivative for a particular function?

CHAPTER 3 REVIEW

Important Terms and Concepts

Acceleration
Antiderivative
Concave downward
Concave upward
Concavity
Critical point
Critical value
Decreasing function
Differentials
Extrema
First derivative test
Increasing function
Inflection point
Linear approximation formula
Maximum
- Absolute
- Local
- Relative

Minimum
- Absolute
- Local
- Relative

Newton's method
Position function
Rectilinear motion
Related rates
Second derivative test
Speed
Velocity
- Average velocity
- Instantaneous velocity

Review Exercises

In Exercises 1–10, find critical values, the extrema, the intervals on which the curve is concave up and concave down, the inflection points, and the asymptotes. Sketch the curve.

1. $f(x) = x^4 - x^2$
2. $g(x) = x^4 - 32x$
3. $h(x) = 18x^2 - x^4$
4. $j(x) = 3x^4 - 4x^3 + 1$
5. $k(x) = \sqrt[5]{x}$
6. $f(x) = x\sqrt[3]{4-x}$
7. $g(x) = \dfrac{x^2+1}{x^2-4}$
8. $h(x) = \dfrac{x-1}{x+2}$
9. $j(x) = \dfrac{x}{x^2+x-2}$
10. $k(x) = \dfrac{x^3}{x^2-9}$

In Exercises 11 and 12, position functions of points moving rectilinearly are given. Find the function for the velocity and acceleration at time *t*. When are the position, velocity, and acceleration maximum and minimum for the given interval?

11. $s(t) = t^3 - 9t^2 + t,\ [-4, 4]$
12. $s(t) = t + \dfrac{4}{t},\ [1, 4]$

In Exercises 13–16, find the antiderivative of each function.

13. $f(x) = 3x^2 - 4x$
14. $g(x) = 2x^{-3} + 5x^6$
15. $h(x) = \sqrt{x} + x^2 - \dfrac{3}{x^2}$
16. $s(t) = 4.9t^2 - 3.6t + 14$

Solve Exercises 17–29.

17. *Physics* An object is thrown directly upward with a velocity of 288 ft/s. Its height $s(t)$ in feet above the ground after t s is given by $s(t) = 288t - 16t^2$. **(a)** What are the velocity and acceleration after t s? **(b)** What are the height, velocity, and acceleration after 4 s? **(c)** What is the maximum height? **(d)** At what time is the height a maximum? **(e)** When does the object strike the ground? **(f)** What is its velocity when it strikes the ground?

18. *Physics* An object rolls down a ramp such that the distance in centimeters after t s is given by $s(t) = 15t^2 + 5$. **(a)** What is its velocity after t s? **(b)** What is its velocity after 2 s? **(c)** When will its velocity be 105 cm/s?

19. *Business* Suppose it costs $1 + 0.00058v^{3/2}$ dollars/mi to operate a truck at v mph. Additional costs, including the driver's salary, amount to $25 per hour. What speed will minimize the total cost of a 1,000 mi trip?

20. *Construction* A cylindrical barrel is to be constructed to hold 246π ft^3 of liquid. The cost per ft^2 of constructing the side of the barrel is three times the cost of constructing the top and the bottom. What are the dimensions of the barrel that will cost the least to construct?

21. *Agriculture* What is the maximum rectangular area that can be enclosed with 1 200 m of fencing?

22. *Petroleum engineering* A storage tank is in the shape of a sphere with a radius of 8 ft. If oil is pumped into the tank at the rate of 20 gal/min, how fast is the oil level rising when it has reached a height of 11 ft? (1 gal ≈ 0.1335805 ft^3.)

23. The radius of a sphere is 1.5 m with a possible error in measurement of 0.01m=1 cm. What is the maximum error you would expect in **(a)** the calculation of the volume? **(b)** the calculation of the surface area?

24. *Sheet metal technology* A gutter is to be formed by bending up equal widths of the opposite sides of a sheet of metal that is 380 mm wide. If the bent sides are perpendicular to the base, what dimensions will give the largest capacity?

25. *Thermodynamics* The voltage of a certain thermocouple as a function of temperature is given by $v = 4.5T + 0.0003T^3$. What is the approximate change in voltage when the temperature changes from 100°C to 101°C?

26. *Navigation* Two ships leave the same port at the same time. One travels west at the rate of 12 km/h and the other travels south at the rate of 5 km/h. At what rate is the distance between them changing 3 h after they leave the port?

27. A right-circular cylinder is to have a volume of 20 kL. What dimensions will make the total surface area a minimum?

28. *Navigation* A motorist is in a desert 8 mi from point A, the nearest point on a long straight road. The motorist wants to get to an aid station at point B on the road 10 mi from A. The car can travel 15 mph in the desert and 55 mph on the road. At what point should the motorist meet the road to get to B in the shortest possible time?

29. *Physics* A ball is dropped from the top of a building. It takes 7 s for the ball to reach the ground. **(a)** How high is the building? **(b)** What is its speed when the ball strikes the ground?

CHAPTER 3 TEST

1. For the function $f(x) = x^4 - 8x^2$, determine **(a)** all critical values, **(b)** all extrema, **(c)** intervals on which f is concave upward, **(d)** intervals on which f is concave downward, and **(e)** sketch the curve.

2. For the function $g(x) = \dfrac{x^2}{9 - x^2}$, determine **(a)** all critical values, **(b)** all extrema, **(c)** intervals on which f is concave upward, **(d)** intervals on which f is concave downward, **(e)** asymptotes, and **(f)** sketch the curve.

3. Use Newton's method to approximate the positive root of $x^2 - 7 = 0$.
4. Determine the antiderivative of $f(x) = 5x^4 - 7x$.
5. The position function of points moving rectilinearly is given by the function $s(t) = t^3 - 12t^2 + 5$. **(a)** Find the function for the velocity and acceleration at time t. **(b)** When are the position, velocity, and acceleration maximum and minimum for the interval $[-2, 4]$?
6. A rectangular page is to contain 24 in.^{2} of print. The margins at the top and bottom of the page are $1\frac{1}{2}$ in., the margin toward the binding is $1\frac{1}{4}$ in., and the margin on the outside edge is $\frac{3}{4}$ in. What should the dimensions of the page be so that the least amount of paper is used?
7. A farm silo consists of a right circular cylinder of radius r and height h topped by a hemisphere. The volume of the silo is $V = \pi r^2 h + \frac{2}{3}\pi r^3$. If $h = 20$ m and $V = 800\,\text{m}^3$, use Newton's method to find r to two decimal places.

CHAPTER

4

Integration

Integration is used to determine area and volume. In Section 4.5, you will learn a numerical method for determining the cross-sectional area of the amount of dirt that must be excavated at a building site.

Courtesy of Mike Nelson

In Chapters 2 and 3, we looked at one of the two main parts of calculus—differential calculus. We have not finished looking at differential calculus, but it is time to take a look at the other main part of calculus—integral calculus.

In Chapter 1, we spent time looking at some of the foundations of calculus. As we discovered, calculus was developed in order to solve two types of problems. One of the problems, called the tangent question, formed the basis for our study of differential calculus. The other type of problem, which we called the area question, gives the foundation for our study of integral calculus.

4.1 THE AREA QUESTION AND THE INTEGRAL

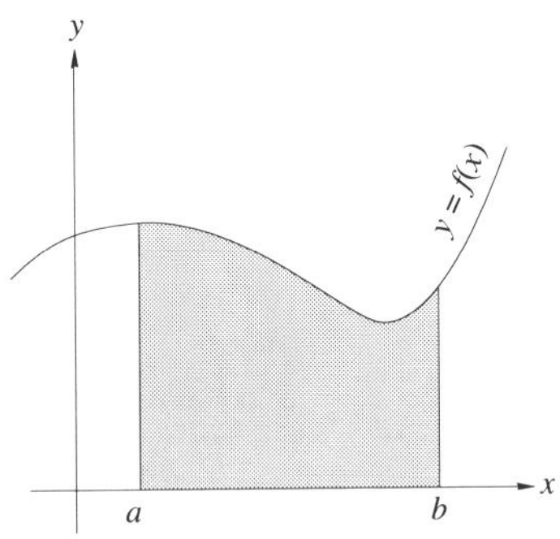

FIGURE 4.1

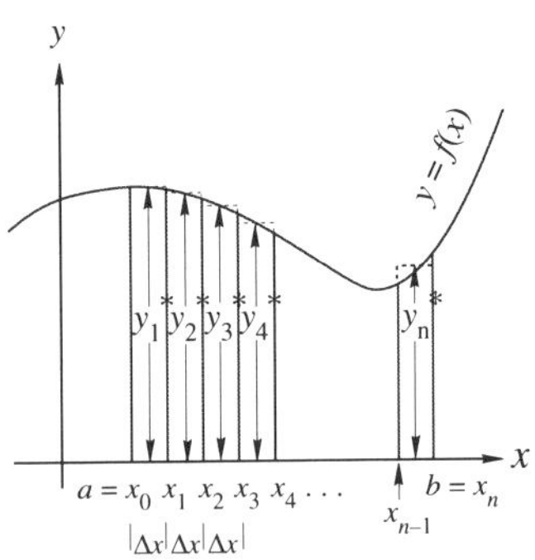

FIGURE 4.2

In Section 1.2, we looked at the area question. We began by looking at the shaded area in Figure 4.1. The area was enclosed by the curve $f(x)$, the x-axis, and the vertical lines $x = a$ and $x = b$.

We began our effort to find the area by dividing the interval $[a, b]$ into n equal segments. We let $a = x_0$ and $b = x_n$ with the first segment $[x_0, x_1]$, the second $[x_1, x_2]$, the third $[x_2, x_3]$, and so on. The midpoint of the ith interval was designated $x_i{}^*$ and we located the height of the curve at that point, labeling it $y_i{}^* = f(x_i{}^*)$. Thus, in the interval $[x_0, x_1]$, the midpoint was $x_1{}^*$ and the height was $y_1{}^* = f(x_1{}^*)$, as shown in Figure 4.2.

These $y_i{}^*$ were used as altitudes of rectangles. The width of each rectangle was $\Delta x = \dfrac{x_n - x_0}{n}$. So, the area of the first rectangle is $(y_1{}^*)(\Delta x)$, the area of the second rectangle is $(y_2{}^*)(\Delta x)$, the area of the third rectangle is $(y_3{}^*)(\Delta x)$, and so on. Thus, the area A under the curve was approximated by

$$\begin{aligned} A &\approx (y_1{}^*)(\Delta x) + (y_2{}^*)(\Delta x) + (y_3{}^*)(\Delta x) + \cdots + (y_n{}^*)(\Delta x) \\ &= (y_1{}^* + y_2{}^* + y_3{}^* + \cdots + y_n{}^*)\Delta x \\ &= [f(x_1{}^*) + f(x_2{}^*) + f(x_3{}^*) + \cdots + f(x_n{}^*)]\Delta x \end{aligned}$$

Using the summation or sigma notation, we have

$$A \approx \sum_{i=1}^{n} f(x_i{}^*)\,\Delta x$$

In this problem, we made things easy by letting each interval be the same size and by selecting the midpoints of the intervals. This did not have to be done. We can let each interval be a different size. The width of the ith interval would be $\Delta x_i = x_i - x_{i-1}$. We can also pick any point in the interval, say $x_i{}'$, and let $f(x_i{}')$ be the altitude of the rectangle at that point. The area would then be approximated by the formula

$$A \approx \sum_{i=1}^{n} f(x_i{}')\,\Delta x_i.$$

In the first method, the more intervals that are selected the closer we will get to the actual area. Taking this further, we can see that

$$A = \lim_{n\to\infty} \sum_{i=1}^{n} f(x_i{}^*)\,\Delta x$$

In the second method, as the widths of the intervals get smaller, we get a closer approximation to the actual area. If we let $\|\Delta x\|$ represent the width of the largest Δx_i, then we can see that

$$A = \lim_{\|\Delta x\|\to 0} \sum_{i=1}^{n} f(x_i{}')\,\Delta x_i$$

In fact, a combination of these two is the easiest to use. We will let the number of equal-sized intervals increase and we will select any convenient point in each interval to get

$$A = \lim_{n\to\infty} \sum_{i=1}^{n} f(x_i{}')\,\Delta x$$

Notice that, since each interval is the same size, increasing the number of intervals has the effect of decreasing the width of the interval.

We will adopt a more convenient notation to represent the above formula:

$$A = \int_a^b f(x)\,dx = \lim_{n\to\infty} \sum_{i=1}^{n} f(x_i{}')\,\Delta x$$

The middle expression, $\int_a^b f(x)\,dx$, is called the **definite integral of f from a to b**. The function $f(x)$ is known as the **integrand** and the numbers a and b are the **limits of integration**. More specifically, a is the **lower limit** and b is the **upper limit**. The symbol $\int$ is an elongated "S" for "sum" and is called the **integral sign**.

The following summation formulas will be helpful in working the examples and exercises in this section.

Summation Formulas

$$\sum_{i=1}^{n} c = nc$$

$$\sum_{i=1}^{n} cf(x) = c\sum_{i=1}^{n} f(x)$$

$$\sum_{i=1}^{n} i = 1+2+3+4+\cdots+n = \frac{n(n+1)}{2}$$

$$\sum_{i=1}^{n} i^2 = 1^2+2^2+3^2+4^2+\cdots+n^2$$

$$= 1+4+9+16+\cdots+n^2 = \frac{n(n+1)(2n+1)}{6}$$

$$\sum_{i=1}^{n} i^3 = 1^3+2^3+3^3+4^3+\cdots+n^3$$

$$= 1+8+27+64+\cdots+n^3 = \left[\frac{n(n+1)}{2}\right]^2$$

EXAMPLE 4.1

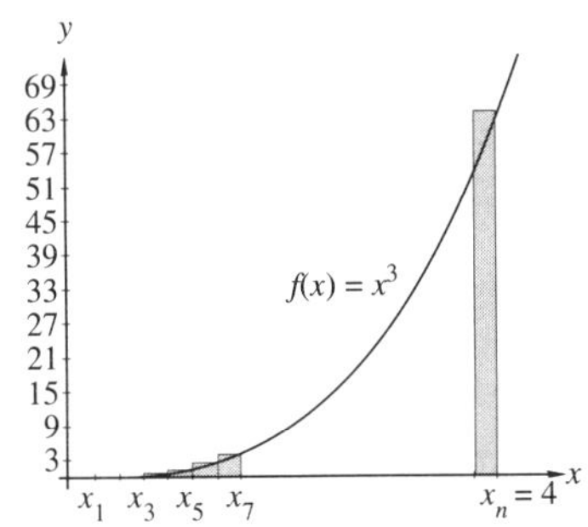

FIGURE 4.3

Evaluate $\int_0^4 x^3\,dx$.

Solution The interval is $[0,4]$ and we will divide it into n subintervals of equal length. Thus, each subinterval has a length $\Delta x = \frac{4}{n}$. While we can select any point in the subinterval, we will select the right-hand endpoint. The values of x_i, as shown in Figure 4.3, are $x_1 = 0 + \frac{4}{n} = \frac{4}{n}$, $\left(x_1 = 0 + \frac{4}{n}\right.$, because 0 is the left endpoint of the interval $\left.[0,4]\right)$, $x_2 = 0 + 2\left(\frac{4}{n}\right) = \frac{8}{n}$, $x_3 = 0 + 3\left(\frac{4}{n}\right) = \frac{12}{n}, \ldots,$ $x_i = 0 + i\left(\frac{4}{n}\right) = \frac{4i}{n}, \ldots, x_n = 0 + n\left(\frac{4}{n}\right) = 4.$

The altitude of each rectangle is computed using $f(x) = x^3$, and so

$$f(x_1) = \left(\frac{4}{n}\right)^3 = \frac{64}{n^3}$$

$$f(x_2) = \left[2\left(\frac{4}{n}\right)\right]^3 = 2^3 \cdot \frac{64}{n^3}$$

$$f(x_3) = \left[3\left(\frac{4}{n}\right)\right]^3 = 3^3 \cdot \frac{64}{n^3}$$

$$\vdots$$

$$f(x_i) = \left[i\left(\frac{4}{n}\right)\right]^3 = i^3 \cdot \frac{64}{n^3}$$

$$\vdots$$

$$\text{and} \qquad f(x_n) = 4^3 = 64$$

The sum of the areas of the rectangles can be written as

$$\sum_{i=1}^{n} f(x_i)\,\Delta x = \sum_{i=1}^{n} i^3 \frac{64}{n^3} \cdot \frac{4}{n} = \frac{256}{n^4} \sum_{i=1}^{n} i^3$$

since $\frac{256}{n^4}$ is a constant. Now $\sum_{i=1}^{n} i^3 = \left[\frac{n(n+1)}{2}\right]^2$, and so

$$\begin{aligned}\frac{256}{n^4}\sum_{i=1}^{n} i^3 &= \frac{256}{n^4}\left[\frac{n(n+1)}{2}\right]^2 \\ &= \frac{256}{n^4}\frac{n^2(n+1)^2}{4} \\ &= \frac{64(n+1)^2}{n^2} \\ &= \frac{64(n^2+2n+1)}{n^2}\end{aligned}$$

EXAMPLE 4.1 (Cont.)

We are now ready to find the integral.

$$\int_0^4 x^3\,dx = \lim_{n\to\infty} \frac{64\left(n^2+2n+1\right)}{n^2}$$
$$= \lim_{n\to\infty} \frac{64\left(1+\dfrac{2}{n}+\dfrac{1}{n^2}\right)}{1}$$
$$= 64$$

Thus, the area of the region is 64 square units.

EXAMPLE 4.2

Evaluate $\int_{-1}^{2} \left(x^2+1\right) dx$.

Solution The interval $[-1, 2]$ has length 3. Using equal subintervals, each subinterval will have length $\Delta x = \dfrac{3}{n}$.

Again, using the right-hand endpoint of each subinterval as shown in Figure 4.4, $x_i = -1 + \dfrac{3i}{n} = \dfrac{3i-n}{n}$. Since $f(x) = x^2+1$, we see that $f(x_i) = \left(\dfrac{3i-n}{n}\right)^2 + 1 = \dfrac{9i^2-6in+n^2}{n^2} + 1 = \dfrac{9i^2}{n^2} - \dfrac{6i}{n} + 2.$

Thus,

$$\sum_{i=1}^{n} f(x_i)\Delta x = \sum_{i=1}^{n} \left(\frac{9i^2}{n^2} - \frac{6i}{n} + 2\right)\left(\frac{3}{n}\right)$$
$$= \sum_{i=1}^{n} \left(\frac{27i^2}{n^3} - \frac{18i}{n^2} + \frac{6}{n}\right)$$
$$= \frac{27}{n^3}\sum_{i=1}^{n} i^2 - \frac{18}{n^2}\sum_{i=1}^{n} i + \frac{6}{n}\sum_{i=1}^{n} 1$$
$$= \frac{27}{n^3}\left[\frac{n(n+1)(2n+1)}{6}\right] - \frac{18}{n^2}\left[\frac{n(n+1)}{2}\right] + \frac{6}{n}\cdot n$$
$$= \frac{27}{n^3}\left(\frac{2n^3+3n^2+n}{6}\right) - \frac{18}{n^2}\left(\frac{n^2+n}{2}\right) + 6$$
$$= \left(9 + \frac{27}{2n} + \frac{9}{2n^2}\right) - \left(9 + \frac{9}{n}\right) + 6$$

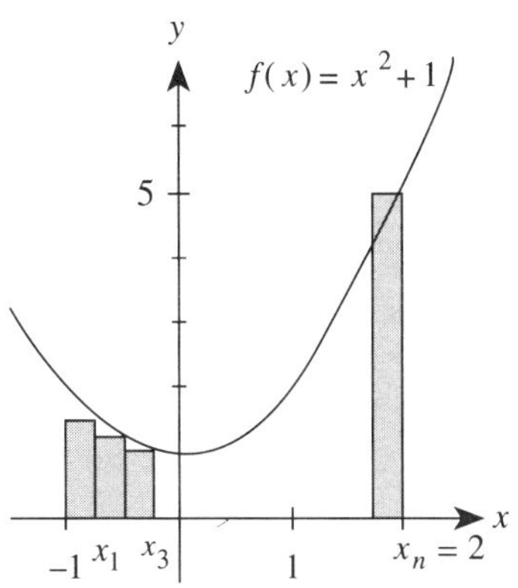

FIGURE 4.4

EXAMPLE 4.2 (Cont.)

We find the integral by taking the limit of this sum as follows:

$$\int_{-1}^{2}(x^2+1)\,dx = \lim_{n\to\infty}\left[\left(9+\frac{27}{2n}+\frac{9}{2n^2}\right)-\left(9+\frac{9}{n}\right)+6\right]$$
$$=(9+0+0)-(9+0)+6=6$$

Thus, the area of the region is 6 square units. ▪

Exercise Set 4.1

In Exercises 1–18, use summation notation and limits to evaluate the definite integrals.

1. $\int_1^3 x\,dx$
2. $\int_0^2 2x\,dx$
3. $\int_0^4 (3x+1)\,dx$
4. $\int_1^3 (2x+2)\,dx$
5. $\int_1^3 (2x-2)\,dx$
6. $\int_0^4 (1-3x)\,dx$
7. $\int_1^4 (3-2x)\,dx$
8. $\int_0^4 (x^2)\,dx$
9. $\int_0^2 (x^2-1)\,dx$
10. $\int_0^4 (x^2+2)\,dx$
11. $\int_0^2 (2x^2+1)\,dx$
12. $\int_0^4 (2-x^2)\,dx$
13. $\int_1^3 x^3\,dx$
14. $\int_1^3 (1-x^3)\,dx$
15. $\int_0^4 (x^3+2)\,dx$
16. $\int_0^3 (2x^3-1)\,dx$
17. $\int_0^2 (x^2+x)\,dx$
18. $\int_0^1 x^4\,dx$

Solve Exercise 19.

19. Use the computer program from Exercise 11 of Exercise Set 1.2 to solve Exercises 1–18. Input different values of n to see how close the computer approximations get to the actual value.

In Your Words

20. In the expression $\int_a^b f(x)\,dx$, what is the integrand, the lower limit of integration, and the upper limit of integration? What does each of them tell you?

21. What are the differences in the meaning between $\sum_1^{10} f(x)\,\Delta x$ and $\int_1^{10} f(x)\,dx$?

4.2 THE FUNDAMENTAL THEOREM OF CALCULUS

In this section, we will examine the relationship between the derivative and the definite integral. We will then look at three properties of integrals. One of the most important tools we will attain in this section will be a general method for evaluating a definite integral.

Figure 4.5 shows the graph of $y=f(x)$. We will assume that $f(x)$ is continuous on the closed interval $[a,b]$ and that its graph does not fall below the x-axis. From Section 4.1, we know that the area of the shaded region is given by $\int_a^b f(x)\,dx$. We

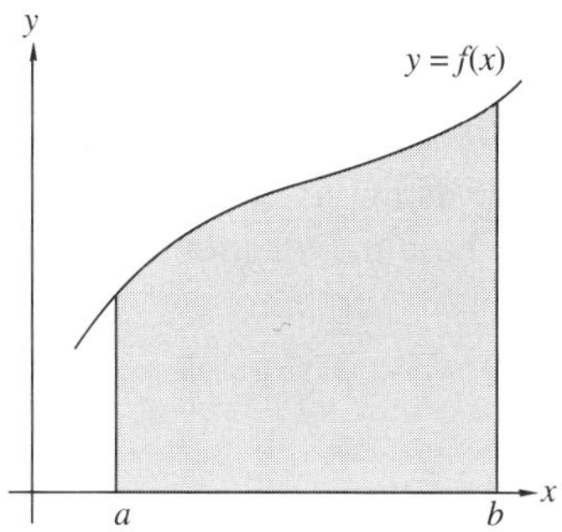

FIGURE 4.5

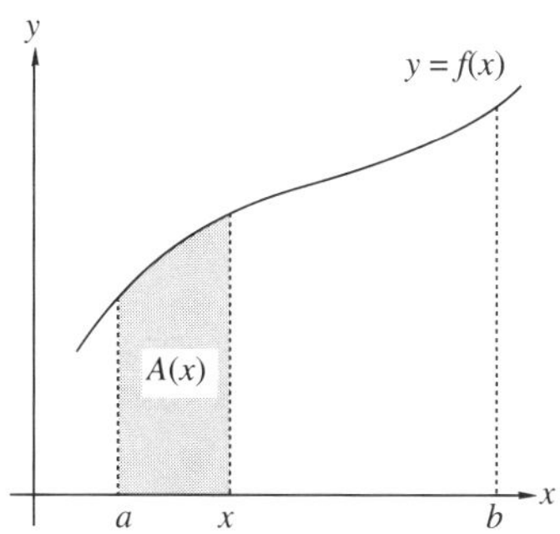

FIGURE 4.6

found that area by dividing the region into smaller and smaller rectangles, finding the sum of the areas of these rectangles, and taking the limit of this sum as the number of rectangles became infinitely large. We will now look at another way to determine the area.

Suppose there is a function $A(x)$, which we will refer to as an "area" function. The function $A(x)$ gives the area of the region below the graph of f and above the x-axis from a to x, where a is a fixed point and $a \leq x \leq b$. The region is shaded in Figure 4.6.

We already know two properties of $A(x)$:

1. $A(a) = 0$, since there is no area from a to a.
2. $A(b) = \int_a^b f(x)\,dx$.

Now, suppose that x is increased by h units as shown in Figure 4.7 and $A(x+h)$ is the area of the new shaded region. The difference in the area of the new region and the area of the original region is shown in Figure 4.8, and is written as $A(x+h) - A(x)$.

The area of the region in Figure 4.9 is the same as the area of a rectangle whose base is h and whose height is some value $\overline{y}$. Since the value of $\overline{y}$ depends on h, $\overline{y}$ is a function of h. The area of the rectangle is $A(x+h) - A(x)$ and is also represented as $\overline{y}h$. Thus, we have the equation

$$A(x+h) - A(x) = \overline{y}h$$

Dividing by h:

$$\frac{A(x+h) - A(x)}{h} = \overline{y}$$

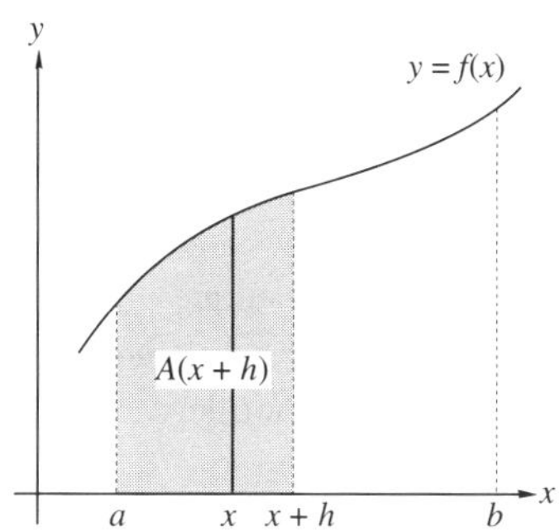

FIGURE 4.7

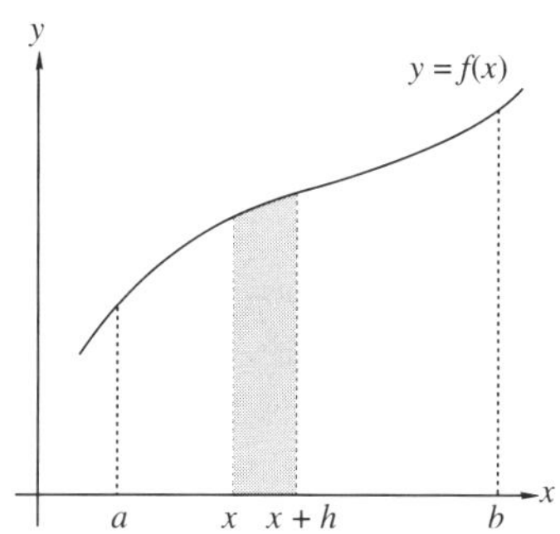

FIGURE 4.8

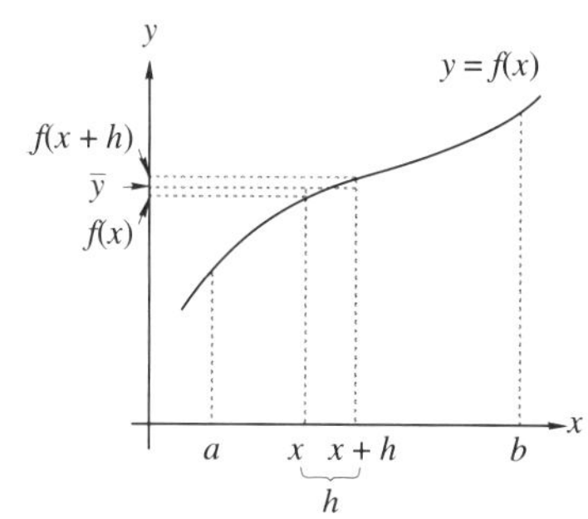

FIGURE 4.9

Now, what happens as the value of h approaches zero? One thing that will happen is that the value of $\overline{y}$ will get closer to the value of $f(x)$, and so

$$\lim_{h \to 0} \frac{A(x+h) - A(x)}{h} = \lim_{h \to 0} \overline{y} = f(x)$$

But, the left-hand side of this equation is the derivative of $A(x)$, thus we have

$$A'(x) = f(x)$$

What this means is that the area function $A(x)$ has the property that its derivative A' is the function f. That means that A is an antiderivative of f.

In Section 3.7, we found that the antiderivative of $f(x)$ was $F(x)+C$, where C is a constant. Thus, we know that $A(x) = F(x)+C$. Now, $A(a) = 0$, so when $x = a$, we have $0 = F(a)+C$ or $C = -F(a)$. Thus, we have the equation

$$A(x) = F(x) - F(a)$$

If $x = b$, then this equation becomes

$$A(b) = F(b) - F(a)$$

But, $A(b) = \int_a^b f(x)\,dx$. Combining these last two equations we get

$$\int_a^b f(x)\,dx = F(b) - F(a)$$

As we can see, there is a relationship between a definite integral and differentiation. All we need to do to find $\int_a^b f(x)\,dx$ is find an antiderivative of f, which we call F, and subtract the value of F at the lower limit from its value at the upper limit. This result is known as the **Fundamental Theorem of Calculus**. It was discovered independently by Newton and Leibnitz.

Fundamental Theorem of Calculus

If f is continuous on the interval $[a, b]$, then

$$\int_a^b f(x)\,dx = F(x)\Big|_a^b = F(b) - F(a)$$

where F is a function such that $F' = f$.

In stating the Fundamental Theorem of Calculus, we introduced some new notation, $F(x)\,\Big|_a^b$. This notation is often used to help keep track of the limits of integration after we have integrated but before the antiderivative is evaluated.

EXAMPLE 4.3

Evaluate $\int_0^3 x^3\,dx$.

Solution This is the same function we evaluated in Example 25.1. We know that the antiderivative of x^3 is $\frac{x^4}{4}+C$, so

$$\begin{aligned}\int_0^3 x^3\,dx &= \left.\frac{x^4}{4}\right|_0^4\\ &= \frac{4^4}{4}-\frac{0^4}{4}=64-0=64\end{aligned}$$

Note

When using antiderivatives with the Fundamental Theorem of Calculus, the constant is not used. If the antiderivative of $f(x)$ is $F(x)+C$, then

$$\begin{aligned}\int_a^b f(x)\,dx &= F(x)+C\,\Big|_a^b\\ &= [F(b)+C]-[F(a)+C]\\ &= F(b)-F(a)\end{aligned}$$

As you can see, the constant C does not have to be included since $C-C=0$. From now on we will call the process of finding an antiderivative by its more common title of **integration**.

EXAMPLE 4.4

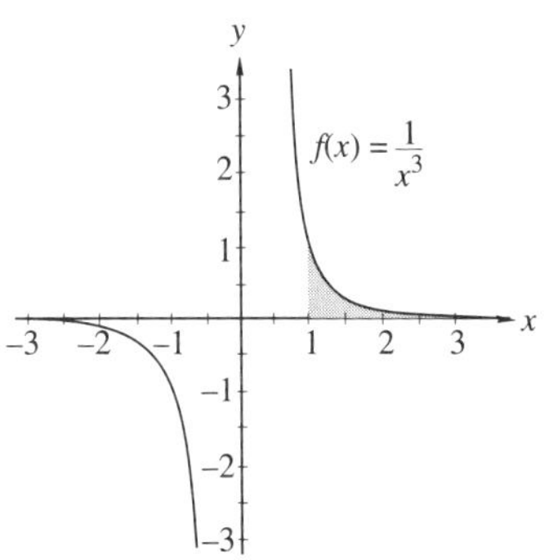

FIGURE 4.10

Find the area under the curve of $f(x)=\frac{1}{x^3}$ from $x=1$ to $x=3$.

Solution The desired area is shown in Figure 4.10. Note that f is continuous from $x=1$ to $x=3$. Since the curve lies above the x-axis in the interval $[-1,3]$, the area can be found by evaluating $\int_1^3 f(x)\,dx$. Using the Fundamental Theorem of Calculus, we have

$$\begin{aligned}\int_1^3 \frac{1}{x^3}\,dx &= \int_1^3 x^{-3}\,dx\\ &= \left.\frac{1}{-2}x^{-2}\right|_1^3\\ &= \left(\frac{1}{-2(3^2)}\right)-\left(\frac{1}{-2\cdot 1^2}\right)\\ &= \frac{-1}{18}+\frac{1}{2}\\ &= \frac{8}{18}=\frac{4}{9}\text{ square units}\end{aligned}$$

At the present time we can only find the area when a curve is above the x-axis.

EXAMPLE 4.5

Evaluate $\int_1^3 (3x-2)\,dx$.

Solution

$$\int_1^3 (3x-2)\,dx = \left(\frac{3}{2}x^2 - 2x\right)\Big|_1^3$$
$$= \left[\frac{3}{2}(3^2) - 2(3)\right] - \left[\frac{3}{2}(1^2) - 2(1)\right]$$
$$= \left(\frac{27}{2} - 6\right) - \left(\frac{3}{2} - 2\right) = 8$$

The following are three properties, or rules, of definite integrals.

Properties of Definite Integrals

Rule 1: $\int_a^b kf(x)\,dx = k\int_a^b f(x)\,dx$, where k is a constant

Rule 2: $\int_a^b [f(x) \pm g(x)]\,dx = \int_a^b f(x)\,dx \pm \int_a^b g(x)\,dx$

Rule 3: If f is continuous on an interval containing a, b, and c, then

$$\int_a^b f(x)\,dx = \int_a^c f(x)\,dx + \int_c^b f(x)\,dx$$

EXAMPLE 4.6

Evaluate $\int_0^2 (3x^2 - \sqrt{x})\,dx$.

Solution

$$\int_0^2 (3x^2 - \sqrt{x})\,dx = \int_0^2 (3x^2 - x^{1/2})\,dx$$
$$= \left(x^3 - \frac{2}{3}x^{3/2}\right)\Big|_0^2$$
$$= \left[2^3 - \frac{2}{3}(2)^{3/2}\right] - \left[0^3 - \frac{2}{3}(0)^{3/2}\right]$$
$$= 8 - \frac{4\sqrt{2}}{3}$$

EXAMPLE 4.7

Evaluate $\int_1^2 4x^2\,dx$.

Solution According to rule 1,

$$\int_1^2 4x^2\,dx = 4\int_1^2 x^2\,dx = 4\left(\frac{1}{3}x^3\Big|_1^2\right)$$
$$= 4\left(\frac{8}{3} - \frac{1}{3}\right)$$
$$= \frac{28}{3}$$

EXAMPLE 4.8

Evaluate $\int_1^2 (x^2 - x)\,dx$.

Solution Using rule 2, we see that

$$\int_1^2 (x^2 - x)\,dx = \int_1^2 (x^2)\,dx - \int_1^2 x\,dx$$
$$= \frac{1}{3}x^3\Big|_1^2 - \frac{1}{2}x^2\Big|_1^2$$
$$= \frac{7}{3} - \frac{3}{2} = \frac{5}{6}$$

EXAMPLE 4.9

Evaluate $\int_{-1}^1 3x^2\,dx$.

Solution Using rule 3, we know that

$$\int_{-1}^1 3x^2\,dx = \int_{-1}^0 3x^2\,dx + \int_0^1 3x^2\,dx$$
$$= x^3\Big|_{-1}^0 + x^3\Big|_0^1$$
$$= (0+1) + (1-0) = 2$$

At present these three rules are not particularly needed. But, we are setting the foundation for later, when they will play an important part in our continued development and use of calculus.

Application

EXAMPLE 4.10

From the top of a building 192 ft high, a ball is thrown upward with an initial velocity of 64 ft/s. **(a)** Find the position function giving the height s as a function of the time t. **(b)** When does the ball strike the ground?

Solutions

(a) We will let $t = 0$ be the initial time. We are given the initial conditions that

$$s(0) = 192 \quad \text{(initial height is 192 ft)}$$
$$s'(0) = 64 \quad \text{(initial velocity is 64 ft/s)}$$

Since acceleration due to gravity is -32 ft/s^2, we have

$$s''(t) = -32$$

and integrating, we obtain

$$\begin{aligned} s'(t) &= \int s''(t)\,dt \\ &= \int -32\,dt = -32t + C_1 \end{aligned}$$

Since $s'(0) = 64 = -32(0) + C_1$, we have $C_1 = 64$. Thus, $s'(t) = -32t + 64$. Now, by integrating $s'(t)$, we have

$$s(t) = \int s'(t)\,dt = \int (-32t + 64)\,dt = -16t^2 + 64t + C_2$$

Using the initial height $s(0) = 192 = -16(0^2) + 64(0) + C_2$, we obtain $C_2 = 192$. We have now found the position function

$$s(t) = -16t^2 + 64t + 192$$

(b) The ball will hit the ground when $s(t) = 0$.

$$\begin{aligned} s(t) = -16t^2 + 64t + 192 &= 0 \\ -16(t^2 - 4t - 12) &= 0 \\ -16(t-6)(t+2) &= 0 \\ t &= -2, 6 \end{aligned}$$

Since t must be positive, we conclude that the ball hits the ground 6 s after it was thrown. ■

Exercise Set 4.2

In Exercises 1–37, use the Fundamental Theorem of Calculus to evaluate the given definite integral.

1. $\int_0^5 6\,dx$

2. $\int_1^2 (x-6)\,dx$

3. $\int_1^5 2x\,dx$

4. $\int_{-1}^2 (6-2x)\,dx$

5. $\int_{-1}^3 \frac{5}{3}x^3\,dx$

6. $\int_0^5 4x^2\,dx$

7. $\int_1^2 \frac{8}{5}x^3\,dx$

8. $\int_{-2}^2 6x\,dx$

9. $\int_0^2 (4x+5)\,dx$

10. $\int_1^3 (-4)\,dt$

11. $\int_2^0 (4x+5)\,dx$

12. $\int_{-1}^2 (2t-t^2)\,dt$

13. $\int_1^3 (-4x)\,dx$

14. $\int_2^5 (5w^2+2w)\,dw$

15. $\int_0^3 (t^2+2t)\,dt$

16. $\int_{-2}^0 (5-t^3)\,dt$

17. $\int_{-1}^3 (3y^2-2y)\,dy$

18. $\int_0^4 (s^2-4s+16)\,ds$

19. $\int_1^4 (4-2w^2)\,dw$

20. $\int_{-2}^0 -dt$

21. $\int_2^3 (x^2+2x+1)\,dx$

22. $\int_{-3}^0 (3y-y^3)\,dy$

23. $\int_3^2 (x^2+2x+1)\,dx$

24. $\int_1^4 \left(\sqrt{x}-\frac{1}{\sqrt{x}}\right)dx$

25. $\int_6^8 dx$

26. $\int_{1/2}^1 \frac{4\,dx}{x^2}$

27. $\int_{-4}^{-2} (3y^2-y-1)\,dy$

28. $\int_1^4 \left(\frac{3}{x^3}-\frac{2}{x^2}\right)dx$

29. $\int_2^4 (x^2-2x+1)\,dx$

30. $\int_0^4 \left(\sqrt{x}\right)^3 dx$

31. $\int_1^2 -3x^{-4}\,dx$

32. $\int_0^1 3x^{5/6}\,dx$

33. $\int_{-2}^{-1} \frac{t^{-2}}{3}\,dt$

34. $\int_0^2 (x^4-x^3)\,dx$

35. $\int_0^2 \sqrt{x}\,dx$

36. $\int_1^3 x\sqrt[3]{x}\,dx$

37. $\int_{1/2}^3 \left(\frac{1}{\sqrt{x}}-2\right)dx$

Solve Exercises 38–48.

38. The Fundamental Theorem of Calculus *seems* to say that

$$\int_{-1}^{1} x^{-2}\,dx = -x^{-1}\Big|_{-1}^{1} = -2$$

But x^{-2} is always positive and we would expect the area under this curve to be positive. Why does the Fundamental Theorem of Calculus appear to give a wrong answer?

39. *Physics* A ball is projected upward along a frictionless inclined plane. The velocity of the ball is given by $v(t) = 4.00t - 1.00t^2$ in meters per second (m/s). What is the change in displacement in the first 4.00 s?

40. *Mechanical technology* The angular velocity in radians per second (rad/s) for a gear is given by $\omega(t) = 1.00t^3 - 27.0t$. What is the change in angular displacement between $t = 1.00$ s and $t = 3.00$ s?

41. *Dynamics* From the top of a building 180 m high, a ball is thrown down with an initial velocity of 10 m/s. **(a)** How fast is the ball moving after 3 s? **(b)** How far does the ball move in the first 3 s? **(c)** How long does it take the ball to hit the ground?

42. *Aeronautics* A hot-air balloon is rising slowly, so the balloonist drops a sandbag from the bottom of the balloon. If the sandbag was dropped at an elevation of 840 ft, what was the speed of the balloon at the time the sandbag was dropped if it took 8 s for the sandbag to hit the ground?

43. *Nuclear technology* A charged particle enters a linear accelerator. Its initial velocity is 500 m/s and its velocity increases with a constant acceleration to a velocity of 10 500 m/s in 0.01 s. **(a)** What is the acceleration? **(b)** How far does the particle travel in that 0.01 s?

44. *Electricity* The current i in amperes (A) to a capacitor is given by $i = 1.00t + 1.00\sqrt{t}$, where t is time in seconds (s). If $i = \frac{dq}{dt}$, where q is the charge in coulombs, find the change in the charge on the capacitor from $t = 1.00$ s to $t = 9.00$ s.

45. *Environmental science* The manager of a wildlife preserve has started a management program to control the preserve's moose population. It is estimated that the population will continue to grow according to the function $N'(t) = 25 - 6t^{1/2}$ per year, where t is the number of years since the plan was implemented. Find the change in the population during the first 9 years of the program.

46. *Industrial engineering* A stamping machine produces computer cases at a rate that varies over the working day according to the equation $\dfrac{dN}{dt} = 175 + 2t - 0.2t^2$, where t is the time in h.
(a) Find an expression for $N(t)$ if $N(0) = 0$.
(b) Find the number of computer cases the machine produces in the first 4 hours of the day.
(c) Find the number of computer cases the machine produces in the last 4 hours of the 8-hour day.

47. *Forestry* A certain type of hardwood tree will grow at the rate of $G'(t) = 0.5 + 4t^{-3}$ ft/yr, where t is time in years after the tree is 20 years old.
(a) How much will the tree grow in the second year after it is 20 years old?
(b) How much will the tree grow in the fifth year after it is 20 years old?
(c) How much will the tree grow in years 1 through 5 after it is 20 years old?

48. *Ecology* Pollution from a factory is entering a lake. The rate of concentration of the pollutant at time t is given by $P'(t) = 91t^{5/2}$, where t is the number of years since the factory began introducing pollutants into the lake. Ecologists estimate that the lake can accept a total level of pollution of 5,720 units before all the fish are killed. How long can the factory pollute the lake before all the fish are killed?

In Your Words

49. Explain the fundamental theorem of calculus.

50. Three properties of definite integrals were given in the text. Explain what each one means and how you can use it.

4.3 THE INDEFINITE INTEGRAL

The Fundamental Theorem of Calculus showed us a relationship between definite integrals and antiderivatives. In fact, we found that we were able to evaluate $\int_a^b f(x)\,dx$ by finding an antiderivative of f. The general form of the antiderivative is called the **indefinite integral**. The integral symbol, $\int$, without any limits is used to indicate an indefinite integral. Thus, the notation $\int f(x)\,dx$ is used to indicate all the antiderivatives of f.

EXAMPLE 4.11

$$\int x^2\,dx = \frac{1}{3}x^3 + C$$

$$\int 4x^3\,dx = x^4 + C$$

$$\int (2x^{-3} + 5)\,dx = -x^{-2} + 5x + C$$

In each of these examples, C represents an arbitrary constant called the **constant of integration**.

Note It is important to point out the difference between a definite integral and an indefinite integral. The definite integral is a number; the indefinite integral is a family of functions.

The value of the definite integral depends on the limits of integration. For example,

$$\int_1^2 3x^2\,dx = x^3\Big|_1^2 = 7 \qquad \int_1^3 3x^2\,dx = x^3\Big|_1^3 = 26$$

Notice that in each case the definite integral is a number.

However, the indefinite integral has no limits to affect the answer. Thus,

$$\int 3x^2\,dx = x^3 + C$$

and C, or the constant of integration, is used to show that this integral represents a family of functions and not just one particular value.

Properties of Indefinite Integrals

At the end of Section 4.2, we gave three rules for definite integrals. We are now going to give several rules for indefinite integrals. Two of these rules are the indefinite integral versions of previous rules. All of them are based on the six rules for derivatives we gave in Section 2.2.

Because an indefinite integral of f is defined as an antiderivative of f, we know that we can differentiate the indefinite integral and get the original function. Thus,

Rule 1

$$\frac{d}{dx}\int f(x)\,dx = f(x)$$

EXAMPLE 4.12

$$\frac{d}{dx}\int \sqrt[3]{x^2-5}\,dx = \sqrt[3]{x^2-5}$$

The next rule is similar to one we had for definite integrals.

Rule 2

$$\int kf(x)\,dx = k\int f(x)\,dx, \text{ where } k \text{ is a real number}$$

This rule states that we can, in effect, factor a constant out of a function before we integrate it.

The next rule is also similar to one we had for definite integrals.

Rule 3

$$\int [f(x) \pm g(x)]\, dx = \int f(x)\, dx \pm \int g(x)\, dx$$

Thus, the indefinite integral of a sum (or difference) is equal to the sum (or difference) of the indefinite integrals.

The next rule is the antiderivative of rule 3 from Section 2.2.

Rule 4

$$\int x^n\, dx = \frac{1}{n+1} x^{n+1} + C, n \neq -1$$

EXAMPLE 4.13

$$\begin{aligned}
\int (3x^2 + 4x - 5)\, dx &= \int 3x^2\, dx + \int 4x\, dx - \int 5\, dx && \text{By rule 3} \\
&= 3\int x^2 dx + 4\int x\, dx - 5\int dx && \text{By rule 2} \\
&= 3\left(\frac{x^3}{3}\right) + 4\left(\frac{x^2}{2}\right) - 5x + C && \text{By rule 4} \\
&= x^3 + 2x^2 - 5x + C
\end{aligned}$$

We used three of the four rules in Example 4.13.

Method of Substitution

All of the rules just given are useful, but they do not show us how to integrate $\int (3x-7)^{10}\, dx$, $\int (x^4+6)^3 4x^3\, dx$, or $\int \sqrt[3]{4x^3 - 8}\, x^2\, dx$. The first integral could be found by expanding $(3x-7)^{10}$ and then integrating, but it would be very long and inefficient. Also, this method will not work on the last two integrals.

Some integrals that cannot be evaluated directly by using the previously stated rules can sometimes be evaluated using the method of substitution. This method involves the introduction of a function that changes the integrand, such that the given rules will work when integrating. Examples 4.14 and 4.15 show how to use the method of substitution on the last two integrals given.

EXAMPLE 4.14

Use the method of substitution to determine

$$\int (x^4+6)^3 4x^3\,dx$$

Solution In this integral, we will first use the substitution $u = x^4+6$. Differentiating with respect to x, we get $\frac{du}{dx} = 4x^3$ or $du = 4x^3\,dx$. We then substitute u for x^4+6 and du for $4x^3\,dx$ in the original integral, with the following result:

$$\begin{aligned}
\int (x^4+6)^3 4x^3\,dx &= \int u^3\,du && \text{Let } x^4+6=u \text{ and } 4x^3\,dx = du.\\
&= \frac{1}{4}u^4 + C && \text{Integrating with respect to } u.\\
&= \frac{1}{4}\left(x^4+6\right)^4 + C && \text{Substituting for } u\text{, we obtain this.}
\end{aligned}$$

EXAMPLE 4.15

Use the method of substitution to find $\int \sqrt[3]{4x^3-8}\,x^2\,dx$.

Solution We will let $u = 4x^3-8$, and so $\frac{du}{dx} = 12x^2$. Thus, $du = 12x^2\,dx$ and $x^2\,dx = \frac{1}{12}\,du$. Substituting these values in the given integral, we get the following solution:

$$\begin{aligned}
\int \sqrt[3]{4x^3-8}\,x^2\,dx &= \int \left(4x^3-8\right)^{1/3} x^2\,dx && \text{Let } 4x^3-8=u \text{ and } x^2\,dx = \frac{1}{12}du\\
&= \int \left(u^{1/3}\right)\frac{1}{12}du\\
&= \frac{1}{12}\int u^{1/3}\,du\\
&= \frac{1}{12}\frac{u^{4/3}}{4/3} + C\\
&= \frac{1}{16}u^{4/3} + C\\
&= \frac{1}{16}\left(4x^3-8\right)^{4/3} + C
\end{aligned}$$

Note We substituted not only for the integrand, but for the differential. The most difficult part is selecting the correct substitution. Once a substitution is chosen, its differential is determined. In all cases, we are attempting to rewrite the integral so as to get a more general form of rule 4:

$$\int u^n\,du = \frac{u^{n+1}}{n+1}, \qquad n \neq -1$$

The substitution we select must allow us to rewrite $f(x)\,dx$ as $u^n\,du$ with the possible exception of a multiplicative constant.

EXAMPLE 4.16

Evaluate $\int \frac{dx}{\sqrt{x^2+5}}$.

Solution If we let $u = x^2+5$, then $du = 2x\,dx$. But $2x$ is not a constant, so we cannot write this integral as a constant multiple of $u^n\,du$. Thus, we cannot use the substitution method and we cannot integrate this problem using the present methods.

EXAMPLE 4.17

Determine $\int (x^3+5)^2\,dx$.

Solution If we let $u = x^3+5$, then $du = 3x^2\,dx$. Again the substitution method cannot be used because the given integrand is not a constant multiple of $u^n\,du$. However, we can expand the given problem and solve it.

$$\begin{aligned}\int (x^3+5)^2\,dx &= \int (x^6+10x^3+25)\,dx \\ &= \frac{1}{7}x^7+\frac{5}{2}x^4+25x+C\end{aligned}$$

Application

EXAMPLE 4.18

The work, W, done by a variable force, F, over a distance, x, is given by $W = \int F\,dx$. What is the formula for the work done if $F = 15(4x+1)^2$ and $W = 4$ ft·lb when $x = 0$ ft?

EXAMPLE 4.18 (Cont.)

Solution Since $W = \int F\,dx$, we have $W = \int 15(4x+1)^2\,dx$. Let $u = 4x+1$ and then $\frac{du}{dx} = 4$ and $4\,dx = du$. Making these substitutions, we obtain the following solution. Notice in the second line of the solution that we rewrite 15 as $\frac{15}{4}\cdot 4$ in order to get $4\,dx$.

$$\begin{aligned} W &= \int 15(4x+1)^2\,dx \\ &= \int \frac{15}{4}(4x+1)^2 4\,dx \qquad \text{Let } 4x+1 = u \text{ and } 4\,dx = du \\ &= \frac{15}{4}\int u^2\,du \\ &= \frac{15}{4}\frac{u^3}{3} + C \\ &= 1.25\,(4x+1)^3 + C \end{aligned}$$

Thus, the formula for the work is $W = 1.25\,(4x+1)^3 + C$. But, we are given some additional information. We are told that $W = 4$ ft·lb when $x = 0$ ft. Substituting $x = 0$ in our formula for W produces $W = 1.25\,(4(0)+1)^3 + C = 1.25(0+1)^3 + C = 1.25 + C$. Since $W = 4$ when $x = 0$, then $1.25 + C = 4$ and $C = 4 - 1.25 = 2.75$. With this value for C we get the desired formula for the work as:

$$W = 1.25\,(4x+1)^3 + 2.75$$

Exercise Set 4.3

In Exercises 1–56, find the given indefinite integral.

1. $\int 9\,dx$

2. $\int 3x\,dx$

3. $\int 6x^2\,dx$

4. $\int \sqrt{x}\,dx$

5. $\int x^2\sqrt{x}\,dx$

6. $\int x^{-2/3}\,dx$

7. $\int (t^3+1)\,dt$

8. $\int (2-4s^3)\,ds$

9. $\int (y^2+4y-3)\,dy$

10. $\int \left(4\sqrt{y}+\frac{1}{4\sqrt{y}}\right)dy$

11. $\int \left(\sqrt[3]{x}-\frac{3}{\sqrt[3]{x^2}}\right)dx$

12. $\int \left(\frac{1}{x^3}-\frac{3}{\sqrt{x}}+\frac{\sqrt{x}}{3}\right)dx$

13. $\int (x^2+3)2x\,dx$

14. $\int (3x^2-5)6x\,dx$

15. $\int (4-2x^2)4x\,dx$

16. $\int (2x^3+1)^4 6x^2\,dx$

17. $\int (3-x^2)^3 2x\,dx$

18. $\int (x^2-3)^4 x\,dx$

19. $\int \sqrt{x^2+4}\,x\,dx$

20. $\int (x^3+1)^5 x^2\,dx$

21. $\int \frac{(\sqrt{x}-1)^3}{\sqrt{x}}\,dx$

22. $\int (x^4-5)^7 x^3\,dx$

23. $\int (3x^3+1)^4 x^2\,dx$

24. $\int (3-6x^2)^{3/2} 2x\,dx$

25. $\int \frac{x\,dx}{\sqrt{x^2+3}}$

26. $\int \frac{x\,dx}{\sqrt{(4x^2-1)^3}}$

27. $\int \frac{x\,dx}{(x^2+3)^3}$

28. $\int (x^2+3)^2\,dx$

29. $\int (3x^2-1)^2\,dx$

30. $\int \frac{x^2-1}{x+1}\,dx$

31. $\int \frac{x^2-x-6}{x+2}\,dx$

32. $\int \frac{3x\,dx}{(x^2+4)^5}$

33. $\int \frac{5x\,dx}{\sqrt[3]{x^2-1}}$

34. $\int \frac{dx}{\sqrt{4-x}}$

35. $\int (1+3x^2)^2 x\,dx$

36. $\int 2x\sqrt{5x^2+3}\,dx$

37. $\int 3x^2(4x^3-5)^{2/3}\,dx$

38. $\int \frac{2x^2\,dx}{(5-3x^3)^{2/3}}$

39. $\int (1+3x^2)^2\,dx$

40. $\int (2x^2+4x)^3(4x+4)\,dx$

41. $\int \sqrt{4x^2+2x}(4x+1)\,dx$

42. $\int (x^2+x-5)^4(2x+1)\,dx$

43. $\int \frac{x-1}{(x-1)^3}\,dx$

44. $\int \frac{2x+1}{(x^2+x-5)^3}\,dx$

45. $\int (2x^4-3)^2 8x\,dx$

46. $\int \frac{x+3}{\sqrt[3]{x^2+6x}}\,dx$

47. $\int (x^3-3x)^{2/3}(x^2-1)\,dx$

48. $\int (x^3-2)\sqrt{x^4-8x}\,dx$

49. $\int (3x+1)^3\,dx$

50. $\int 7x(x^3-2)^2\,dx$

51. $\int (2x^3-1)\sqrt[4]{x^4-2x}\,dx$

52. $\int \frac{x^2}{(4-x^3)^{3/2}}\,dx$

53. $\int (t+7)^{1/2}\,dt$

54. $\int \frac{x^4-1}{x-1}\,dx$

55. $\int \frac{y^3-8}{y-2}\,dy$

56. $\int (6x-7)^8\,dx$

In Exercises 57–66, evaluate the given definite integral.

57. $\int_0^2 (x^2-4)2x\,dx$

58. $\int_0^4 x(4x^2+2)^3\,dx$

59. $\int_1^2 6x(3x^2-7)\,dx$

60. $\int_1^3 2x^2(3x^3-1)^2\,dx$

61. $\int_0^2 2x(20-3x^2)\,dx$

62. $\int_0^5 x\sqrt{x^2+144}\,dx$

63. $\int_0^3 (x^3+2)x^2\,dx$

64. $\int_1^2 \frac{dx}{(x+4)^2}$

65. $\int_1^4 (3x^2-2)^4 x\,dx$

66. $\int_{-1}^0 (6x^2-1)^3 x\,dx$

Solve Exercises 67–72.

67. *Environmental science* The air quality control office estimates that for a population of x thousand people, the level of pollutant in the air is increasing at a rate of $L'(x) = 0.3 + 0.002x$ parts per million (ppm) per thousand people. Find a function, $L(x)$, that estimates the level of the pollutants if the level is 7.2 ppm when the population is 20,000 people.

68. *Ecology* Biologists are treating a stream contaminated with bacteria. The level of contamination is changing at the rate of $N'(t) = -\frac{750}{t^2} - 120$ bacteria/cm^3/day, where t is the number of days since the treatment began. Find a function, $N(t)$, to estimate the level of contamination if the level after 1 day was about 7,320 bacteria/cm^3.

69. *Business* The marginal profit of a fast-food restaurant is given by $P'(x) = 4x + 35$, where x is the sales volume in thousands of hamburgers. The "profit" is $\$-150$ when no hamburgers are sold. Determine the profit function, $P(x)$.

70. *Medical technology* A medical laboratory estimates that t hours after some certain bacteria are introduced into a culture, the population will be increasing at the rate of $P'(t) = \dfrac{1800}{(18-0.5t)^{1/2}}$ bacteria per hour. Find the increase in the population during the first 8 hours.

71. *Thermodynamics* The temperature, T, in °C, in an industrial furnace varies from its center to a point outside. The rate of temperature change is given by the equation

$$\frac{dT}{dx} = -\frac{5750}{(x+1)^3}$$

Find an expression for T at a distance of x m from the center if $T(0) = 2900$°C.

72. *Electronics* The current, I, and the charge, q, are related by $I = \dfrac{dq}{dt}$. At $t = 0$ s, the current to a discharged capacitor is $I = 0.005$ A. If the current remains constant, what is the charge, q, in coulombs on the capacitor?

In Your Words

73. What are the differences in the meaning of $\int f(x)\,dx$ and $\int_a^b f(x)\,dx$?

74. Describe how to use the method of substitution.

4.4 THE AREA BETWEEN TWO CURVES

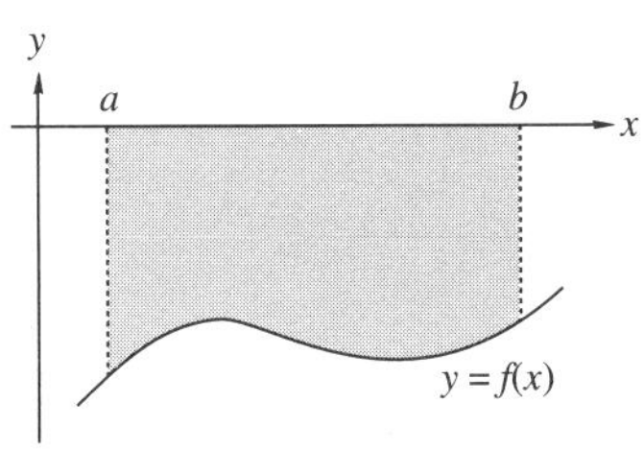

FIGURE 4.11

We introduced the definite integral as an answer to one of the two basic questions that lead to the development of calculus—the area question. In this section, we will continue our study of area.

Until now, we have only been concerned with finding the area of a curve that is above the x-axis. Suppose we want to find the area for a curve that goes below the x-axis, such as the one shown in Figure 4.11. If we were to compute its area, since $f(x) < 0$, we would get $\int_a^b f(x)\,dx < 0$ or a negative area. This does not make sense. The idea of area means that the area of something is a nonnegative number. Thus, we need a better idea for finding the area. This will become clearer as we look at the next example.

EXAMPLE 4.19

Find the area between the curve $f(x) = x^3$ and the x-axis from $x = -2$ to $x = 2$.

Solution The graph of the desired region is shaded in Figure 4.12. Using the definite integral we see that

$$\begin{aligned}\int_{-2}^{2} x^3\,dx = \frac{1}{4}x^4\Big|_{-2}^{2} &= \frac{1}{4}(2^4) - \frac{1}{4}(-2)^4 \\ &= \frac{1}{4}(16) - \frac{1}{4}(16) \\ &= 4 - 4 \\ &= 0\end{aligned}$$

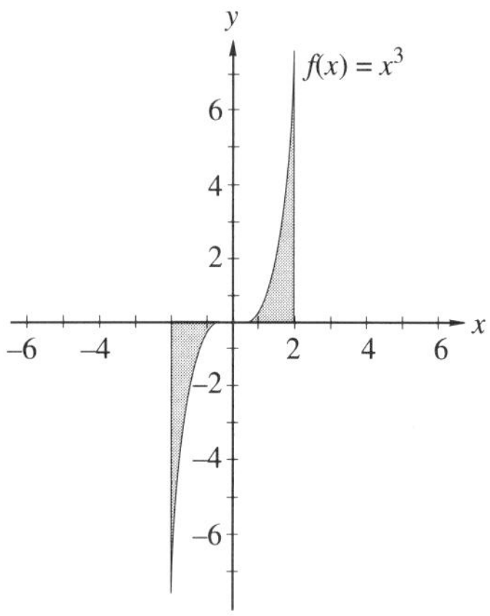

FIGURE 4.12

This does not make sense. As we can see, the shaded area of Figure 4.12 is not zero. Because this shaded area is symmetric about the origin, we can see that the area above the graph is the same as the area below the graph. Thus, we can see that the total area is $2\int_{0}^{2} x^3\,dx = 2(4) = 8$.

The integral $\int_a^b f(x)\,dx$ defines the area between the curve $y = f(x)$ and the x-axis only when $f(x)$ is a nonnegative continuous function for all x in the interval $[a, b]$. If $f(x)$ is continuous but $f(x) < 0$ for all x in the interval $[a, b]$, then $\int_a^b f(x)\,dx < 0$ and we define the area to be $\left|\int_a^b f(x)\,dx\right|$. If $f(x)$ is continuous and assumes both positive and negative values on the interval $[a, b]$, then the area is found by separating the integral into two or more parts.

One of the rules for definite integrals states that if a, b, and c are three points in an interval and if f is continuous on that interval, then

$$\int_a^b f(x)\,dx = \int_a^c f(x)\,dx + \int_c^b f(x)\,dx$$

Until now we have used this rule only once, in Example 4.9. Here, in Example 4.19, we have the opportunity to use it again. Since the curve crosses the x-axis at $x = 0$, we can compute the area as

$$\begin{aligned}\int_{-2}^{2} x^3\,dx &= \left|\int_{-2}^{0} x^3\,dx\right| + \int_{0}^{2} x^3\,dx \\ &= \left|\frac{1}{4}x^4\Big|_{-2}^{0}\right| + \left.\frac{1}{4}x^4\right]_0^2 \\ &= \left|\frac{1}{4}0^4 - \frac{1}{4}(-2)^4\right| + \left[\frac{1}{4}2^4 - \frac{1}{4}0^4\right] \\ &= |0 - 4| + (4 - 0) \\ &= 4 + 4 = 8\end{aligned}$$

Caution

As you can see, it is very important that you sketch the graph when you are doing an area problem.

EXAMPLE 4.20

Find the area between the curve $g(x) = x^4 - 1$ and the x-axis from $x = -1$ to $x = 2$.

Solution The desired area is shown in Figure 4.13. The curve lies below the x-axis for the interval $[-1, 1]$ and above the x-axis for $[1, 2]$. The total area is the sum of these two separate areas. Thus,

$$\text{Area} = \left| \int_{-1}^{1} (x^4 - 1)\, dx \right| + \int_{1}^{2} (x^4 - 1)\, dx$$

Now,

$$\int_{-1}^{1} (x^4 - 1)\, dx = \left[\frac{1}{5}x^5 - x\right]_{-1}^{1}$$
$$= \left(\frac{1}{5} - 1\right) - \left(-\frac{1}{5} + 1\right) = -\frac{8}{5}$$

and

$$\int_{1}^{2} (x^4 - 1)\, dx = \left[\frac{1}{5}x^5 - x\right]_{1}^{2}$$
$$= \left(\frac{32}{5} - 2\right) - \left(\frac{1}{5} - 1\right) = \frac{26}{5}$$

Thus, the total area of the region is $\left|-\frac{8}{5}\right| + \frac{26}{5} = \frac{34}{5} = 6.8$ square units.

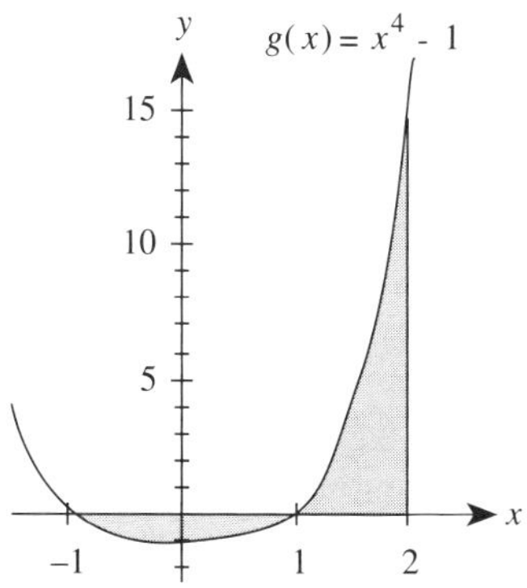

FIGURE 4.13

EXAMPLE 4.21

Find the area between the curve $f(x) = x^{1/3}$ and the x-axis from $x = -1$ to $x = 8$.

Solution The desired area is the shaded region of Figure 4.14. The graph of $f(x) = x^{1/3}$ crosses the x-axis at $x = 0$, so the area A will be

$$A = \left| \int_{-1}^{0} x^{1/3}\, dx \right| + \int_{0}^{8} x^{1/3}\, dx$$
$$= \left| \frac{3}{4}x^{4/3} \Big|_{-1}^{0} \right| + \frac{3}{4}x^{4/3} \Big|_{0}^{8}$$
$$= \left| \frac{3}{4}(0)^{4/3} - \frac{3}{4}(-1)^{4/3} \right| + \left[\frac{3}{4}(8)^{4/3} - \frac{3}{4}(0)^{4/3}\right]$$
$$= \left| 0 - \frac{3}{4} \right| + (12 - 0)$$
$$= \frac{3}{4} + 12 = 12\frac{3}{4}$$

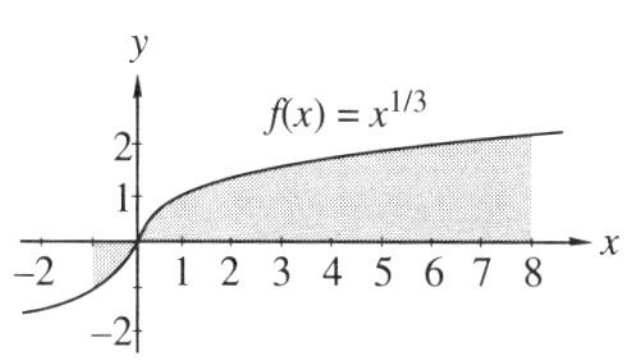

FIGURE 4.14

Note The direct evaluation of $\int_{-1}^{8} x^{1/3}\,dx = \frac{3}{4}x^{4/3}\Big|_{-1}^{8} = 12 - \frac{3}{4} = 11\frac{1}{4}$. This is not the area of the shaded region in Figure 4.14. It is, however, the value of the definite integral $\int_{-1}^{8} x^{1/3}\,dx$. You will need to be careful when evaluating a definite integral to proceed using the method that is correct for the desired meaning.

Let's expand our thinking to include a different type of area. Until now, each area has been for some region between a curve and the x-axis. Now we want to find the area between two curves.

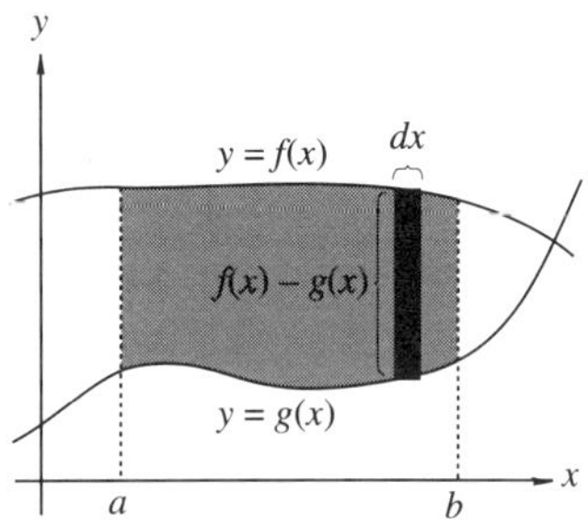

FIGURE 4.15

Consider the functions f and g as shown in Figure 4.15. We want to find the area between the curves $y = f(x)$ and $y = g(x)$ and the lines $x = a$ and $x = b$. We will return to finding the area by dividing the region into rectangular regions of thickness dx. The height of cach of these rectangles is $f(x) - g(x)$, regardless of the position of f and g. As a result, the area is given by

$$\int_a^b [f(x) - g(x)]\,dx$$

The result works whether the graphs lie above the x-axis, partially above and partially below the x-axis, or both lie below the x-axis.

EXAMPLE 4.22

Find the area enclosed by the graphs of $f(x) = 12x - 3x^2$ and $g(x) = 3x - 12$.

Solution The graph of the two functions is shown in Figure 4.16. Before we can compute the area, we need to locate the points of intersection of the two graphs. Thus, we need to determine when $f(x) = g(x)$ or when

$$12x - 3x^2 = 3x - 12$$
$$3x^2 - 9x - 12 = 0$$
$$3(x-4)(x+1) = 0$$

and so the graphs intersect when $x = -1$ and $x = 4$.

From Figure 4.16, we can see that $f(x) \geq g(x)$ on $[-1, 4]$, so the area we want is

$$\begin{aligned}\int_{-1}^{4} [f(x) - g(x)]\,dx &= \int_{-1}^{4} [(12x - 3x^2) - (3x - 12)]\,dx \\ &= \int_{-1}^{4} (9x - 3x^2 + 12)\,dx \\ &= \left(\frac{9}{2}x^2 - x^3 + 12x\right)\Big|_{-1}^{4} \\ &= (72 - 64 + 48) - \left(\frac{9}{2} + 1 - 12\right) \\ &= 56 - (-6.5) = 62.5\end{aligned}$$

The desired area is 62.5 square units.

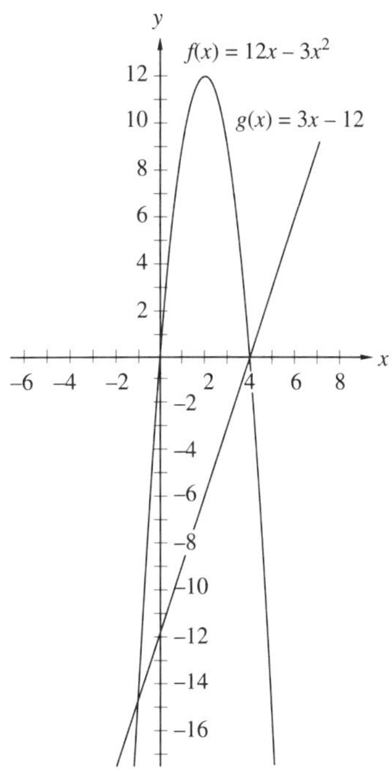

FIGURE 4.16

In the next example, we will combine the techniques that we used in the previous two examples.

EXAMPLE 4.23

Find the area enclosed by the graphs of $f(x) = x^3$ and $g(x) = x$.

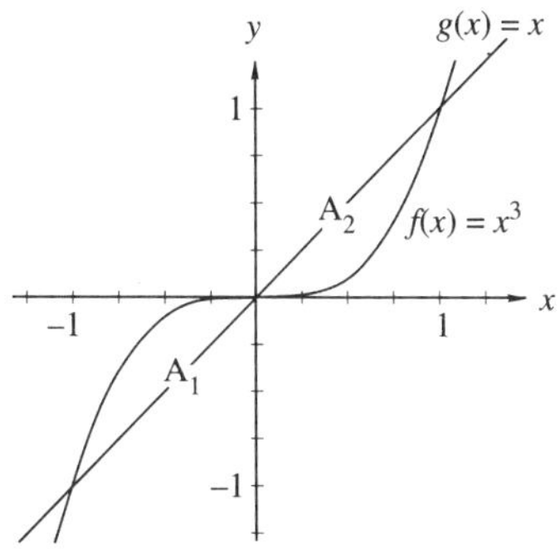

FIGURE 4.17

Solution The graph of the two functions is shown in Figure 4.17. The points of intersection occur when

$$f(x) = g(x)$$

or when

$$x^3 = x$$
$$x^3 - x = 0$$
$$x(x^2 - 1) = x(x-1)(x+1) = 0$$

The two graphs intersect when $x = 0$, $x = 1$, and $x = -1$.

Study the graph of the two curves. On the interval $[-1, 0]$, $f(x) \geq g(x)$, and so the area A_1 enclosed by the curves in this interval is

$$\begin{aligned} A_1 = \int_{-1}^{0} [f(x) - g(x)]\,dx &= \int_{-1}^{0} (x^3 - x)\,dx \\ &= \left(\frac{1}{4}x^4 - \frac{1}{2}x^2\right)\Bigg|_{-1}^{0} \\ &= 0 - \left(\frac{1}{4} - \frac{1}{2}\right) = \frac{1}{4} \end{aligned}$$

On the interval $[0, 1]$, $g(x) \geq f(x)$, and so the area A_2 enclosed by the curves in this interval is

$$\begin{aligned} A_2 = \int_{0}^{1} [g(x) - f(x)]\,dx &= \int_{0}^{1} (x - x^3)\,dx \\ &= \left(\frac{1}{2}x^2 - \frac{1}{4}x^4\right)\Bigg|_{0}^{1} \\ &= \left(\frac{1}{2} - \frac{1}{4}\right) - 0 = \frac{1}{4} \end{aligned}$$

The total area is $A_1 + A_2 = \frac{1}{4} + \frac{1}{4} = \frac{1}{2}$.

Note Again, we need to point out the importance of graphing. If you had applied the rule for the area between two curves without graphing, you could easily have reached the incorrect result shown below:

$$\begin{aligned}\int_{-1}^{1}[f(x)-g(x)]\,dx &= \int_{-1}^{1}(x^3-x)\,dx\\ &= \left(\frac{1}{4}x^4-\frac{1}{2}x^2\right)\Big|_{-1}^{1}\\ &= \left(\frac{1}{4}-\frac{1}{2}\right)-\left(\frac{1}{4}-\frac{1}{2}\right)\\ &= 0\end{aligned}$$

EXAMPLE 4.24

Find the area enclosed by the graphs of $f(x)=\sqrt{9-9x}$, $g(x)=\sqrt{9-x}$, and the x-axis.

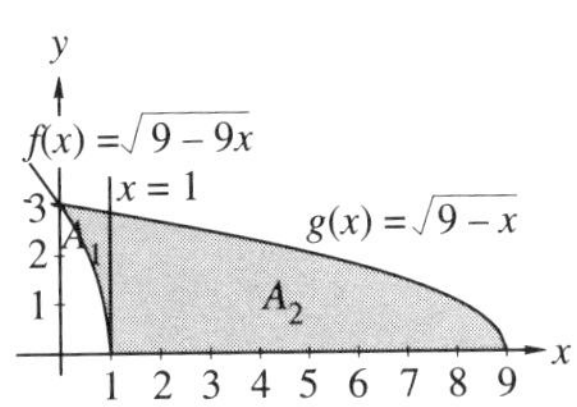

FIGURE 4.18

Solution The graphs of these functions are shown in Figure 4.18. The shading indicates the area we want to measure.

Again, we need to exercise caution. In the interval $[0, 1]$, the shaded region A_1 is between $f(x)$ and $g(x)$ with $g(x) \geq f(x)$. Thus

$$\begin{aligned}A_1 &= \int_0^1 [g(x)-f(x)]\,dx\\ &= \int_0^1\left(\sqrt{9-x}-\sqrt{9-9x}\right)dx\\ &= \int_0^1 \sqrt{9-x}\,dx-\int_0^1\sqrt{9-9x}\,dx\\ &= -\frac{2}{3}(9-x)^{3/2}\Big|_0^1-\frac{-2}{27}(9-9x)^{3/2}\Big|_0^1\\ &= \left[-\frac{2}{3}(8)^{3/2}+\frac{2}{3}(9)^{3/2}\right]-\left[\frac{-2}{27}(0)+\frac{2}{27}(9)^{3/2}\right]\\ &= \left(-\frac{16}{3}\sqrt{8}+18\right)-(0+2)\\ &= 16-\frac{16}{3}\sqrt{8}\end{aligned}$$

In the interval $[1, 9]$, the shaded region A_2 is between $g(x)$ and the x-axis, with $g(x) \geq 0$, and so

$$A_2 = \int_1^9 \sqrt{9-x}\,dx = -\frac{2}{3}(9-x)^{3/2}\Big|_1^9 = \frac{16}{3}\sqrt{8}$$

The total area is $A_1 + A_2 = 16 - \frac{16}{3}\sqrt{8} + \frac{16}{3}\sqrt{8} = 16$. ■

Partitions Along the *y*-axis

Until now, we have always integrated with respect to x. We have done this because the functions have always been given in terms of x and the x-axis. Let's look again at Example 4.24 and rewrite the functions in terms of y.

We had $f(x)=\sqrt{9-9x}$ or $y=\sqrt{9-9x}$. If we solve this for x, we get $y^2=9-9x$ or $x=\frac{1}{9}(9-y^2)$. Let's call this function $h(y)=\frac{1}{9}(9-y^2)$. In the same way, if we solve $y=g(x)$ for x, we get $y=\sqrt{9-x}$ or $x=9-y^2$. We will call this function $j(y)$, and so $j(y)=9-y^2$.

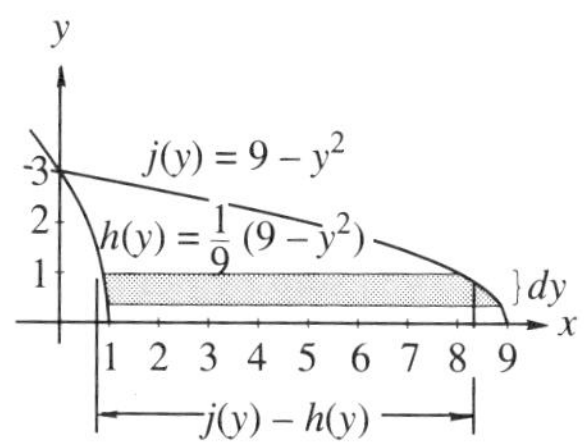

FIGURE 4.19

Look at Figure 4.19. This time we have partitioned the y-axis and we get rectangles of width dy and length $j(y)-h(y)$. We want the area A between these curves and the x-axis (or when $y=0$). Thus,

$$\begin{aligned}A &= \int_0^3 [j(y)-h(y)]dy\\ &= \int_0^3 \left[(9-y^2)-\frac{1}{9}(9-y^2)\right]dy\\ &= \int_0^3 \left(8-\frac{8}{9}y^2\right)dy\\ &= \left(8y-\frac{8}{27}y^3\right)\Big|_0^3\\ &= 8(3)-\frac{8}{27}(3)^3 = 24-8=16\end{aligned}$$

As you can see in this case, a partition of the y-axis made the integration simpler. There is no clear method to tell which way to partition the graph. Usually, the graph will give you some indication.

Improper Integrals

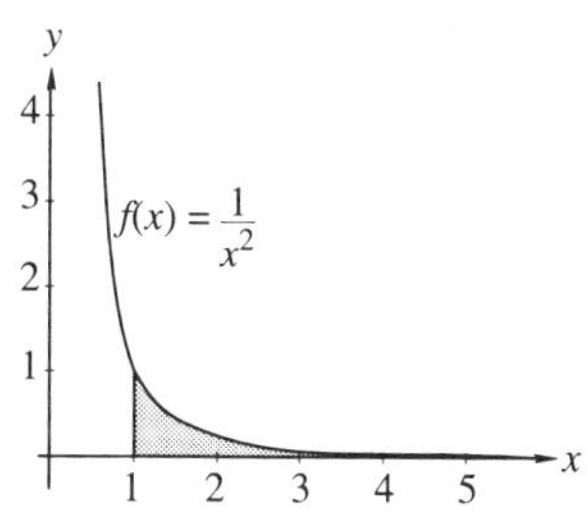

FIGURE 4.20

In the next example, we want to determine the area between a curve and the x-axis. In this case, we want the area between $f(x)=\dfrac{1}{x^2}$, the x-axis, and to the right of $x=1$, as shown in Figure 4.20. We know that the x-axis is an asymptote of $y=\dfrac{1}{x^2}$, and the curve never intersects the x-axis. Thus, it looks as if the area is determined by

$$\int_1^\infty \frac{1}{x^2}\,dx$$

When we defined the definite integral at the beginning of this chapter, it was over a closed interval $[a,b]$. This integral is on the half-open interval $[a,\infty)$. To get around this problem, we will define

$$\int_1^\infty \frac{1}{x^2}\,dx = \lim_{b\to\infty}\int_1^b \frac{1}{x^2}\,dx, \text{ if the limit exists}$$

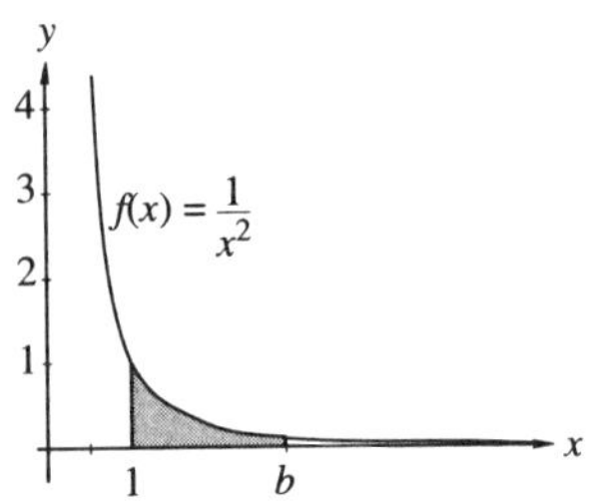

FIGURE 4.21

This is an example of an **improper integral**. Geometrically, we are finding the area under the graph on the interval $[1, b]$, as shown in Figure 4.21, and then we let $b \to \infty$. If the limit exists, the limit is defined to be the area. For this problem,

$$\begin{aligned}\int_1^\infty \frac{1}{x^2}\,dx &= \lim_{b\to\infty}\int_1^b \frac{1}{x^2}\,dx \\ &= \lim_{b\to\infty}\left(-\frac{1}{x}\right)\Bigg|_1^b \\ &= \lim_{b\to\infty}\left(-\frac{1}{b}+1\right) = 1\end{aligned}$$

Exercise Set 4.4

In Exercises 1–42, find the area of the region bounded by the given curves. Sketch the graphs and shade the desired area first.

1. $f(x) = x$, $x = 4$, x-axis
2. $f(x) = -2x$, $x = -2$, x-axis
3. $f(x) = x^2$, $x = 4$, x-axis
4. $h(x) = x^2 + 4$, $x = -1$, $x = 1$, x-axis
5. $g(x) = 3x + 1$, $x = 10$, x-axis
6. $h(x) = 10 - 2x$, $x = 0$, x-axis
7. $k(x) = 2x + 5$, $x = -2$, $x = 1$, x-axis
8. $j(x) = 2x^2$, x-axis, $x = 1$, $x = 2$
9. $k(x) = 2x^2 - x$, x-axis, $x = -2$
10. $j(x) = x^3 + 1$, $x = 1$, x-axis
11. $m(x) = x^2 - 4x$, x-axis
12. $n(x) = x^2 - 4$, x-axis
13. $f(x) = 9 - x^2$, x-axis
14. $h(x) = 6 - x - x^2$, x-axis
15. $g(x) = x^3 - 4x$, x-axis
16. $f(x) = x^3 - x^2$, x-axis
17. $h(x) = \sqrt{x+9}$, x-axis, $x = 0$
18. $j(x) = \sqrt{x}$, $x = 9$, x-axis
19. $f(x) = x^2$, $g(x) = 2x$
20. $g(x) = 4 - x^2$, $h(x) = x + 2$
21. $f(x) = x^2 + 3$, $g(x) = 9$
22. $f(x) = x^2 + 4$, $x = -2$, $x = 2$, $y = -1$
23. $f(x) = x - 4$, $g(x) = \sqrt{2x}$
24. $f(x) = \sqrt{x}$, $x = 0$, $y = 4$
25. $g(x) = 4 - x^2$, $h(x) = -3x$
26. $f(x) = x^2 - 4$, $g(x) = 8 - 2x^2$
27. $h(x) = 2x - \frac{1}{2}x^2$, $j(x) = \frac{1}{2}x - 2$
28. $m(x) = x^4$, $n(x) = x^2$
29. $j(x) = 8 - x^2$, $k(x) = x^2$, $x = -2$, $x = 1$
30. $f(x) = 2\sqrt{x}$, $y = 4$, $x = -2$, x-axis
31. $f(x) = x^3$, $g(x) = 4x$
32. $h(x) = x + 3$, $k(x) = 9 - x^2$, $x = 3$
33. $g(x) = 4x - x^2 + 8$, $h(x) = x^2 - 2x$
34. $f(x) = \sqrt{x}$, $g(x) = -x$, $h(x) = x - 2$
35. $f(x) = \sqrt{4x}$, $y = 3$, $x = 0$
36. $h(x) = x^3 - 3x + 2$, $k(x) = x + 2$
37. $y = \sqrt{16 - 2x}$, $y = \sqrt{16 - 4x}$, x-axis
38. $f(x) = x^{1/3}$, $g(x) = -2$, $x = 8$
39. $f(x) = x^{-5/3}$, x-axis, to the right of $x = 8$
40. $f(x) = \frac{1}{x^2}$, $g(x) = 4$, $x = -2$, $x = 2$, x-axis
41. $x^4 y = 1$, x-axis, to the right of $x = 1$
42. $g(x) = x^4 - x^2$, x-axis

Solve Exercises 43–50.

43. *Product design* A tile manufacturing company plans to produce enamel tiles with the design and color scheme shown in Figure 4.22. Will more blue or white enamel be needed?

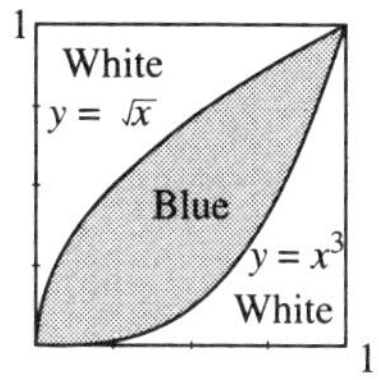

FIGURE 4.22

44. *Environmental science* A new smog-control device will reduce the output of sulfur oxides from automobile exhausts but increase the production of nitrous oxides. It is estimated that using this device will produce a rate of savings to the community of $S(x) = -x^2 + 11x + 15$, in millions of dollars after x years of using the device. The increase in the production of nitrous oxides will result in a rate of additional costs to the community of $C(x) = \frac{5}{4}x^2$, in millions of dollars. If $S(x) - C(x)$ represents the net savings, **(a)** For how many years will it pay to use this new device? **(b)** What will be the net total savings for this period of time?

45. *Industrial design* A double-channel trough used on an assembly line is constructed with end plates defined by the curve $y = x^4 - 2x^2 + 1$ and the line $y = 9$, where x and y are measured in cm. What is the volume of a trough that is 10 m long and has these end plates?

46. *Machine technology* Find the area of the cam outlined by the curves $y = 10x - x^2$ and $4y = x^2$, where x and y are measured in cm.

47. *Automotive engineering* The work cycle, or **Otto cycle**, of an internal combustion engine may be approximated by the pressure-volume (or pV) diagram shown in Figure 4.23. The area of the region between the curves is proportional to the work done by the engine during one cycle. The curves represent adiabatic, or zero, heat input/output processes, where $pV^\gamma = k$, where k is a constant and $\gamma \approx 1.33$ for gasoline vapor. Find the work done by the cycle shown in Figure 4.23. (Note: the answer will be in British thermal units, Btu.)

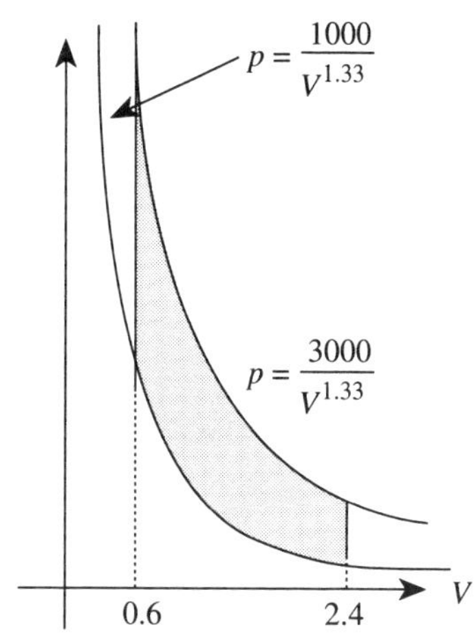

FIGURE 4.23

48. *Automotive engineering* The work cycle, or **Diesel cycle**, of a diesel engine may be approximated by the pV-diagram shown in Figure 4.24. The area of the region between the curves is proportional to the work done by the engine during one cycle. The curves represent adiabatic, or zero, heat input/output processes, where $pV^\gamma = k$, where k is a constant, and $\gamma \approx 1.40$ for diesel vapor. Find the work done by the cycle shown in Figure 4.24. (Note: the answer will be in British thermal units, Btu.)

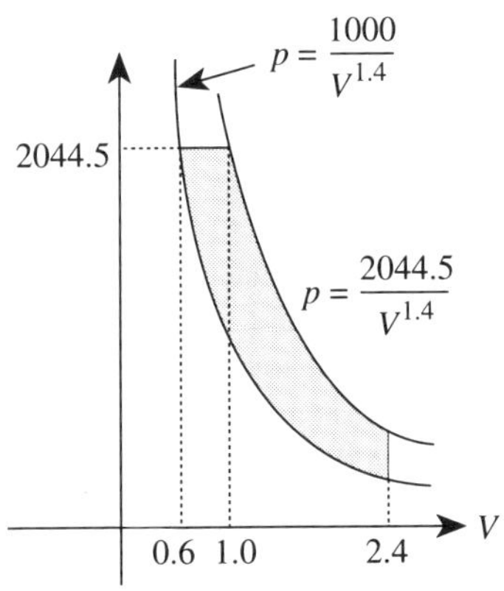

FIGURE 4.24

49. *Mechanical engineering* A heavy flywheel initially rotating on its axis is slowing because of friction in its bearings. If its angular acceleration, α, is given by $\alpha = \frac{d\omega}{dt} = \frac{100}{\sqrt{(t+1)^3}}$, find its angular velocity, ω, at time $t = 10$ s, if $\omega = 0$ at $t = 0$.

50. *Electronics* The magnetic potential, V, at a point on the axis of a certain coil is given by an equation of the form $V = l\int_a^{\infty} \frac{x}{(R^2+x^2)^{3/2}}\,dx$, where k and R are constants. Integrate in order to find an expression for V.

In Your Words

51. Describe the process used to find the area between two curves.

52. What is an indefinite integral?

4.5 NUMERICAL INTEGRATION

The Fundamental Theorem of Calculus provides a useful method for finding the area under a curve. It is not always an easy task to find the antiderivative of a function. It is also possible that we do not know how to find the antiderivative of a certain function. In this section, we will look at two methods for approximating the value of a definite integral to as many decimal places as desired. All techniques of this type come under the heading of **numerical integration**. The two methods that we will examine are called the **trapezoidal rule** and **Simpson's rule**. As with most numerical techniques, your work with these can be eased by the use of a calculator or a computer.

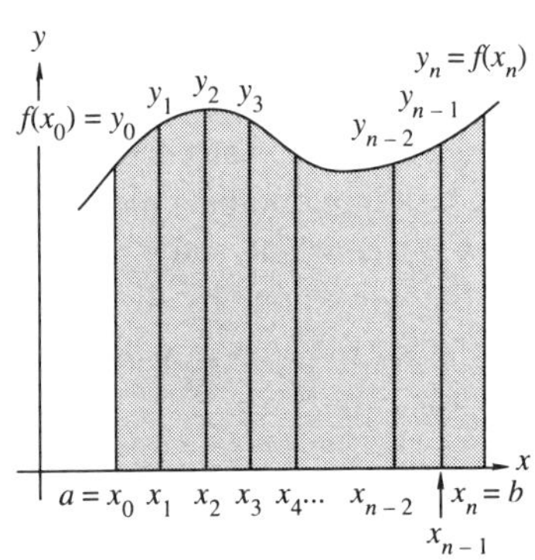

FIGURE 4.25

The Trapezoidal Rule

In both the trapezoidal rule and Simpson's rule, we will use some of the work we did in developing the definite integral back in Section 4.1. First we will assume that we want to find the area under some curve, which we will describe by $y = f(x)$. We will also assume that we want to find the area under the curve $f(x)$, above the x-axis, and between the lines $x = a$ and $x = b$, with $a < b$.

We will divide the interval $[a, b]$ into n equal parts of length $dx = \frac{b-a}{n}$, where $a = x_0 < x_1 < x_2 < x_3 < \cdots < x_{n-1} < x_n = b$. The corresponding points on the curve are $y_0 = f(x_0)$, $y_1 = f(x_1)$, $y_2 = f(x_2)$, $y_3 = f(x_3)$, $\ldots$, $y_{n-1} = f(x_{n-1})$, $y_n = f(x_n)$. If we connect these points, we obtain n trapezoids as shown in Figure 4.25.

The area of a trapezoid is equal to one-half the product of the altitude and the sum of its bases. A typical trapezoid is shown in Figure 4.26a. If you draw the trapezoid as shown in Figure 4.26b, you will find it easier to see that the altitude of

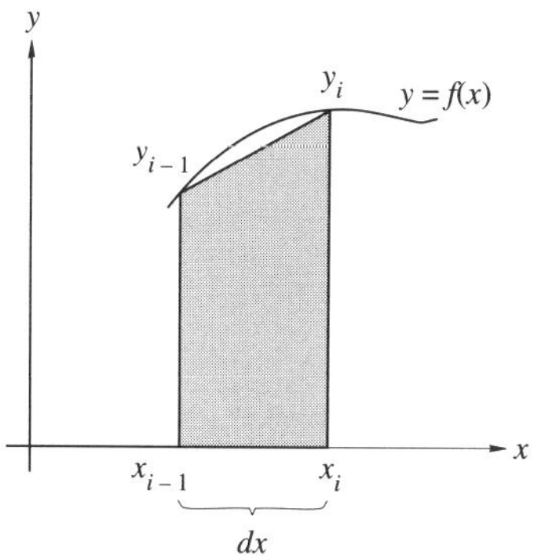

FIGURE 4.26a

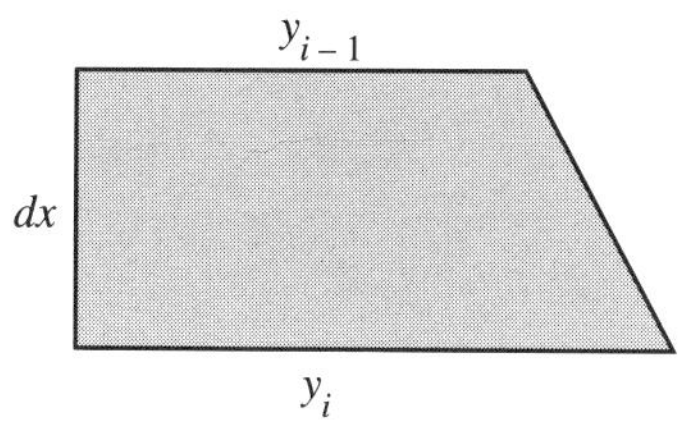

FIGURE 4.26b

this trapezoid is dx and the lengths of its bases are y_{i-1} and y_i. Thus, the area of the trapezoid in Figure 4.26a is $\frac{1}{2}(y_{i-1}+y_i)\,dx$.

The area under the curve in Figure 4.25 can be approximated by adding the areas of all the trapezoids. Thus,

$$\begin{aligned}\int_a^b f(x)\,dx &\approx \frac{1}{2}(y_0+y_1)\,dx + \frac{1}{2}(y_1+y_2)\,dx + \frac{1}{2}(y_2+y_3)\,dx + \cdots \\ &\quad + \frac{1}{2}(y_{n-1}+y_n)\,dx \\ &= \frac{1}{2}(y_0+2y_1+2y_2+2y_3+\cdots+2y_{x-1}+y_n)\,dx\end{aligned}$$

Using the more conventional function notation, $y = f(x)$, we can rewrite the formula as follows:

> **Trapezoidal Rule** $$\int_a^b f(x)\,dx \approx \frac{dx}{2}[f(x_0)+2f(x_1)+2f(x_2)+2f(x_3)+\cdots +2f(x_{n-1})+f(x_n)]$$
>
> where $dx = \dfrac{b-a}{n}$.

EXAMPLE 4.25

Use the trapezoidal rule with $n = 6$ to approximate the value of $\int_0^2 x^3\,dx$.

Solution For $n = 6$, we have $dx = \frac{2-0}{6} = \frac{1}{3}$. Thus, $x_0 = 0$, $x_1 = \frac{1}{3}$, $x_2 = \frac{2}{3}$, $x_3 = 1$, $x_4 = \frac{4}{3}$, $x_5 = \frac{5}{3}$, and $x_6 = 2$. Using the trapezoidal rule, we have

$$\begin{aligned}\int_0^2 x^3\,dx &\approx \frac{1/3}{2}\left[0^3+2\left(\frac{1}{3}\right)^3+2\left(\frac{2}{3}\right)^3+2(1)^3+2\left(\frac{4}{3}\right)^3+2\left(\frac{5}{3}\right)^3+2^3\right] \\ &= \frac{1}{6}\left(0+\frac{2}{27}+\frac{16}{27}+2+\frac{128}{27}+\frac{250}{27}+8\right) \\ &= \frac{37}{9} \\ &\approx 4.111\end{aligned}$$

The exact value obtained by integration is 4. ▪

Simpson's Rule

For the second approximation technique, we will use the same curve. Again, the interval $[a, b]$ is divided into n equal parts, but this time n must be an even number. Instead of the straight lines we drew connecting y_0, y_1, y_2, y_3, and so on, we are

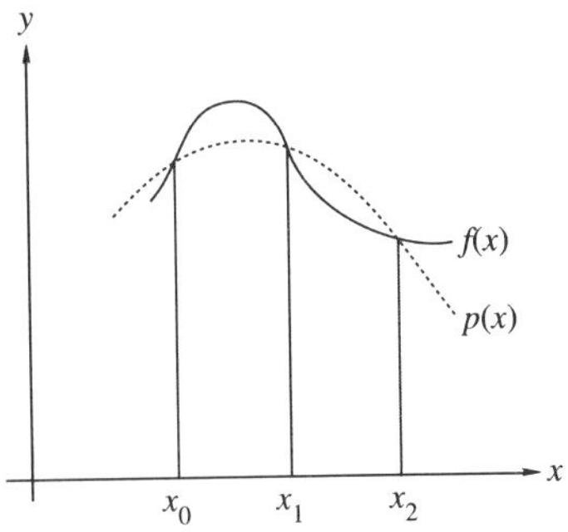

FIGURE 4.27

going to draw parabolas through each group of three consecutive points, as shown in Figure 4.27. A lengthy algebraic manipulation will show that

$$\int_{x_0}^{x_2} f(x)\,dx = \frac{dx}{3}[f(x_0)+4f(x_1)+f(x_2)]$$

Now, if we did this for each subinterval $[x_0, x_2], [x_2, x_4], [x_4, x_6], \ldots, [x_{n-2}, x_n]$, we would get the result known as Simpson's rule.

Simpson's Rule
$$\int_a^b f(x)\,dx \approx \frac{dx}{3}[f(x_0)+4f(x_1)+2f(x_2)+4f(x_3)+2f(x_4) + \cdots + 2f(x_{n-2})+4f(x_{n-1})+f(x_n)]$$

where n is an even number and $dx = \dfrac{b-a}{n}$.

EXAMPLE 4.26

Use Simpson's rule with $n = 6$ to approximate the value of $\int_0^2 x^3\,dx$.

Solution Since $n = 6$, n is an even number. This is the same integral we approximated in Example 4.25, so the values of x and dx are the same.

$$\begin{aligned}\int_0^2 x^3\,dx &\approx \frac{1/3}{3}\left[0^3+4\left(\frac{1}{3}\right)^3+2\left(\frac{2}{3}\right)^3+4(1)^3+2\left(\frac{4}{3}\right)^3+4\left(\frac{5}{3}\right)^3+2^3\right]\\ &= \frac{1}{9}\left(0+\frac{4}{27}+\frac{16}{27}+4+\frac{128}{27}+\frac{500}{27}+8\right)\\ &= \frac{1}{9}(36)\\ &= 4\end{aligned}$$

This happens to be a better approximation than we got using the trapezoidal rule. In fact, for this function it gives the exact value of the integral. ∎

Both the trapezoidal rule and Simpson's rule are much easier to use if one can take advantage of a calculator or computer.

Some scientific calculators use Simpson's rule to evaluate numeric integrals of functions. The next five examples demonstrate how to do this. Examples 4.27 and 4.28 use a Casio fx-7700G and Examples 4.29, 4.30, and 4.31 use a TI-82.

EXAMPLE 4.27

Use a Casio fx-7700G to evaluate $\int_0^2 x^3\,dx$.

Solution Press the following sequence of keys:

[SHIFT] [$\int dx$] [X,θ,T] [x^y] 3 [SHIFT] [,] 0 [SHIFT] [,] 2 [ )] [EXE]

After a slight pause, the calculator will return the solution 4, the same as that which we got in Example 4.26. ▪

EXAMPLE 4.28

Use a Casio fx-7700G to evaluate $\int_1^5 (2x^5 - 3x)\,dx$.

Solution Press:

[SHIFT] [$\int dx$] 2 [X,θ,T] [x^y] 5 [−] 3 [X,θ,T] [SHIFT] [,] 1 [SHIFT] [,] 5 [ )] [EXE]

After a fairly long pause, we get the result 5,172. So, according to the calculator, $\int_1^5 (2x^5 - 3x)\,dx = 5{,}172$. How does this compare with the actual value?

$$\begin{aligned}\int_1^5 (2x^5 - 3x)\,dx &= \frac{1}{3}x^6 - \frac{3}{2}x^2 \Big|_1^5 \\ &= \left(\frac{15{,}625}{3} - \frac{75}{2}\right) - \left(\frac{1}{3} - \frac{3}{2}\right) \\ &= 5{,}172\end{aligned}$$

▪

EXAMPLE 4.29

Use a TI-82 to evaluate $\int_0^2 x^3\,dx$.

Solution There are two ways to use a TI-82 to evaluate a definite integral. We will show the first method in this example and the other in the next example.

In the first method you press the following sequence of keys:

[MATH] 9 [X,T,θ] [∧] 3 [,] [X,T,θ] [,] 0 [,] 2 [ )] [ENTER]

After a slight pause, the calculator will return the solution 4, the same as we got in Example 4.27. (Note that pressing [MATH] 9 accesses the "`fnInt(`" option on the calculator.) ▪

EXAMPLE 4.30

Use a TI-82 to evaluate $\int_1^5 (2x^5 - 3x)\,dx$.

Solution In this method you begin as if your are going to graph the integrand. So, for this integral begin by entering the integrad as $y_1 = 2x^5 - 3x$. Then, after you access the `fnInt(` option on the calculator, you recall the name of the function, in this case, y_1. The remainder of the procedure is exactly the same as shown in

EXAMPLE 4.30 (Cont.)

Example 4.28. Press:

MATH 9 2nd VARS 1 1 , X,T,θ , 1 , 5) ENTER

After a brief pause, we get the result `5172`.

According to the calculator, $\int_1^5 (2x^5 - 3x)\,dx = 5172$. How does this compare with the actual value? As we showed in Example 25.27, $\int_1^5 (2x^5 - 3x)\,dx = 5{,}172$, so the calculator gave the exact value of this definite integral. ■

It is also possible to get the calculator to both shade the region being studied and evaluate the integral. Since an integral is a measure of the area under a curve, this can provide a graphical and numerical interpretation of an integral.

EXAMPLE 4.31

Use a TI-82 to graph the region, and determine the area under the curve of $f(x) = x^2 + \frac{1}{x^3} - 4x + 6$ from $x = 1$ to $x = 4.2$.

Solution As in the previous method, you begin as if you are going to graph the integrand. So, for this integral begin by entering the integrand as $y_1 = x^2 + \frac{1}{x^3} - 4x + 6$. You may want to graph the curve, in order to select a viewing window that shows the graph over the entire interval from $x = 1$ to $x = 4.2$. In this example, a viewing window of $[0, 4.7] \times [-1, 8]$ was selected.

Now access 2nd CALC. (On a TI-82, the CALC key is above the TRACE key.) Select the `∫f(x)dx` option on the calculator, by pressing 7. The result in Figure 4.28a shows the graph of the function, puts a blinking cursor on the curve at the x-value in the middle of the window, asks for a `Lower Limit?`, and gives the coordinates of the cursor. Use the left or right arrow key, ◀ or ▶, to move the cursor to the lower limit, in this case $x = 1$, and press the ENTER key. The result is shown in Figure 4.28b. Notice that the calculator is now asking for the `Upper Limit?`. Use the ▶ key to move the cursor to the x-value that indicates the upper limit as shown in Figure 4.28c. Press ENTER. You should see the calculator shade the area of the region being intergrated and then, at the bottom of the screen, give an approximation of the area as 10.754322, as shown in Figure 4.28d.

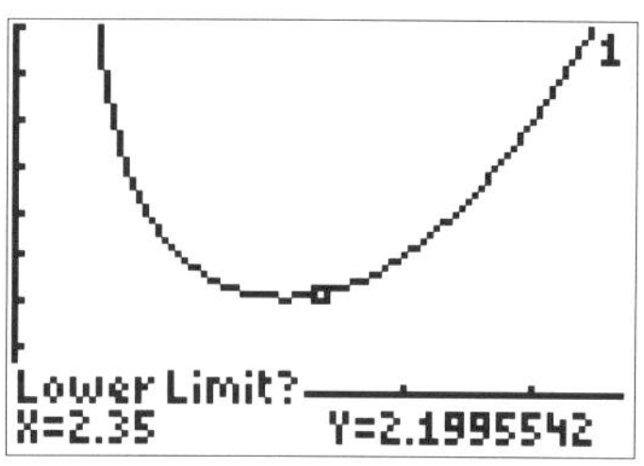

[0, 4.7, 1] x [–1, 8, 1]

FIGURE 4.28a

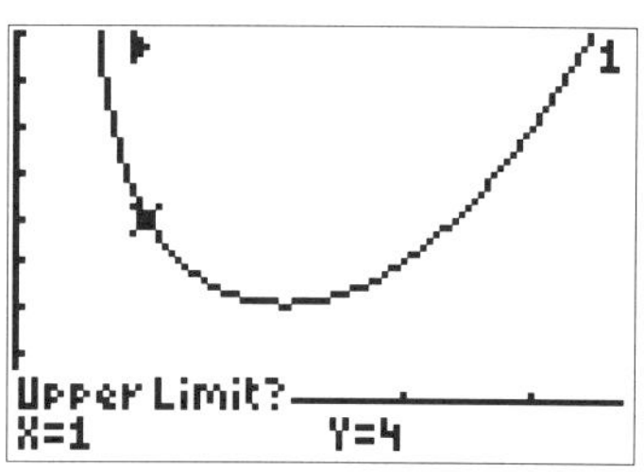

[0, 4.7, 1] x [–1, 8, 1]

FIGURE 4.28b

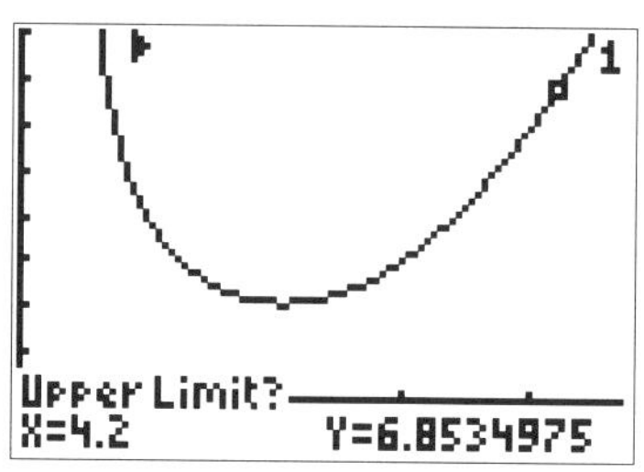

[0, 4.7, 1] x [–1, 8, 1]

FIGURE 4.28c

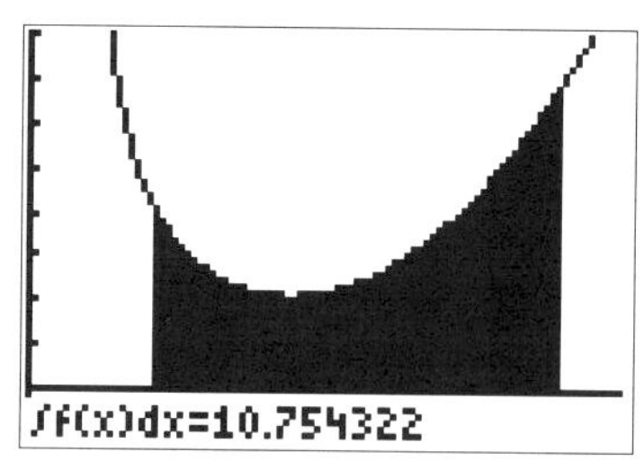

[0, 4.7, 1] x [–1, 8, 1]

FIGURE 4.28d ■

It is also possible to use both of these formulas to obtain integrals of functions that are available only in graphical or tabular form, as shown by the next example.

Application

EXAMPLE 4.32

The following table gives the results of a series of drillings meant to determine the depth of the bedrock at a building site. These drillings were taken along a straight line down the middle of the lot where the building will be placed. In the table, x is the distance from the front of the parking lot and y is the corresponding depth. Both x and y are given in feet.

x	0	20	40	60	80	100	120	140	160
y	33	35	40	45	42	38	46	40	48

Approximate the area of this cross-section by using Simpson's rule.

Solution We first determine that $n = 8$, an even number, and $dx = 20$. By Simpson's rule,

$$\begin{aligned}\int_0^{160} f(x)\,dx &\approx \frac{20}{3}[33+4(35)+2(40)+4(45)+2(42)+4(38)\\&\quad +2(46)+4(40)+48]\\&= \frac{20}{3}(969)\\&= 6{,}460\,\text{ft}^2\end{aligned}$$

Exercise Set 4.5

In Exercises 1–16, find the approximate value of the given integrals with the specified value of n by (a) the trapezoidal rule and (b) Simpson's rule.

1. $\int_1^2 x^4\,dx$, $n = 6$ (Check by integration.)
2. $\int_0^2 2x^5\,dx$, $n = 8$ (Check by integration.)
3. $\int_0^1 \sqrt{x}\,dx$, $n = 4$ (Check by integration.)
4. $\int_1^2 x^{1/3}\,dx$, $n = 4$ (Check by integration.)
5. $\int_1^5 \frac{dx}{x}$, $n = 8$
6. $\int_1^4 \frac{dx}{x^2}$, $n = 6$ (Check by integration.)
7. $\int_1^5 \frac{dx}{1+x}$, $n = 8$
8. $\int_0^2 x^{\pi}\,dx$, $n = 4$ (Check by integration.)
9. $\int_0^1 \frac{4}{1+x^2}$, $n = 10$
10. $\int_0^2 e^x\,dx$, $n = 4$
11. $\int_0^2 \sqrt{1+x^3}\,dx$, $n = 8$
12. $\int_0^{0.2} \sin x\,dx$, $n = 4$ (x in radian measure)
13. $\int_{-2}^2 \sqrt{x+4}\,dx$, $n = 8$
14. $\int_1^3 \ln x\,dx$, $n = 8$

15. $\int_0^4 \frac{\sqrt{x}}{1+\sqrt{x}}\,dx, n = 8$

16. $\int_0^1 \frac{e^x + e^{-x}}{2}\,dx, n = 4$

In Exercises 17–19, use the trapezoidal rule and Simpson's rule to find the approximate area under the curve defined by the given sets of experimental data.

17.

x	5.0	6.0	7.0	8.0	9.0	10.0	11.0	12.0	13.0
y	4.2	3.9	3.8	4.0	3.5	3.4	3.9	4.1	4.3

18.

x	2.00	2.10	2.20	2.30	2.40	2.50	2.60
y	4.32	4.57	5.14	5.78	6.84	6.62	6.51

19.

x	1.00	1.25	1.50	1.75	2.00	2.25	2.50
y	16.32	16.48	16.73	16.42	16.38	16.29	16.25

Solve Exercises 20–26.

20. *Environmental science* Find the area of the pond in Figure 4.29.

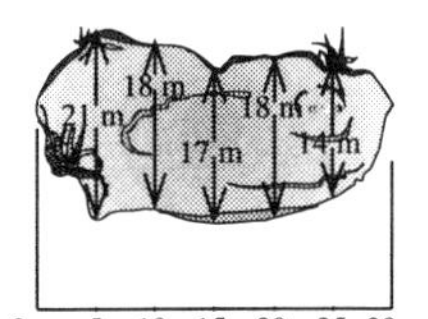

FIGURE 4.29

21. *Electronics* The current i in amperes (A) to a capacitor is given by $i = 4.0e^{-10t} - 4.0e^{-20t}$, where t is time in seconds (s). If $i = \frac{dq}{dt}$ where q is the charge in coulombs (C), use **(a)** the trapezoid rule and **(b)** Simpson's rule to find the approximate change in charge on the capacitor from $t = 1.0$ s to $t = 2.0$ s, using four intervals.

22. *Light* The graph in Figure 4.30 shows the light in lumens given off by the flash of a flashbulb. A typical focal plane shutter is open from 10 milliseconds to 60 milliseconds after the camera button is pressed. What is the total amount of light given by the flashbulb during this interval?

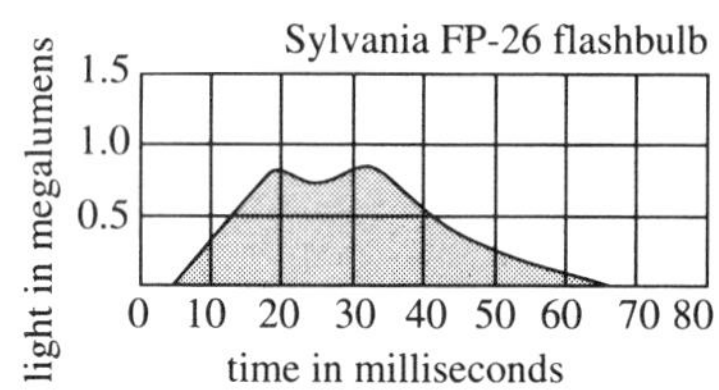

FIGURE 4.30

23. Write a computer program to determine the approximate area under a curve. The program should allow you to select whether to approximate the area by **(a)** the trapezoidal rule, **(b)** Simpson's rule, or **(c)** both rules. Your program should also allow for the input of the endpoints and the number of intervals. If Simpson's rule is used, the program should check to make sure that the number of intervals is even. The program should also allow for either a function $y = f(x)$ to be given or a set of observed data points to be used. Check your program by comparing results with Examples 4.25 and 4.26.

24. *Business* The marginal revenue for the sale of computers by a store is given by the function $R'(x) = 25\sqrt{x}e^{-0.004x}$, where $0 \le x \le 600$. Use the trapezoidal rule with $n = 10$ to determine the approximate revenue from the sale of the first 600 computers sold by the store.

25. *Agriculture* To determine the amount of fertilizer needed for an irregularly shaped field, a farmer needed to estimate the number of acres in the field. The farmer drew a line across the field and then, at 100 ft intervals, measured across the land perpendicular to the farmer's line. The farmer's measurements are shown in the following table:

x	0	100	200	300	400	500	600	700	800	900	1000
y	300	300	350	400	500	600	500	600	550	650	600

Use both the trapezoidal rule and Simpson's rule to estimate the number of acres if 1 acre = 43,500 ft^2.

26. *Pharmacology* The reaction rate to a new drug is given by $r = e^{-t^2} + \frac{1}{t}$ where t is the time measured in hours after the drug is administered. Use both the trapezoidal rule and Simpson's rule with $n = 8$ to estimate the total reaction to the drug from $t = 1$ to $t = 9$.

In Your Words

27. Explain the purpose of the trapezoidal rule and Simpson's rule.

28. Write an application in your technology area of interest that requires you the use of either the trapezoidal rule or Simpson's rule for its solution. Give your problem to a classmate and see if he or she understands and can solve your problem. Rewrite the problem as necessary to remove any difficulties encountered by your classmate.

CHAPTER 4 REVIEW

Important Terms and Concepts

Area between curves
Definite integral
  Properties of
Fundamental Theorem of Calculus
Improper integrals
Indefinite integrals
  Properties of
Numerical integration
Simpson's rule
Substitution method
Summation formulas
Trapezoidal rule

Review Exercises

In Exercises 1–30, evaluate the given definite or indefinite integral.

1. $\int x^5\,dx$

2. $\int (\sqrt[3]{x}+4)\,dx$

3. $\int_0^3 (x^2+4x-6)\,dx$

4. $\int_1^4 (x+4)^2\,dx$

5. $\int_1^8 \frac{dt}{\sqrt[3]{t}}$

6. $\int_1^4 (\sqrt{x}+4)\,dx$

7. $\int \frac{dt}{(t+5)^4}$

8. $\int \frac{x\,dx}{\sqrt[3]{x^2+4}}$

9. $\int_1^8 x\sqrt[3]{x}\,dx$

10. $\int_1^4 \frac{4\,dx}{3\sqrt{x^3}}$

11. $\int (u^4-u^{-4})\,du$

12. $\int \frac{x^4-1}{x^3}\,dx$

13. $\int 2x\sqrt{x^2-5}\,dx$

14. $\int_0^4 x\sqrt{x^2+9}\,dx$

15. $\int x\sqrt[3]{6-x^2}\,dx$

16. $\int 3x\left(4-x^2\right)^3\,dx$

17. $\int_{-1}^{2} (x^2+4)(x^3-3)\,dx$

18. $\int_1^2 x\,(x+2)^2\,dx$

19. $\int \frac{y^2\,dy}{(y^3-5)^{2/3}}$

20. $\int \frac{5-6x+x^2}{1-x}\,dx$

21. $\int_0^1 (1-x^2)^3\,dx$

22. $\int \frac{x+2}{(x^2+4x)^2}\,dx$

23. $\int \left(\sqrt{2x}-\frac{1}{\sqrt{2x}}\right)\,dx$

24. $\int \left(x^3-4x\right)\left(3x^2-4\right)\,dx$

25. $\int_1^4 \frac{9-t^3}{3t^2}\,dt$

26. $\int_0^1 \sqrt{x}\left(3\sqrt{x}-5\right)\,dx$

27. $\int_{-1}^{8} \frac{4\sqrt[3]{u}\,du}{(1+u^{4/3})^3}$

28. $\int_1^2 \frac{3x^3-2x^{-3}}{x^2}\,dx$

29. $\int_1^{\infty} \frac{dx}{x^4}$

30. $\int_0^{\infty} \frac{4x\,dx}{(3x^2+1)^5}$

In Exercises 31–38, find the areas bounded by the given curves.

31. $f(x)=4x-6$, x-axis, $x=-3$, $x=4$

32. $g(x)=2-x-x^2$, x-axis

33. $f(x)=x^2-x-12$, $g(x)=-6$

34. $h(x)=\sqrt{x+2}$, x-axis, $x=7$

35. $f(x)=4x^3$, $g(x)=x$

36. $f(x)=x^3-3x^2$, $g(x)=x-3$

37. $y^2+x=0$, $y=2x+1$

38. $f(x)=2x^3$, $g(x)=x^4$

In Exercises 39–44, find the approximate value of the given integrals with the specified values of *n* by (a) the trapezoidal rule and (b) Simpson's rule.

39. $\int_0^1 \sqrt{1-x^2}\,dx$, $n=10$

40. $\int_1^4 \frac{1}{3x}\,dx$, $n=6$

41. $\int_2^{2.5} \sqrt{x^3-1}\,dx$, $n=10$

42. $\int_{-1}^{4} \sqrt{1+x^5}\,dx$, $n=10$

43. $\int_0^1 \left(e^x-e^{-x}\right)\,dx$, $n=4$

44.

x	1.00	1.25	1.50	1.75	2.00
y	4.25	5.72	5.13	3.19	2.10

x	2.25	2.50	2.75	3.00
y	0.15	1.65	3.10	3.70

Solve Exercises 45–48.

45. *Dynamics* An object is moving rectilinearly with a velocity $v(t)=t+\frac{1}{\sqrt{1+t}}$ m/s. **(a)** If the object began at the origin, what is its position when $t=15$? **(b)** What is its acceleration when $t=15$?

46. *Electricity* The power P in watts (W) in a circuit is given by $P=\frac{d\omega}{dt}$, where ω is dissipated energy in joules (J) and t is time in seconds (s). If the power in a circuit is given by $P=10t^3$, find the energy dissipated from $t=3.0$ s to $t=5.0$ s.

47. *Electricity* The induced voltage v in volts (V) in an inductor is given by $v=-L\frac{di}{dt}$, where i is the current in amperes (A) and L is the inductance in henries (H). If v is given by $v=2.0-1.0t^2$, where t is time in seconds (s) and $L=2$ H, find the change in the current to the inductor from $t=1.0$ s to $t=6.0$ s.

48. *Highway safety* A car moving initially at 88 ft/s, decelerates at the constant rate of 4 ft/s^2. **(a)** How long does it take to stop the car? **(b)** How far does it travel from the time the brakes are applied until it stops?

CHAPTER 4 TEST

In Exercises 1–7, evaluate the given integral.

1. $\int_0^1 2x\,dx$

2. $\int_1^4 (x+1)(x^2+1)\,dx$

3. $\int (x^2+3)^2\,dx$

4. $\int 4(x^4+7)^5 x^3\,dx$

5. $\int \frac{x^2+2}{x^2}\,dx$

6. $\int (x^2+8x)^3(x+4)\,dx$

7. $\int_1^2 (3x^2\sqrt{x^3+x}+\sqrt{x^3+x})\,dx$

Solve Exercises 8–11.

8. Find the total area of the region between the curve $y=4x^3-4$ and the x-axis over the interval $[-1,2]$.

9. Find the area of the region enclosed by the curve $y=x^2-2$ and the line $x+y=4$.

10. Use the trapezoidal and Simpson's rules to determine $\int_2^4 \frac{1}{x}\,dx$ with $n=8$.

11. For a particular circuit, the current, in amperes, after time t, in seconds, at a certain point, P, is given by $i=0.0004(t^{-1/2}+t^{1/2})$. Find the charge, in C, that passes point P during the first second.

CHAPTER

5

Applications of Integration

A tank truck, like the milk truck shown here, is used to carry liquids. In Section 5.7, you will learn how to determine the force on one end of this truck when it is full of milk.

Courtesy of H.P. Hood Inc.

We introduced the integral as a method for determining the area under a curve. Later, we showed that integrals could be used to determine the velocity and position functions from the acceleration. We also used integrals to find the area between two curves. In this chapter, we will see some of the many other applications for the integral.

We begin with the average value and root mean square of a function. The average value will be used to determine the amount of heating oil that will be used during one winter day. Examples will use the root mean square to determine the effective current and voltage in an electric circuit.

Next, we look at methods for determining the volume of a three-dimensional object. Two different methods will be examined to allow you to select the better method for a particular problem. Then we will look at methods for finding the length of a curve and the area of a surface formed by revolving that curve around a line.

We will end the chapter with three sections that look at several physical applications—an object's mass, its center of gravity, its moment of inertia, the amount of work needed to move something, and the pressure exerted by a liquid.

5.1 AVERAGE VALUES AND OTHER ANTIDERIVATIVE APPLICATIONS

In this section, we will look at two ideas: One is the idea of the average value of a function; the other idea looks at some additional applications of the antiderivative.

Average Value of a Function

The concept of an average is a common one to most of us. Sometimes the term is applied to a finite number of items such as the average grade on a test by a group of students. Sometimes it is based on a continuous function. There are times when that function is known and times when it is not.

One example that most of us hear every day is the reference to the average temperature. The weather bureau assumes that there is some function f that will give the temperature T at some time t. Thus, $T = f(t)$. Normally, what is done is to take a temperature reading every hour. The average or mean temperature $\bar{T}$ is the sum of these temperatures divided by the number of hours. Thus, for one day, we have

$$\bar{T} = \frac{1}{24}\sum_{t=1}^{24} f(t)$$

If we divide the day into smaller intervals and take a temperature reading at each of those intervals, we get a more accurate reading. So, if we take n temperature readings we have

$$\bar{T} = \frac{1}{n}\sum_{i=1}^{n} f(t_i)$$

where t_i represents the time of the ith temperature reading.

Naturally, the larger that n becomes the more accurate is the average. If we wanted the true average, we would take the limit of $\frac{1}{n}\sum_{i=1}^{n} f(t_i)$ as n approaches infinity or

$$\bar{T} = \lim_{n\to\infty}\frac{1}{n}\sum_{i=1}^{n} f(t_i)$$

The right-hand side of this equation looks familiar. If dt represents the width of a time interval, $\frac{b-a}{n}$, where a is the beginning of the first interval and b the end of the last interval, and if we write $\frac{1}{n}$ as $\frac{b-a}{n}\cdot\frac{1}{b-a}$, then

$$\begin{aligned}\bar{T} &= \lim_{n\to\infty}\frac{1}{b-a}\sum_{n=1}^{n} f(t_i)\frac{b-a}{n}\\ &= \frac{1}{b-a}\lim_{n\to\infty}\sum_{i=1}^{n} f(t_i)dt\\ &= \frac{1}{b-a}\int_a^b f(t)dt\end{aligned}$$

If we assume that f is continuous, then we can use this idea for the following definition.

Average Value

If f is a continuous function over the interval $[a,b]$, then the **average value** $\bar{y}$ is

$$\bar{y} = \frac{1}{b-a}\int_a^b f(x)dx$$

Application

EXAMPLE 5.1

A ball is dropped from a height of 490 m. Find its average velocity between the time it is dropped and the time it hits the ground.

Solution If acceleration is $-9.8\ \text{m/s}^2$ and, since $v(0)=0$, then we have $v(t) = -9.8t$. We also have $s(0)=490$, and so $s(t) = -4.9t^2+490$. Solving for t, we see that it takes 10 s to hit the ground. The average velocity is

$$\begin{aligned}\bar{v} &= \frac{1}{10-0}\int_0^{10} v(t)dt \\ &= \frac{1}{10}[s(t)]\Big|_0^{10} \\ &= \frac{1}{10}(-4.9t^2+490)\Big|_0^{10} \\ &= \frac{1}{10}(-490) = -49\end{aligned}$$

The average velocity is 49 m/s downward.

Application

EXAMPLE 5.2

The amount of heating oil required to heat a house for one day during the winter can be approximated by multiplying the difference between 20°C and the average daily temperature by 0.8 gal. During one particular cold spell, the temperature T, at any time of the day t h after midnight, was given by $T(t) = 2-\frac{1}{15}(t-15)^2$. Determine the level of oil consumption during one of these days.

Solution If we let A represent the amount of oil required and $\bar{T}$ the average temperature during a day, then we have

$$A = (20^\circ\text{C} - \bar{T})(0.8\,\text{gal}/^\circ\text{C})$$

We will first compute the average temperature during one of these days. We will let t range from midnight at the start of the day, $t=0$, to midnight at the end of the

EXAMPLE 5.2 (Cont.)

day, $t = 24$. Thus, for our formula, $a = 0$ and $b = 24$

$$\begin{aligned}\bar{T} &= \frac{1}{24-0}\int_0^{24} T(t)dt \\ &= \frac{1}{24-0}\int_0^{24}\left[2-\frac{1}{15}(t-15)^2\right]dt \\ &= \frac{1}{24}\left[2t-\frac{1}{45}(t-15)^3\right]_0^{24} \\ &= \frac{1}{24}(31.80-75) = \frac{1}{24}(-43.2) = -1.8\end{aligned}$$

So, the average temperature during this particular 24-hour period was $-1.8°C$.

We can now determine the amount of fuel oil consumed during this period.

$$\begin{aligned}A &= (20°C - \bar{T})(0.8\,\text{gal/°C}) \\ &= [20°C - (-1.8°C)](0.8\,\text{gal/°C}) \\ &= (21.8°C)(0.8\,\text{gal/°C}) \\ &= 17.44\,\text{gal}\end{aligned}$$

A total of 17.44 gal of heating oil were used during this 24-h period.

Root Mean Square

The **root mean square**, rms, of a function $y = f(x)$ over the interval $[a, b]$ is defined to be the square root of the average value of y^2 on the interval.

Root Mean Square

If a function f is continuous on $[a, b]$, then the **root mean square**, f_{rms}, is

$$f_{\text{rms}} = \sqrt{\frac{1}{b-a}\int_a^b [f(x)]^2\,dx}$$

EXAMPLE 5.3

Find the root mean square of $f(x) = x^4$ on $[-1, 3]$.

Solution

$$\begin{aligned}f_{\text{rms}} &= \sqrt{\frac{1}{3-(-1)}\int_{-1}^{3}(x^4)^2\,dx} \\ &= \sqrt{\frac{1}{4}\int_{-1}^{3}x^8\,dx}\end{aligned}$$

EXAMPLE 5.3 (Cont.)

$$= \sqrt{\frac{1}{4}\cdot\frac{1}{9}x^9\Big|_{-1}^{3}}$$

$$= \sqrt{\frac{1}{36}[19{,}683-(-1)]}$$

$$= \sqrt{\frac{19{,}684}{36}} \approx 23.38$$

Because an alternating current changes continuously, its maximum value, $\pm I_{\max}$, does not indicate its ability to do work or to produce heat as does the magnitude of a direct current. For this reason, it is customary to refer to the **effective current**. The same is true for the voltage, where it is more common to refer to the **effective voltage** rather than $V_{\max}$.

Note The effective current and the effective voltage are the same as the root mean square of the current and voltage. In symbols, this means that $i_{\text{eff}} = i_{\text{rms}}$ and $V_{\text{eff}} = V_{\text{rms}}$.

If the resistance is constant, then the average or mean power, in watts, produced by the current is

$$P = i_{\text{rms}}^2 \cdot R$$

Application

EXAMPLE 5.4

If a current $i = t - 2t^2$ A flows through a resistor of $20\,\Omega$ from $t = 0$ to $t = 3$ s, find i_{eff} and the power generated.

Solution

$$i_{\text{eff}} = i_{\text{rms}} = \sqrt{\frac{1}{3}\int_0^3 (t-2t^2)^2\,dt}$$

$$= \sqrt{\frac{1}{3}\int_0^3 (t^2-4t^3+4t^4)dt}$$

$$= \sqrt{\frac{1}{3}\left[\frac{1}{3}t^3 - t^4 + \frac{4}{5}t^5\right]_0^3}$$

$$= \sqrt{\frac{1}{3}(9-81+194.4)}$$

$$= \sqrt{40.8} \approx 6.39\,\text{A}$$

and

$$P = \left(\sqrt{40.8}\right)^2 \cdot 20$$

$$\approx 816\,\text{W}$$

Current and Power

The current i, in amperes, in an electric circuit is defined as the time rate of change of the charge q, in coulombs, which passes a given point in the circuit. This is written as

$$i = \frac{dq}{dt}$$

When this is placed in the differential form, we get $dq = i\,dt$. Integrating both side of this equation, we get

$$q = \int i\,dt$$

Thus we can find the total charge transmitted in a circuit.

The charge in a capacitor is given by

$$q = CV_c$$

where C is the capacitance in farads and V_c represents the voltage between the capacitor terminals. Substituting for q from the previous formula and solving for V_c, we get

$$V_c = \frac{1}{C}q$$

$$\text{or} \quad V_c = \frac{1}{C}\int i\,dt$$

Application

EXAMPLE 5.5

The current in a circuit is given by the equation $i = 4t^3 + 2$ A. How many coulombs are transmitted in 3 s?

Solution We have $q = \int i\,dt$ and from the information given we can see that

$$\begin{aligned} q &= \int_0^3 (4t^3 + 2)dt \\ &= (t^4 + 2t)\big|_0^3 \\ &= 87\,\text{C} \end{aligned}$$

Note The C at the end of Example 5.5 stands for coulombs. Do not confuse it with the C for capacitance, with the constant of integration, or with °C for degrees Celsius.

Application

EXAMPLE 5.6

A 1-μF capacitor receives a charge such that it has a terminal-to-terminal voltage of 33 V. We connect this capacitor at $t = 0$ s to a source that sends a current $i = t^2$ A into it. What is the voltage across the capacitor when $t = 0.2$ s?

EXAMPLE 5.6 (Cont.)

Solution We will need the formula $V_c = \frac{1}{C}\int i\,dt$. We are given $C = 1\,\mu\text{F}$ $= 10^{-6}\,\text{F}$ and $i = t^2$, so

$$V_c = \frac{1}{10^{-6}}\int t^2\,dt = 10^6 \cdot \frac{1}{3}t^3 + k$$

(We will let k be the constant of integration.) Now when $t = 0$, $V_c = 33$ and $k = 33$ and so

$$V_c = 10^6 \cdot \frac{1}{3}t^3 + 33$$

When $t = 0.2$, we get

$$\begin{aligned} V_c &= 10^6 \cdot \frac{1}{3}(0.2)^3 + 33 \\ &\approx 2{,}699.67\,\text{V} \\ &\approx 2{,}700\,\text{V} \end{aligned}$$

Application

EXAMPLE 5.7

A certain capacitor has a voltage of 150 V across it. A current given by $i = 0.08\sqrt{t}$ is sent through the circuit. After 0.36 s, the voltage across the capacitor is measured at 200 V. What is the capacitance of this capacitor?

Solution We will use the equation $V_c = \frac{1}{C}\int i\,dt$. Since $i = 0.08\sqrt{t}$, we have

$$\begin{aligned} V_c &= \frac{1}{C}\int 0.08t^{1/2}\,dt \\ &= \frac{0.08}{C}\int t^{1/2}\,dt \\ &= \frac{0.16}{3C}t^{3/2} + k \end{aligned}$$

When $t = 0$, we know that $V_c = 150$ and so $k = 150$. Thus,

$$V_c = \frac{0.16}{3C}t^{3/2} + 150$$

Now $V_c = 200\,\text{V}$ when $t = 0.36\,\text{s}$, and so

$$\begin{aligned} 200 &= \frac{0.16}{3C}(0.36)^{3/2} + 150 \\ 200 &= \frac{0.01152}{C} + 150 \end{aligned}$$

Solving for C, we get

$$\begin{aligned} C &= 0.0002304\,\text{F} \\ &= 2.304 \times 10^{-4}\,\text{F} \\ &= 230.4\,\mu\text{F} \end{aligned}$$

As a last application in this section, the amount of energy (**work**), W, expended in an electrical system is given by

$$W = \int P\,dt$$

Here W is in joules, the power P is in watts, and t is in seconds.

Application

EXAMPLE 5.8

The power in a certain electrical system is changed according to $P = 5t^{1/4}$ W. What is the formula for energy in this circuit?

Solution Using the formula $W = \int P\,dt$ with $P = 5t^{1/4}$, we have

$$\begin{aligned} W &= \int 5t^{1/4}\,dt \\ &= 4t^{5/4} + k \text{ J} \end{aligned}$$

The **instantaneous power** in a circuit can be represented by $P = Vi$, $P = i^2R$, or $P = V^2/R$, and any of these can be substituted for P in the formula $W = \int P\,dt$.

Exercise Set 5.1

In Exercises 1–12, find the average value and the root mean square of the given function over the indicated interval.

1. $f(x) = x^2$, over $[0, 1]$
2. $g(x) = 3x^2$, over $[2, 4]$
3. $h(x) = x^2 - 1$, over $[1, 3]$
4. $j(x) = 9 - x^2$, over $[-3, 3]$
5. $f(x) = \sqrt{x+4}$, over $[0, 5]$
6. $k(x) = x^3 - 1$, over $[1, 3]$
7. $f(t) = t^2 + 4$, over $[0, 4]$
8. $g(t) = 16t^2 - 25$, over $[2, 4]$
9. $f(x) = 6x - x^2$, over $[1, 2]$
10. $h(t) = 16t^2 - 8t + 1$, over $[0, 3]$
11. $j(t) = 4.9t^2 - 2.8t + 4$, over $[0, 2.5]$
12. $k(t) = 4.9t^2 - 4.2t + 9$, over $[0, 1.6]$

Solve Exercises 13–32.

13. *Meteorology* The rainfall per day, measured in centimeters, x days after the beginning of the year is $0.0002(4991 + 366x - x^2)$. What is the average rainfall for the first 90 days? For the year (365 days)?

14. *Material science* The mass density of a metal bar of length 4 m is given by $\rho(x) = 1 + x - \sqrt{x}$ kg/m^3, where x is the distance in meters from one end of the bar. What is the average mass density over the length of the bar?

15. *Physics* If a ball is dropped from a height of 360 ft, find its average height and average velocity between the time it is dropped and the time it hits the ground.

16. A 20 000-L water tank takes 10 min to drain. After t min, the amount of water left in the tank is given by $V(t) = 200(100 - t^2)$. What is the average amount of water in the tank while it is draining?

17. *Meteorology* One day the temperature of the air t h after noon was given by $60+4t-t^{2/3}$°F.
    (**a**) What was the average temperature between noon and 6 p.m.?
    (**b**) What was the average temperature between 3 p.m. and 9 p.m.?
18. *Business* A laboratory receives a shipment of 900 cases of test tubes every 30 days. The number of cases on hand t days after the shipment arrives is given by $I(t)=900-15\sqrt{120t}$. What is the average daily inventory?
19. *Electricity* Find the average current from $t=0$ to $t=3$ s, if $i=1.0t+\sqrt{t}$ A.
20. *Electricity* Find the effective current for the current in Exercise 19.
21. *Electricity* Find the average current from $t=0$ to $t=4$ s, if $i=4t\sqrt{t^2+1}$ A.
22. *Electricity* What is the effective current for the current in Exercise 21?
23. *Electricity* If the current in Exercise 21 flows through a 30-Ω resistor, what is its average power?
24. *Electricity* A current $i=t^3-2t^2$ A flows through a resistance of 5.0 Ω from $t=0.0$ to $t=0.25$ s. Find the average power developed over that interval.
25. *Electricity* The voltage across a 10-Ω resistor is given by $V=4\sqrt{t}-2t$. (**a**) What is the effective voltage during the time interval from $t=0$ to $t=6$ s?
    (**b**) What is the average power generated during this time interval?
26. *Electricity* A voltage of $v=8.0t^{1/2}-16t^{3/2}$ V is applied across a resistance of 9.75 Ω. Find the average power developed during the period when v is greater than 0?
27. *Electricity* A capacitor contains a charge of 0.01 C. If it is supplied with a current of $i=2t$ A, what charge resides in the capacitor after 0.1 s?
28. *Electricity* An 80-μF capacitor is charged to 100 V. The capacitor is then supplied a current $i=0.04t^3$ A. After what time interval does the capacitor voltage reach 225 V?
29. *Electricity* The voltage applied to a circuit was $V=2t+1$ V. If the current was $i=0.03t$ A, what was the energy delivered from $t=0$ to $t=50$ s?
30. *Electricity* The voltage across a 90-μF capacitor is zero. What is the voltage after 0.001 s, if a current of 0.20 A charges the capacitor?
31. *Electricity* The voltage across a 7.5-μF capacitor is zero. What is the voltage after 0.005 s, if a current of $i=0.4t$ A charges the capacitor?
32. *Electricity* The current in a certain wire changes with time according to the relation $i=\sqrt[3]{1+5t}$. How many coulombs of charge pass a certain point in the first 3 s?

In Your Words

33. Describe how to calculate the average value of a function.
34. What is the root mean square? Give three applications of the root mean square.

≡ 5.2 VOLUMES OF REVOLUTION: DISK AND WASHER METHODS

Consider the area under the graph of a continuous nonnegative function, $y=f(x)$, over the interval $[a,b]$. If this area is revolved around the x-axis, then we get a solid of revolution such as the one shown in Figure 5.1. Common examples of solids of revolution include a cone, a cylinder, a soft drink bottle, and a table leg formed by turning a piece of wood on a lathe.

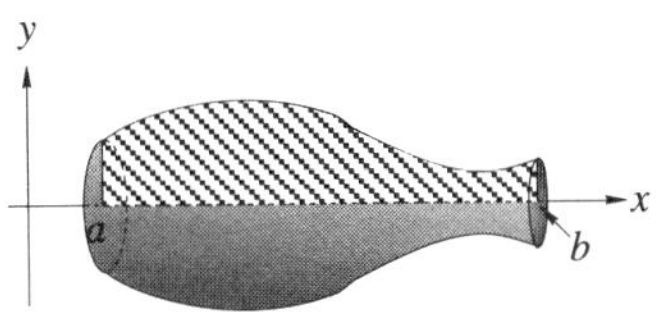

FIGURE 5.1

A cone is formed by revolving the area under a line of the form $f(x) = mx$ from 0 to h around the x-axis, as shown in Figure 5.2a. A cylinder is formed when the area under a horizontal line from a to b is revolved around the x-axis as shown in Figure 5.2b. A soft drink bottle is formed by revolving the area under a curve, such as the one shown in Figure 5.2c.

Disks

In this section and in Section 5.3, we will develop ways for finding the volume of solids of revolution such as these. As when we developed the definite integral, we will let $y = f(x)$ be a continuous function with $f(x) \geq 0$ on the interval $[a, b]$. If we rotate the area under this curve around the x-axis, we generate a solid of revolution. If we cut this solid with parallel planes all perpendicular to the x-axis, we will see that a typical cross-section is a circle and a typical slice is a disk as shown in Figure 5.3a.

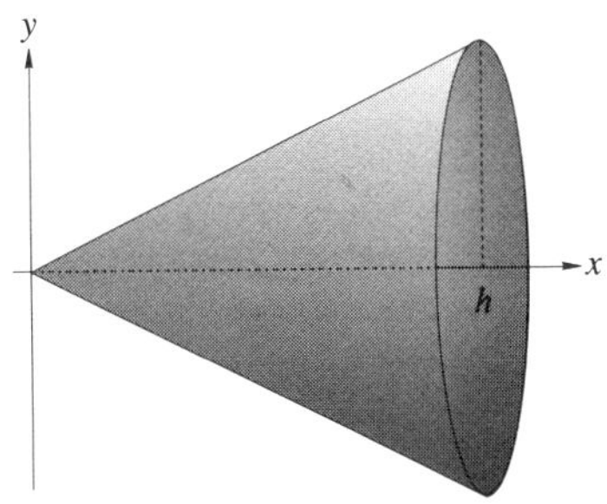

FIGURE 5.2a

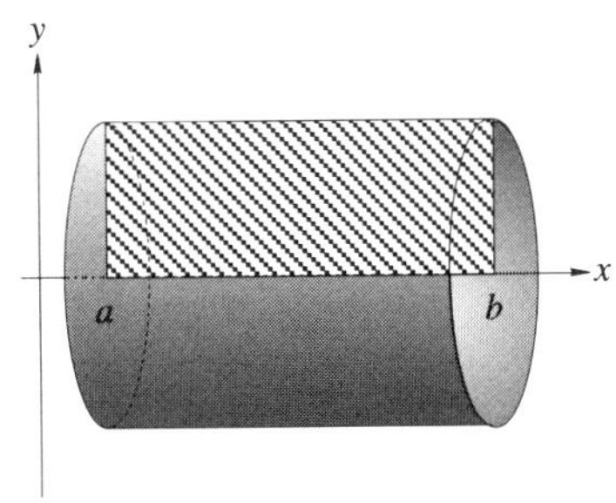

FIGURE 5.2b

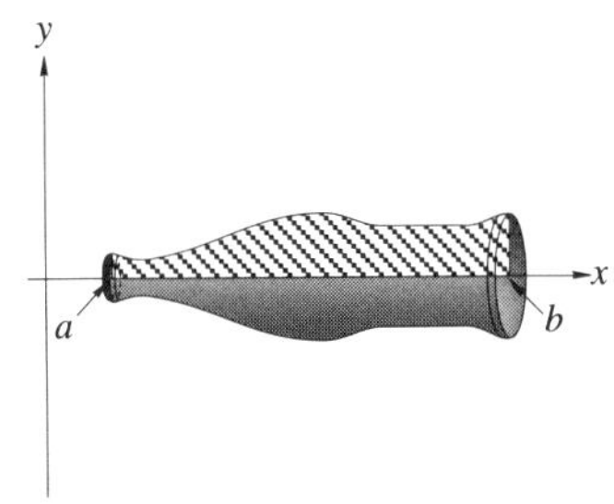

FIGURE 5.2c

If we divide $[a, b]$ into subintervals and let x_i be a point in one of the subintervals, then a typical disk is a cylinder with height dx and radius $f(x_i)$ as shown in Figure 5.3b. The volume of a cylinder is $\pi r^2 h$, so the volume of this disk is $\pi[f(x_i)]^2\,dx$. Summing the volume of all of these disks over the interval $[a, b]$, we get the total volume as shown in the following box.

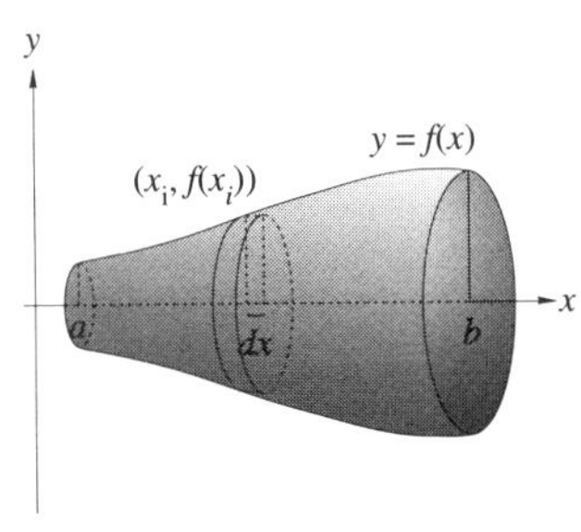

FIGURE 5.3a

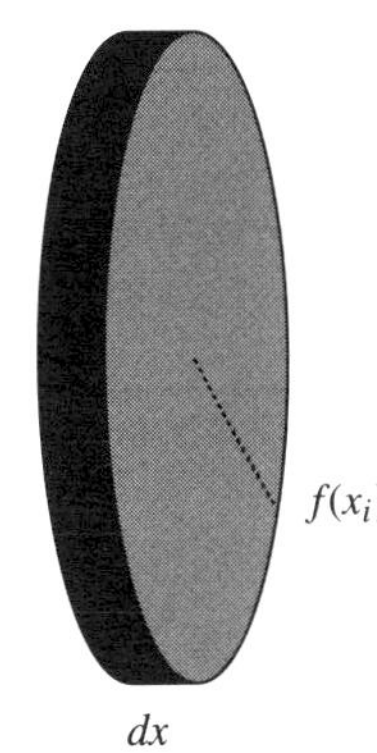

FIGURE 5.3b

Disk Method for Finding a Volume of Revolution (x-Axis)

The volume V of the solid generated by revolving the area between the graph of $y = f(x)$ and the x-axis from $x = a$ to $x = b$ around the x-axis is

$$V = \pi \int_a^b [f(x)]^2 \, dx$$

Hint

It is often easier to remember the disk method by using the following guidelines:

1. Sketch the graph over the limits of integration.
2. Draw a typical disk by slicing the solid perpendicular to the axis of revolution.
3. Determine the radius and thickness of the disk.
4. Determine the volume of a typical disk by using $\pi\,(\text{radius})^2 \cdot (\text{thickness})$.
5. Use the following integration formula to find the total volume:

$$V = \pi \int_a^b (\text{radius})^2 \cdot (\text{thickness})$$

EXAMPLE 5.9

Find the volume of the solid of revolution generated by revolving the area under the graph of $f(x) = \sqrt{x}$ from $x = 0$ to $x = 4$ around the x-axis.

Solution The graph of $f(x) = \sqrt{x}$ on $[0, 4]$ is given in Figure 5.4a. A rectangle has been drawn to show a typical element.

The solid that results from revolving the graph of f around the x-axis is shown in Figure 5.4b, with a disk formed by the typical element. The radius of the disk is $y = \sqrt{x}$, and its thickness is dx. A typical disk is shown in Figure 5.4c. Its volume is

$$\pi\,(\text{radius})^2 \cdot (\text{thickness}) = \pi \left(\sqrt{x}\right)^2 dx$$

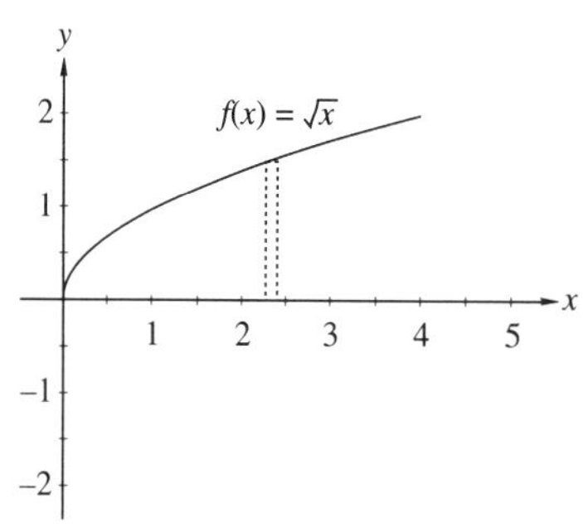

FIGURE 5.4a

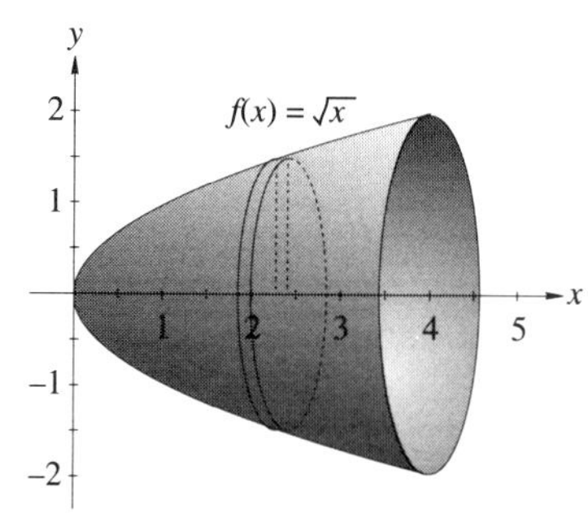

FIGURE 5.4b

EXAMPLE 5.9 (Cont.)

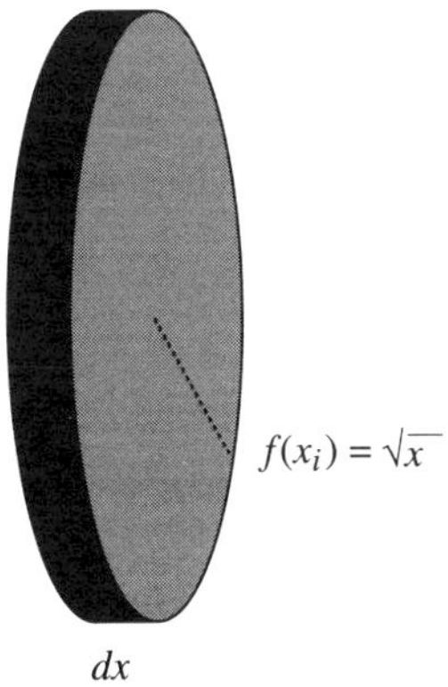

FIGURE 5.4c

Thus, we get the following integral.

$$
\begin{aligned}
V &= \pi \int_0^4 (\sqrt{x})^2\, dx \\
&= \pi \int_0^4 x\, dx \\
&= \pi \left(\frac{1}{2}x^2\right)\Bigg|_0^4 = 8\pi
\end{aligned}
$$

The volume is 8π units^3.

Application

EXAMPLE 5.10

A wing tank for an airplane is a solid of revolution formed by rotating the region bounded by the graph of $y = \frac{1}{16}x\sqrt{4-x}$ and the x-axis from $x = 0$ to $x = 4$ around the x-axis, where x and y are measured in meters, as shown in Figure 5.5a. Find the volume of the tank.

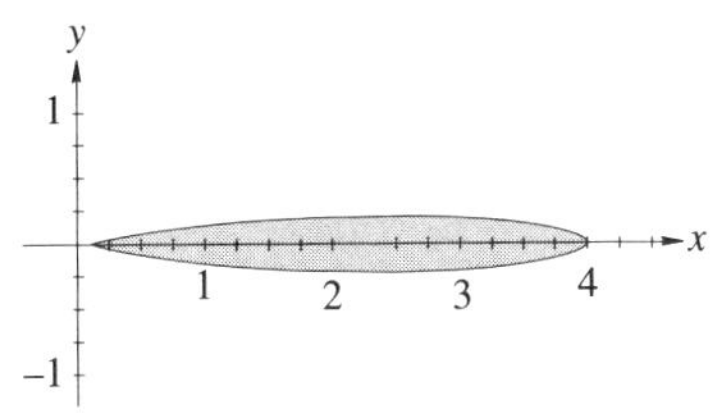

FIGURE 5.5a

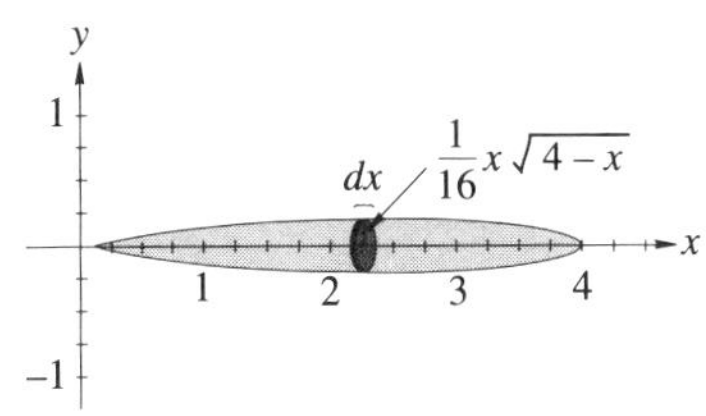

FIGURE 5.5b

Solution A disk formed from a typical element is shown in Figure 5.5b. The radius of the disk is $\frac{1}{16}x\sqrt{4-x}$ and its thickness is dx. Thus the volume of a typical disk is

$$\pi\,(\text{radius})^2 \cdot (\text{thickness}) = \pi \left(\frac{1}{16}x\sqrt{4-x^2}\right)^2 dx$$

and the total volume is

$$
\begin{aligned}
V &= \pi \int_0^4 \left(\frac{1}{16}x\sqrt{4-x}\right)^2 dx \\
&= \pi \int_0^4 \frac{1}{256}x^2(4-x)\, dx \\
&= \frac{\pi}{256}\int_0^4 (4x^2 - x^3)\, dx
\end{aligned}
$$

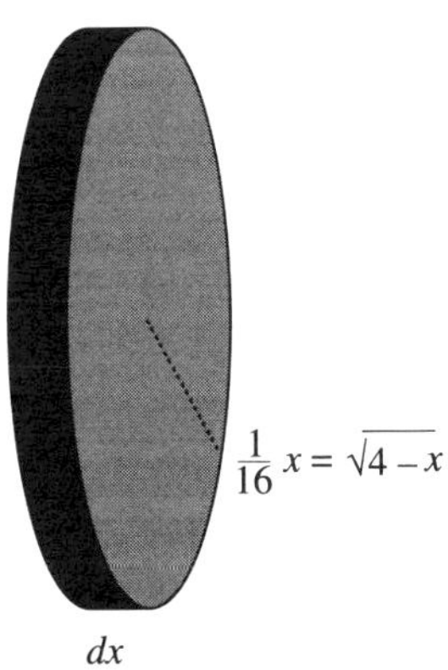

FIGURE 5.5c

EXAMPLE 5.10 (Cont.)

$$= \frac{\pi}{256}\left(\frac{4}{3}x^3 - \frac{1}{4}x^4\right)\Bigg|_0^4$$

$$= \frac{\pi}{256}\left(\frac{256}{3} - 64\right) = \frac{\pi}{12} \approx 0.262$$

The volume of this wing tank is approximately $0.262\,\text{m}^3$.

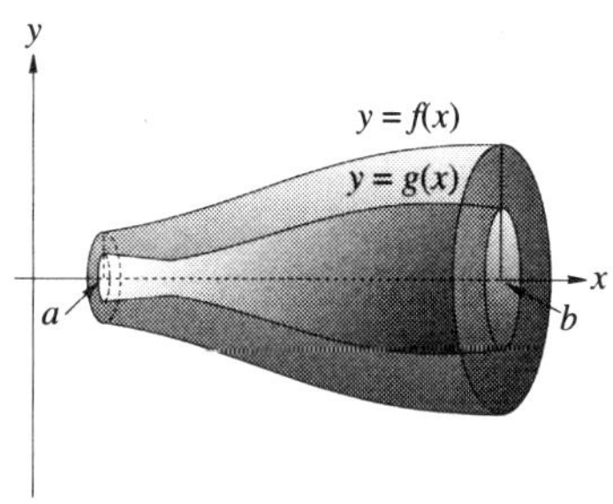

FIGURE 5.6a

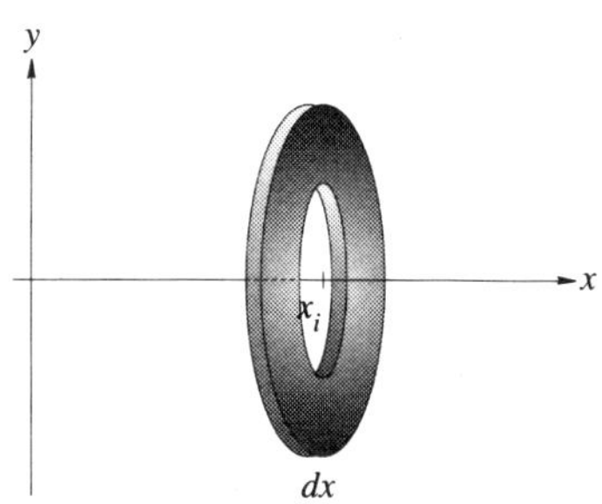

FIGURE 5.6b

Washers

In the disk method, we revolved the area between one curve and the x-axis around the x-axis. But, what happens if we revolve the area between two curves around the x-axis? In Figure 5.6a, we revolved the area enclosed by $y = f(x)$ and $y = g(x)$, where $f(x) \geq g(x)$, and between the lines $x = a$ and $x = b$.

If we sliced this as we did in the disk method, we would get a disk with a hole in the middle. Thus, the slice would look like a washer, as shown in Figure 5.6b. As before, the height of this washer is dx. Its volume could be determined by finding the volume of the solid disk formed by $y = f(x)$ and subtracting the disk that forms the hole in the washer, $y = g(x)$. The area of the entire solid disk is $\pi[f(x_i)]^2\,dx$ and of the hole is $\pi[g(x_i)]^2\,dx$. Thus, the area of the washer is

$$\pi[f(x)]^2\,dx - \pi[g(x)]^2\,dx = \pi\left\{[f(x)]^2 - [g(x)]^2\right\}\,dx$$

Summing this over the entire interval, we get a second method for finding a volume of revolution.

Washer Method for Finding a Volume of Revolution

The volume V of the solid generated by revolving the area between the graph of $y = f(x)$ and the graph of $y = g(x)$ from $x = a$ to $x = b$ around the x-axis is

$$V = \pi\int_a^b \left\{[f(x)]^2 - [g(x)]^2\right\}dx$$

where $f(x)$ = the outer radius and $g(x)$ = the inner radius of the washer.

EXAMPLE 5.11

Find the volume of the solid of revolution generated by revolving the area enclosed by $y = x^2$ and $y = 4x$ around the x-axis.

EXAMPLE 5.11 (Cont.)

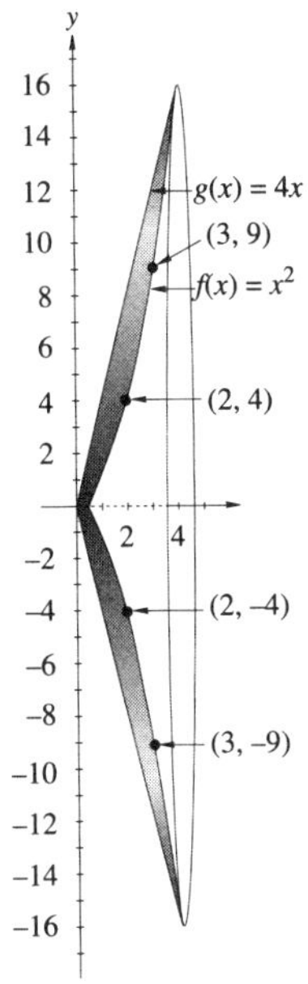

FIGURE 5.7

Solution Let $f(x) = x^2$ and $g(x) = 4x$. The graph of these two functions is shown in Figure 5.7. The two graphs intersect when $f(x) = g(x)$ or when $x^2 = 4x$, which is at $x = 0$ and $x = 4$. On the interval $[0, 4]$, $4x \geq x^2$, the outer radius is $f(x) = x^2$, and the inner radius is $g(x) = 4x$. We determine the volume as

$$\begin{aligned} V &= \pi \int_0^4 \left[(4x)^2 - (x^2)^2\right] dx \\ &= \pi \int_0^4 (16x^2 - x^4)dx \\ &= \pi \left(\frac{16}{3}x^3 - \frac{1}{5}x^5\right)\Bigg|_0^4 \\ &= \pi \left(\frac{2{,}048}{15}\right) \approx 136.53\pi \end{aligned}$$

The volume is about 136.53π units^3.

Hint

The following guidelines are a variation of those for the disk method. Following them should make it easier to remember the washer method.

1. Sketch the graphs of the two functions and find the limits of integration.
2. Draw a typical washer, as in Figure 5.8, and remember to slice the solid perpendicular to the axis of revolution.
3. Find the outer and inner radii for the washer.
4. Determine the thickness of the washer.
5. The volume of a typical washer is

$$\pi \left[(\text{outer radius})^2 - (\text{inner radius})^2\right] \cdot (\text{thickness})$$

6. Use the following integration formula to find the total volume:

$$V = \pi \int_a^b \left[(\text{outer radius})^2 - (\text{inner radius})^2\right] \cdot (\text{thickness})$$

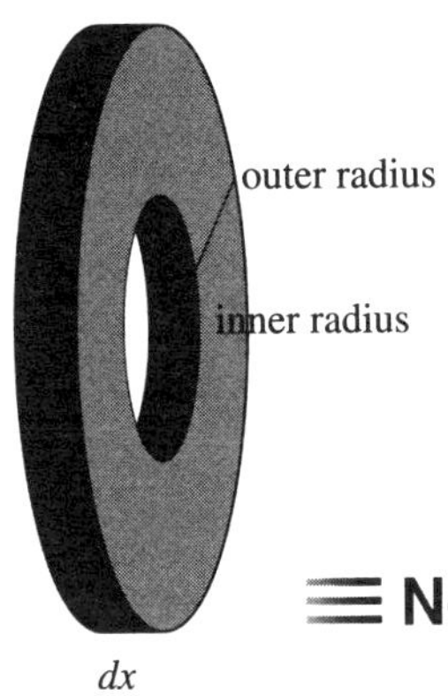

FIGURE 5.8

Note

Be sure to label the volume in appropriate cubic units.

Application

EXAMPLE 5.12

A drafter is designing a plastic bead that can be strung to help form a necklace. The bead is in the shape of a sphere with a hole drilled through the center, as shown in Figure 5.9a. If the sphere has a radius of 0.25 in. and the hole a radius of 0.07 in., how much plastic will be needed to make one bead?

EXAMPLE 5.12 (Cont.)

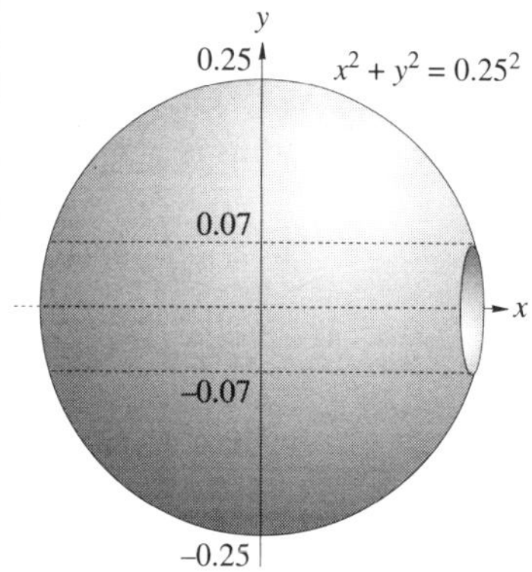

FIGURE 5.9a

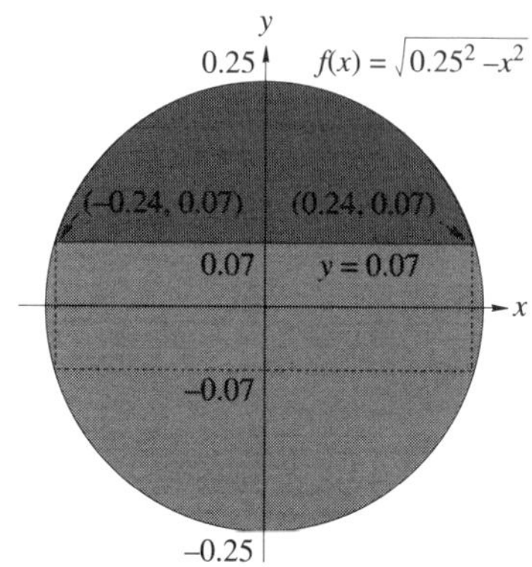

FIGURE 5.9b

Solution We will sketch this so the origin is at the center of the sphere. A cross-section of the bead, as in Figure 5.9b, shows that the sphere is formed by rotating the top half of the circle $x^2 + y^2 = 0.25^2$, so the outer radius is $f(x) = \sqrt{0.25^2 - x^2}$. The hole is formed by rotating the region formed by the line $y = 0.07$, so the inner radius is $g(x) = 0.07$. If we let $y = 0.7$ and we solve the equation $x^2 + y^2 = 0.25^2$, we get the points of intersection at $x = \pm 0.24$ in. Since the thickness is dx, we have all the information we need, and the volume is given by

$$\begin{aligned} V &= \pi \int_{-0.24}^{0.24} \left[\left(\sqrt{0.25^2 - x^2} \right)^2 - (0.07)^2 \right] dx \\ &= \pi \int_{-0.24}^{0.24} (0.0625 - x^2 - 0.0049)\, dx \\ &= \pi \int_{-0.24}^{0.24} (0.0576 - x^2)\, dx \\ &= \pi \left[0.0576x - \frac{x^3}{3} \right]_{-0.24}^{0.24} \\ &= \frac{\pi}{3}(0.055296) = 0.018432\pi \approx 0.0579 \end{aligned}$$

It will take about 0.0579 in.^{3} of plastic to make each bead.

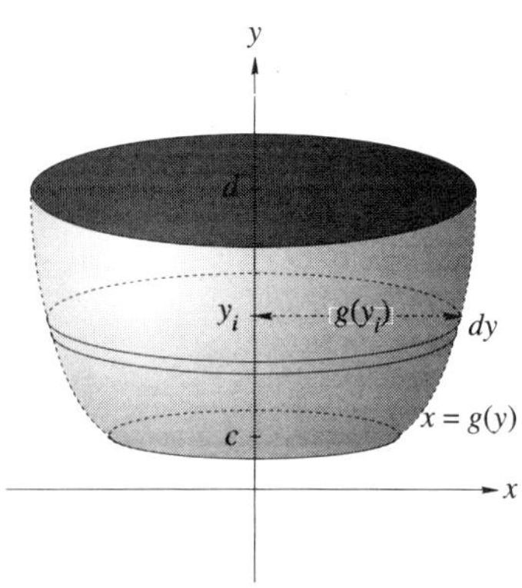

FIGURE 5.10

Revolving Around the y-Axis

Until now we have found the volumes by revolving regions around the x-axis. It is possible to revolve a region around any line. We will look at revolving the area enclosed by $x = g(y)$, $y = c$, and $y = d$, where $g(y) \geq 0$, around the y-axis, as shown in Figure 5.10. The process is the same as when a region is revolved around the x-axis. Again, slices are taken, only this time they are perpendicular to the y-axis. A typical disk is shown in Figure 5.10. The height of the disk is dy and its radius is $g(y_i)$, so the volume of the disk is $\pi[g(y_i)]^2\,dy$ and the total volume is given in the following box.

Disk Method for Finding a Volume of Revolution (y-Axis)

The volume V of the solid generated by revolving the area between the graph of $x = g(y)$ and the y-axis from $y = c$ to $y = d$ around the y-axis is

$$V = \pi \int_c^d (\text{radius})^2 \cdot (\text{thickness}) = \pi \int_c^d [g(y)]^2 \, dy$$

EXAMPLE 5.13

Find the volume of revolution generated by revolving the area enclosed by $y = \frac{1}{2}x^2$, $y = 0$, and $y = 8$ around the y-axis.

Solution Since the region is to be revolved around the y-axis, we will draw, horizontally, the rectangle that forms a typical disk as shown in Figure 5.11. Since this is revolved around the y-axis, we need to describe the curve as a function of y. Solving $y = \frac{1}{2}x^2$ for x, we get $x = g(y) = \sqrt{2y}$. Thus, the volume is

$$V = \pi \int_0^8 (\sqrt{2y})^2 \, dy = \pi \int_0^8 2y \, dy = \pi y^2 \big|_0^8 = 64\pi$$

The volume is 64π units^3. ▪

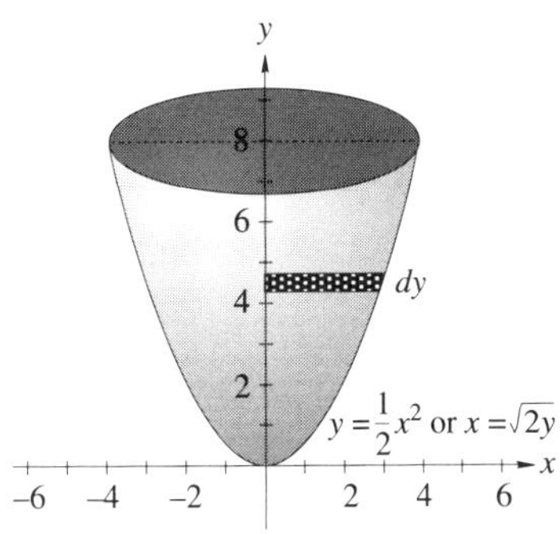

FIGURE 5.11

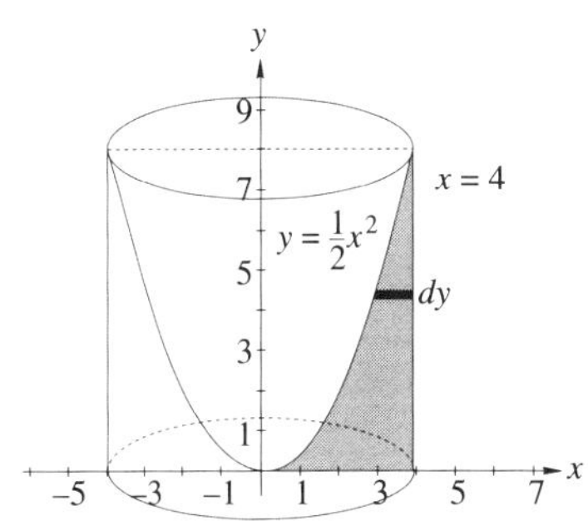

FIGURE 5.12

EXAMPLE 5.14

Find the volume of revolution generated by revolving the area enclosed by $y = \frac{1}{2}x^2$, $x = 0$, $x = 4$, and the x-axis around the y-axis.

Solution The graph of the region that is being revolved is shown in Figure 5.12. The point of intersection of $x = 4$ and $y = \frac{1}{2}x^2$ is $y = 8$. A typical element would

EXAMPLE 5.14 (Cont.)

form a washer, so its volume is $[4^2 - (\sqrt{2y})^2]dy$ and the total volume is

$$V = \pi \int_0^8 \left[(4)^2 - (\sqrt{2y})^2\right] dy = \pi \int_0^8 (16 - 2y)dy$$
$$= \pi \left(16y - y^2\right)\Big|_0^8 = 64\pi$$

The volume is 64π units^3.

Application

EXAMPLE 5.15

A hemispherical water tank, as shown in Figure 5.13a, has a radius of 4 m. The water is 1.5 m deep at the center. How much water is in the tank?

Solution We will sketch a cross-section of the tank with the center at the origin of the quarter-circle used to generate the hemisphere, as shown in Figure 5.13b. A typical element has been sketched in the figure. The length of this element is determined by solving the equation for the circle, $x^2 + y^2 = 4^2$ for x, with the result $x = \sqrt{4^2 - y^2}$. The thickness of the element is dy. Since the elements will move from the bottom of the tank, when $y = -4$, to the top of the water, when $y = -2.5$, we have the limits of integration. We see that the volume is

$$V = \pi \int_{-4}^{-2.5} \left(\sqrt{4^2 - y^2}\right)^2 dy$$
$$= \pi \int_{-4}^{-2.5} \left(16 - y^2\right) dy$$
$$= \pi \left[16y - \frac{y^3}{3}\right]_{-4}^{-2.5}$$
$$= \pi \left(-\frac{835}{24} + \frac{128}{3}\right) = \left(\frac{189}{24}\right)\pi = \frac{63}{8}\pi \approx 24.74$$

There is about 24.74 m^3 of water in the tank.

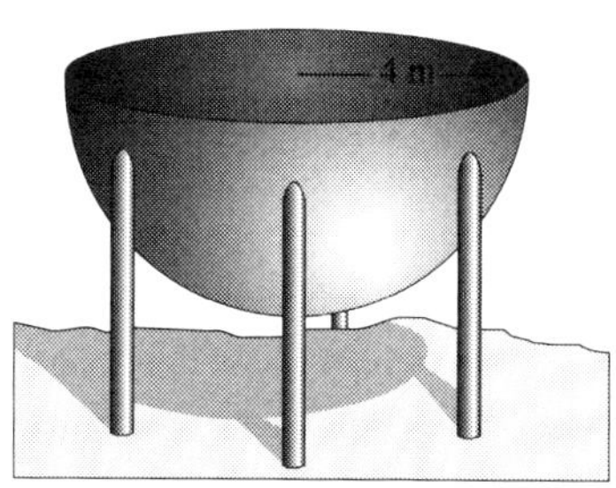

FIGURE 5.13a

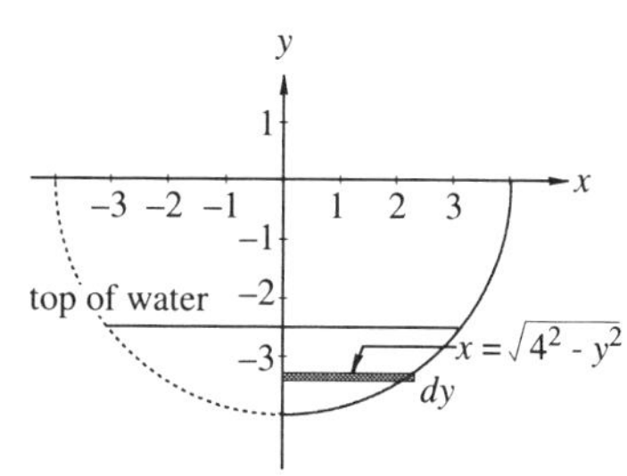

FIGURE 5.13b

Exercise Set 5.2

In Exercises 1–19, find the volume of the solid of revolution generated by revolving the enclosed region about the indicated axis.

1. $y=4x$, $x=1$, $x=5$, $y=0$, about the x-axis
2. $y=x^2$, $x=0$, $x=2$, $y=0$, about the x-axis
3. $y=x$, $x=4$, $y=0$, about the x-axis
4. $y=6-x$, $x=0$, $y=0$, about the x-axis
5. $y=\sqrt{x+1}$, $x=1$, $x=7$, $y=0$, about the x-axis
6. $y=x^3$, $x=2$, $y=1$, about the x-axis
7. $y=x^2$, $y=4$, $x=0$, about the y-axis
8. $x+y=5$, $y=0$, $x=0$, about the y-axis
9. $y=x$, $y=4-x$, $x=0$, about the x-axis
10. $y=\sqrt[3]{x}$, $x=0$, $y=8$, about the y-axis
11. $y=\sqrt[3]{x}$, $x=8$, $y=0$, about the y-axis
12. $y=\sqrt{x}$, $y=x^2$, about the y-axis
13. $y=4-x^2$, $y=8-2x^2$, about the x-axis
14. $y=x^2$, $y=x^3$, about the y-axis
15. $y=x^2$, $y=x^3$, about the x-axis
16. $y=x^2$, $y=4x$, about the x-axis
17. $y=x^2$, $y=4x$, about the y-axis
18. $y=\sqrt{8-x^3}$, $x=0$, $y=0$, about the x-axis
19. $y=\sqrt{8-x^2}$, $x=0$, $y=0$, about the y-axis

Solve Exercises 20–28.

20. Derive the formula for the volume of a cone of radius r and height h.
21. Derive the formula for the volume of a sphere of radius r.
22. The equation of an ellipse centered at the origin is
$$\frac{x^2}{a^2}+\frac{y^2}{b^2}=1$$
Calculate the volume of the solid generated by revolving this ellipse around the x-axis.
23. *Electronics* A parabolic reflector is formed by rotating a parabola around its axis. Suppose a satellite television antenna is a parabolic reflector with a diameter of 3 m and a depth of 1 m. **(a)** Find the equation of this parabola. **(b)** What is the volume contained by the reflector?
24. *Electronics* If the "skin" of the parabolic reflector in Exercise 19 has a constant 0.4 cm thickness, what is the volume of the skin? (Assume the dimensions given in Exercise 18 are the interior dimensions.)
25. *Nuclear engineering* The cooling towers at a nuclear power plant are designed in the shape of the hyperbola $g(y)=\sqrt{(147)^2+0.16y^2}$. The base of the tower is 442 ft below the vertex and the top is 123 ft above the vertex as shown in Figure 5.14. If we assume the hyperbola is rotated around the y-axis, what is the volume of the cooling tower?
26. *Nuclear engineering* If the walls of the cooling tower in Exercise 25 are a constant 6 in. thick, then the interior walls are in the shape of a hyperbola with the equation $x=\sqrt{146.5^2+0.16y^2}$ that is rotated around the y-axis. What is the volume of the walls of the cooling tower?
27. *Wastewater technology* The tank on a water tower is a sphere of radius 18 m.
    (a) Determine the volume of water when the water is 4 m deep at the center.
    (b) Determine the volume of water when the water is 24 m deep at the center.
28. *Mechanical technology* A flange is formed in the shape of a solid of revolution generated by rotating about the x-axis the region bounded by a graph of $y=x^2+5$, where x and y are in centimeters; the y-axis; and the lines $y=2$ and $x=4$. Determine the mass of this flange if it is made of a material of density $0.016\,\text{kg/cm}^3$. (Remember: mass is the product of the density and volume.)

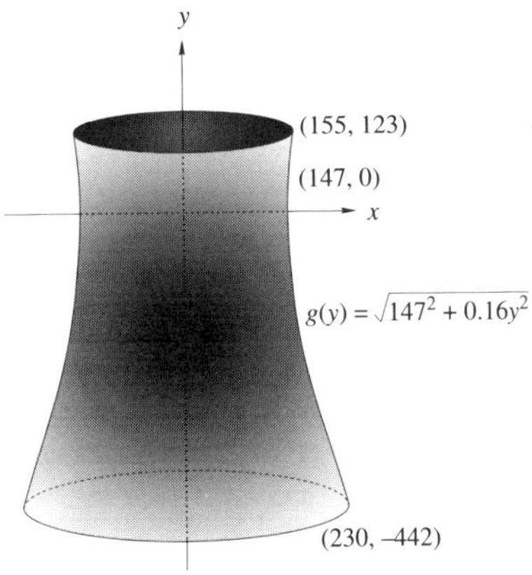

FIGURE 5.14

In Your Words

29. Describe how to find the volume of a solid of revolution by using the disk method.
30. Describe how to find the volume of a solid of revolution by using the washer method.
31. Describe the changes that have to be made when a volume of revolution is formed by revolving a region around the y-axis rather than around the x-axis.

5.3 VOLUMES OF REVOLUTION: SHELL METHOD

In Section 5.2, we found the volume of a solid of revolution by the disk and washer methods. In this section, we will learn how to find these volumes by another method—the shell method. The shell method is based on finding the volume of cylindrical shells and adding those volumes to get the total volume.

An example of a cylindrical shell is a water pipe. A cylindrical shell has some thickness to it. In order to find its volume, we need to find the inner radius r_1 of the shell and the outer radius r_2, as shown in Figure 5.15. If the shell has height h, then its volume is

$$V = \pi(r_2)^2h - \pi(r_1)^2h = \pi h({r_2}^2 - {r_1}^2)$$

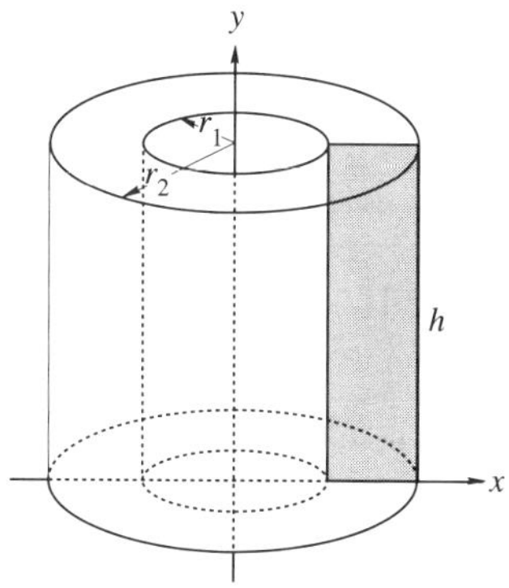

FIGURE 5.15

Some extra algebraic work will change this to a more useful formula.

$$V = \pi h(r_2{}^2 - r_1{}^2) = \pi h(r_2 + r_1)(r_2 - r_1) = 2\pi h\left(\frac{r_2 + r_1}{2}\right)(r_2 - r_1)$$

This formula is very involved, but we will show you a way to remember it after we work Example 5.16. In this last statement, $\frac{r_2 + r_1}{2}$ represents the average radius of the shell and $r_2 - r_1$ represents the thickness.

If we have two curves, $y = f(x)$ and $y = g(x)$, with $f(x) \geq g(x)$, that are continuous over the interval $[a, b]$ with $a \geq 0$, and we revolve the area between these curves around the y-axis, we get a solid of revolution similar to the one in Figure 5.16. Next we divide the interval $[a, b]$ into subintervals. Each subinterval will have the width $r_2 - r_1 = dx$, height $f(x_i) - g(x_i)$, and average radius $\frac{x_{i-1} + x_i}{2}$. Thus, when this subinterval is revolved around the y-axis, it has a volume of

$$2\pi[f(x_i) - g(x_i)]\left(\frac{x_{i-1} + x_i}{2}\right)dx$$

If dx is small enough $\left(\frac{x_{i-1} + x_i}{2} \approx x_i\right)$, and if we sum all the partitions over the interval $[a, b]$, we get the following formula.

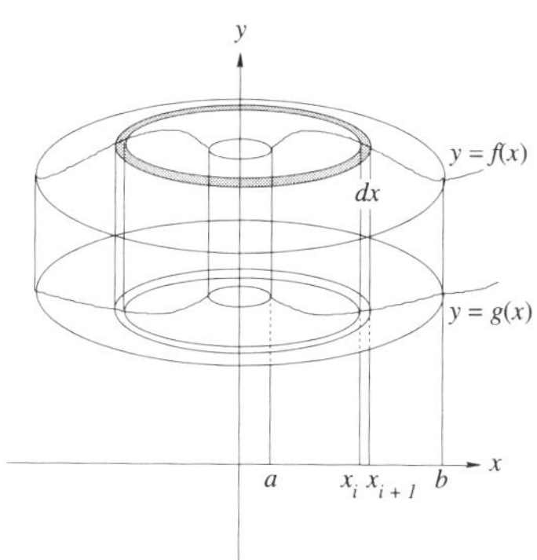

FIGURE 5.16

Shell Method for Finding a Volume of Revolution (y-Axis)

The volume V of the solid generated by revolving the area between the graph of $y = f(x)$ and $y = g(x)$ from $x = a$ to $x = b$, where a and b are both on the same side of the y-axis around the y-axis, is

$$V = 2\pi \int_a^b x[f(x) - g(x)]\,dx$$

EXAMPLE 5.16

Find the volume of the solid generated by revolving around the y-axis the region between $y = 5x^3 - x^4$ and the x-axis.

Solution The graph of this region is shown in Figure 5.17. (Notice that the scale is not the same on the x- and y-axes.) These two curves, $y = 0$ and $y = 5x^3 - x^4$,

EXAMPLE 5.16 (Cont.)

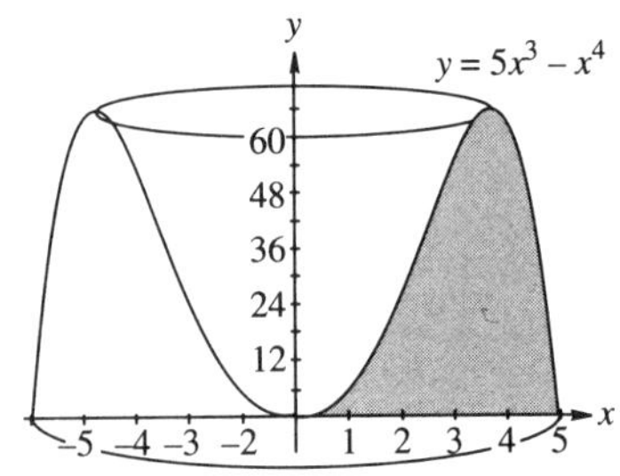

FIGURE 5.17

intersect at $x = 0$ and at $x = 5$; so, $a = 0$ and $b = 5$. Using the new formula, we see that

$$\begin{aligned} V &= 2\pi \int_0^5 x[(5x^3 - x^4) - 0]\,dx \\ &= 2\pi \int_0^5 (5x^4 - x^5)dx = 2\pi \left(x^5 - \frac{1}{6}x^6 \right)\Big|_0^5 \\ &= 2\pi \left(3{,}125 - \frac{15{,}625}{6} \right) = 1{,}041\frac{2}{3}\pi \end{aligned}$$

The volume is $1{,}041\frac{2}{3}\pi$ units^3.

It would have been possible to solve Example 5.16 by using the washer method. But, that would have meant solving the equation $y = 5x^3 - x^4$ for x in order to get a function of y. This would have been a very difficult and complicated task. The shell method provides a much easier alternative.

Hint

It is easier to remember the shell method by using the following guidelines:

1. Sketch the graph over the limits of integration.
2. Draw a typical shell parallel to the axis of revolution.
3. Determine the radius, height, and thickness of the shell.
4. The volume of a typical shell is $2\pi(\text{radius}) \cdot (\text{height}) \cdot (\text{thickness})$.
5. Use the following integration formula to determine the total volume:

$$V = 2\pi \int_a^b (\text{radius}) \cdot (\text{height}) \cdot (\text{thickness})$$

EXAMPLE 5.17

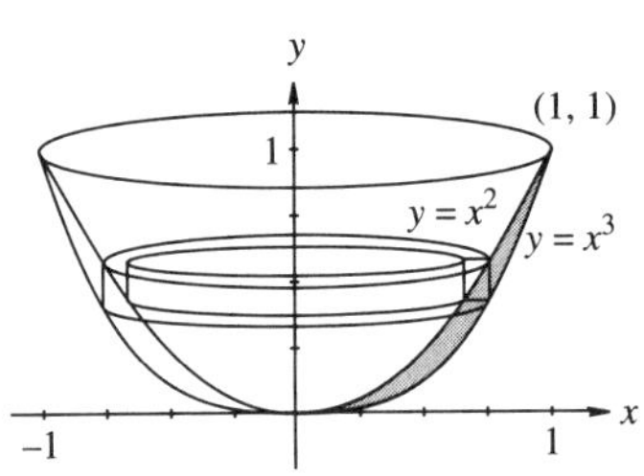

FIGURE 5.18

Find the volume of the solid generated by revolving around the y-axis the region in the first quadrant that is between $y = x^2$ and $y = x^3$.

Solution These two curves intersect at $(0, 0)$ and $(1, 1)$ as shown in Figure 5.18, so $a = 0$ and $b = 1$. On this interval, $x^2 \geq x^3$ and so, using the shell method, the volume is

$$\begin{aligned} V &= 2\pi \int_0^1 x(x^2 - x^3)dx \\ &= 2\pi \int_0^1 (x^3 - x^4)dx \\ &= 2\pi \left(\frac{1}{4}x^4 - \frac{1}{5}x^5 \right)\Big|_0^1 \\ &= 2\pi \left(\frac{1}{4} - \frac{1}{5} \right) = \frac{\pi}{10} \end{aligned}$$

The volume is $\frac{\pi}{10}$ units^3.

We can just as easily use the shell method to revolve a region around the x-axis. When this is done, the function will have to be written to ensure that it is a function of y.

EXAMPLE 5.18

Find the volume of the solid generated by revolving around the x-axis the region bounded by $y=\sqrt{x}$, $y=0$, and $x=4$.

Solution The graph of this region is shown in Figure 5.19. In this instance, we have two curves that must be written as functions of y. The first is $x=4$ and the other, $y=\sqrt{x}$, is $x=y^2$. Thus, the height of each shell is $4-y^2$ and the radius is y. Then the volume is

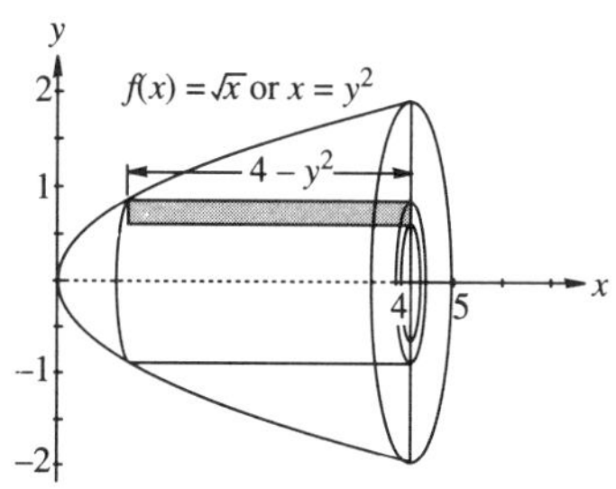

FIGURE 5.19

$$\begin{aligned} V &= 2\pi\int_0^2 y(4-y^2)dy \\ &= 2\pi\int_0^2 (4y-y^3)dy \\ &= 2\pi\left(2y^2-\frac{1}{4}y^4\right)\Big|_0^2 \\ &= 2\pi(8-4)=8\pi \end{aligned}$$

The volume is 8π units^3.

Compare the result of Example 5.18 to that of Example 5.9, where we found the volume of the same solid by using the disk method.

Rotation about a Noncoordinate Axis

As we said, it is possible to rotate a region around any line in order to get a desired volume of revolution. In the next example, we will rotate a region around a line parallel to the y-axis. The formula for the shell method will not work. Instead, we need to rely on the basic concept behind the shell method. In that concept we saw that

$$V=2\pi(\text{radius})(\text{height})(\text{thickness})$$

EXAMPLE 5.19

Find the volume of the solid formed by revolving the region bounded by $y=x^2+2$ and $y=x+8$ around the line $x=4$.

Solution The region is shown in Figure 5.20. Since this region is revolved around the line $x=4$, the radius of a typical shell is $4-x$. The two curves, $y=x+8$ and $y=x^2+2$, intersect at $(-2,6)$ and $(3,11)$. In the interval $[-2,3]$ we can see that

EXAMPLE 5.19 (Cont.)

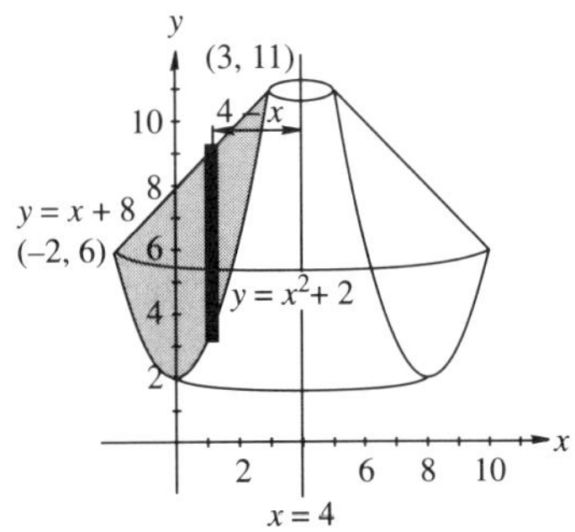

FIGURE 5.20

$x+8 \geq x^2+2$ and so the height of a typical shell is $(x+8)-(x^2+2)=6+x-x^2$. Thus, the volume of a typical shell is

$$\begin{aligned} V &= 2\pi \int_{-2}^{3} (6+x-x^2)(4-x)dx \\ &= 2\pi \int_{-2}^{3} (24-2x-5x^2+x^3)dx \\ &= 2\pi \left(24x - x^2 - \frac{5}{3}x^3 + \frac{1}{4}x^4 \right)\Bigg|_{-2}^{3} \\ &\approx 2\pi\,[38.25-(-34.67)] = 145.84\pi \end{aligned}$$

The volume is 145.84π units^3.

Summary of Disk and Shell Methods

In Sections 5.2 and 5.3, we have shown two methods for finding the volume of a solid of revolution. These two methods are summarized in Table 26.1.

TABLE 5.1 Summary of Disk and Shell Methods

	Disk or Washer Method	Shell Method
Revolution about x-axis	Partition the x-axis. Use vertical strips of width dx. Integrate with respect to x.	Partition the y-axis. Use horizontal strips of width dy. Integrate with respect to y.
Revolution about y-axis	Partition the y-axis. Use horizontal strips of width dy. Integrate with respect to y.	Partition the x-axis. Use vertical strips of width dx. Integrate with respect to x.

Exercise Set 5.3

In Exercises 1–18, use the shell method to obtain the volume of the solid of revolution generated by revolving the enclosed region about the indicated axis.

1. $y=x^2$, $y=0$, $x=2$, about the y-axis
2. $x=y^3$, $x=8$, about the y-axis
3. $y=x^2-1$, $x=1$, $y=8$, about the y-axis
4. $y=x^3+x$, $x=0$, $x=2$, about the y-axis
5. $y=25-x^2$, $y=0$, about the y-axis
6. $y=x^2$, $y=8-x^2$, about the y-axis
7. $y=x^3$, $x=0$, $y=8$, about the x-axis
8. $y=x^4$, $y=x^3$, about the x-axis
9. $y=\sqrt[3]{x}$, $x=0$, $y=2$, about the x-axis
10. $x=\sqrt[3]{y^2}$, $x=0$, $y=8$, about the x-axis
11. $y=x^{1/2}-x^{3/2}$, $y=0$, about the y-axis
12. $x=y$, $x+2y=3$, $y=0$, about the x-axis
13. $y=2x^3$, $y^2=4x$, about the x-axis
14. $x=4y-y^3$, $x=0$, about the x-axis

15. $y = x^2$, $y = 2x$, about the line $y = 6$

16. $y = 2 - x$, $y = x^2$, $y = 1$, about the x-axis (the area above $y = 1$)

17. $y = 2 - x$, $y = x^2$, about $x = 1$

18. $y = 4 - x^2$, $y = 0$, about the line $x = 4$

Solve Exercises 19–22.

19. A hole of radius 2 is drilled vertically through the center of a sphere of radius 5. Find the volume removed by the drilling.

20. *Civil engineering* A water tank is in the shape of a sphere with a radius of 10 m. **(a)** What is the volume of water in the tank if it is filled to a depth of 3 m? **(b)** If the density of water is 1 000 kg/m^3, what is the mass of the water in the tank?

21. *Sheet metal technology* The inside of a bundt pan, used for baking, is designed by revolving the region between $y = -(x - 3.25)^4 + \left(\frac{5}{4}\right)^4$ and the x-axis around the y-axis. Assume that the measurements are in inches, and find the volume of this pan.

22. *Industrial design* The glass reflector for a flashlight is made in the shape of a solid of revolution. It is obtained by revolving the region between the curves $x = 0.5y^2$, $x = 0.5y^2 + 0.1$, the x-axis, and $y = 2$ for $0 \le x \le 2$ around the x-axis. If all measurements are in centimeters, what is the volume of the reflector?

In Your Words

23. Describe how to find the volume of a solid of revolution by using the shell method.

24. Describe the changes that have to be made when a volume of revolution is formed using the shell method rather than either the disk or washer method.

25. Draw an example of a region for which the washer method is more appropriate than the shell method when the region is revolved around the y-axis. Explain why the washer method is more appropriate.

26. Draw an example of a region for which the shell method is more appropriate than the washer method when the region is revolved around the y-axis. Explain why the shell method is more appropriate.

5.4 ARC LENGTH AND SURFACE AREA

In this section, we will look at two related ideas. One is the length of a curve; the second is the area of the surface that is formed when that curve is rotated around an axis.

One useful application for the length of a curve is in determining the lengths of the cables that are used to support a suspension bridge. The cables between the towers are parabolic in shape and so this results in finding the length of a parabola between two values.

Consider a curve such as $y = f(x)$, as shown in Figure 5.21. We would like to find the length of this curve from $x = a$ to $x = b$. As before, we will partition the interval $[a, b]$ into n subintervals. With each subinterval $x_0, x_1, x_2, \ldots, x_n$ there is a corresponding point on the graph $P_0, P_1, P_2, \ldots, P_n$. Each point P_i has the coordinates $(x_i, f(x_i))$. If we connect these points in succession, we get a series of line segments that approximate the curve. The sum of the lengths of these segments

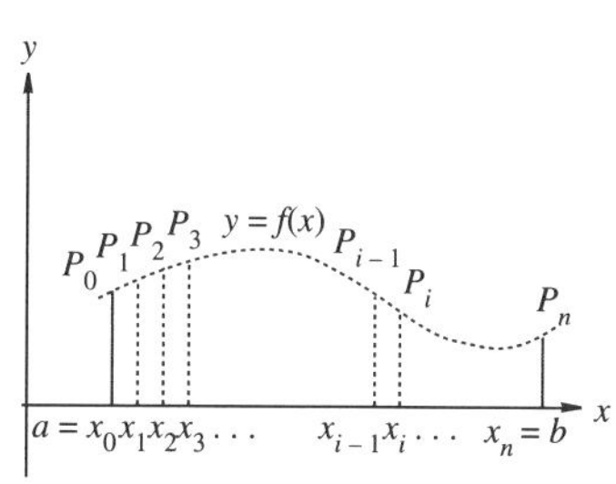

FIGURE 5.21

will approximate the length of the curve. This sum L can be written as

$$L = d(P_0, P_1) + d(P_1, P_2) + d(P_2, P_3) + \cdots + d(P_{n-1}, P_n)$$
$$= \sum_{i=1}^{n} d(P_{i-1}, P_i)$$

where $d(P_{i-1}, P_i)$ is the length of the segment from P_{i-1} to P_i.

Using the distance formula, we get

$$d(P_{i-1}, P_i) = \sqrt{(x_i - x_{i-1})^2 + [f(x_i) - f(x_{i-1})]^2}$$
$$= \sqrt{(\Delta x_i)^2 + (\Delta y_i)^2}$$

Factoring out $(\Delta x_i)^2$ produces

$$= \sqrt{1 + \left(\frac{\Delta y_i}{\Delta x_i}\right)^2}\,\Delta x_i$$

where $\Delta x_i = x_i - x_{i-1}$, and $\Delta y_i = f(x_i) - f(x_{i-1})$. Thus, we can rewrite the formula for L as

$$L = \sum_{i=1}^{n} \sqrt{1 + \left(\frac{\Delta y_i}{\Delta x_i}\right)^2}\,\Delta x_i$$

We are going to assume that f has a derivative; then $\lim\limits_{\Delta x_i \to 0} \dfrac{\Delta y_i}{\Delta x_i} = \dfrac{dy_i}{dx_i} = f'(x_i)$. With this change, the length of the curve L becomes

$$L = \sum_{i=1}^{n} \sqrt{1 + [f'(x_i)]^2}\,dx_i$$

Now, as the length of each subinterval gets smaller and smaller, we get the desired formula as a limit.

Length of a Graph

If f is a function with a continuous derivative on $[a, b]$, then the length of the graph of $y = f(x)$ on $[a, b]$ is

$$L = \int_a^b \sqrt{1 + [f'(x)]^2}\,dx$$

EXAMPLE 5.20

Find the length of $y = 3 + x^{2/3}$ from $x = 1$ to $x = 8$.

Solution If $y = f(x) = 3 + x^{2/3}$, then $f'(x) = \dfrac{2}{3x^{1/3}}$. Thus,

$$\begin{aligned} L &= \int_1^8 \sqrt{1 + \left[\frac{2}{3x^{1/3}}\right]^2}\, dx \\ &= \int_1^8 \sqrt{1 + \frac{4}{9x^{2/3}}}\, dx \\ &= \int_1^8 \sqrt{\frac{9x^{2/3} + 4}{9x^{2/3}}}\, dx \\ &= \int_1^8 \frac{\sqrt{9x^{2/3} + 4}}{3x^{1/3}}\, dx \end{aligned}$$

We will use the substitution method to solve this. If $u = 9x^{2/3} + 4$, then $du = 6x^{-1/3} dx$ and so $\frac{1}{3}x^{-1/3}\, dx = \frac{1}{18}\, du$. Thus,

$$\begin{aligned} L &= \int \frac{\sqrt{u}\, du}{18} \\ &= \frac{1}{18} \int \sqrt{u}\, du \\ &= \frac{1}{18} \cdot \frac{2}{3} u^{3/2} = \frac{1}{27} u^{3/2} \end{aligned}$$

Since $u = 9x^{2/3} + 4$, we have

$$\begin{aligned} L &= \frac{1}{27} \left(9x^{2/3} + 4\right)^{3/2} \Big|_1^8 \\ &= \frac{1}{27} \left(40^{3/2} - 13^{3/2}\right) \\ &\approx 7.63 \end{aligned}$$

We should point out that if the curve is a function of y, such as $x = g(y)$, where g' is continuous on $[c, d]$, then the length of L, of the graph of $x = g(y)$ from c to d, is given by

$$L = \int_c^d \sqrt{1 + [g'(y)]^2}\, dy$$

EXAMPLE 5.21

Find the length of $x = y^{3/2}$ from $y = 1$ to $y = 4$.

Solution Since $x' = \frac{3}{2}y^{1/2}dy$,

$$\begin{aligned} L &= \int_1^4 \sqrt{1 + \left(\frac{3}{2}y^{1/2}\right)^2}\, dy \\ &= \int_1^4 \sqrt{1 + \frac{9}{4}y}\, dy \\ &= \frac{4}{9}\int_1^4 \frac{9}{4}\sqrt{1 + \frac{9}{4}y}\, dy \\ &= \frac{2}{3}\cdot\frac{4}{9}\left(1 + \frac{9}{4}y\right)^{3/2}\Bigg|_1^4 \\ &= \frac{8}{27}\left[(10)^{3/2} - \left(\frac{13}{4}\right)^{3/2}\right] \approx 7.63 \end{aligned}$$

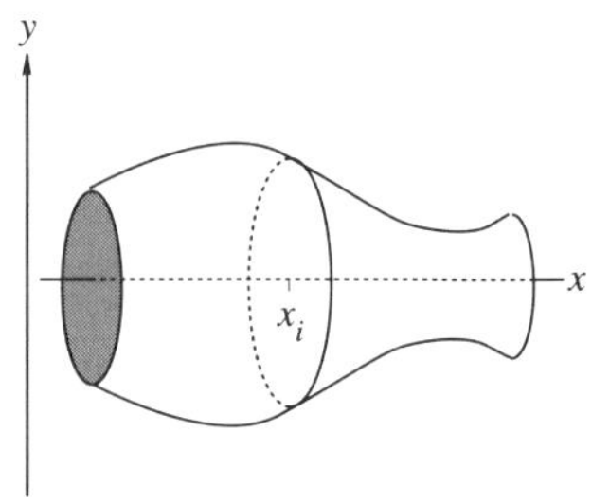

FIGURE 5.22

Surface Area

Now suppose that the curve in Figure 5.21 is rotated around the x-axis. Then a surface is formed. If we want to find the surface area of this figure, we need to think of it as a series of circles. In Figure 5.22, the circle with center at x_i has a circumference of $2\pi f(x_i)$, or $2\pi y_i$, where $y_i = f(x_i)$. If we add the circumferences of all the circles over the entire length of the curve from a to b, we get a total surface area, S.

Surface Area

If f is a function with a continuous derivative on $[a, b]$, then the area of the surface generated by revolving the graph of $y = f(x)$ about the x-axis is

$$S = \int_a^b 2\pi y\sqrt{1 + [f'(x)]^2}\, dx$$

EXAMPLE 5.22

Find the surface area of the solid generated by rotating $y = \frac{1}{3}x^3$ from $x = 0$ to $x = 2$ around the x-axis.

Solution The formula for the surface is

$$S = \int_a^b 2\pi y\sqrt{1 + [f'(x)]^2}\, dx$$

EXAMPLE 5.22 (Cont.)

Here, $y = \frac{1}{3}x^3$ and $f'(x) = x^2$, so we have

$$S = \int_0^2 \frac{2}{3}\pi x^3 \sqrt{1+(x^2)^2}\,dx = \frac{2}{3}\pi \int_0^2 x^3\sqrt{1+x^4}\,dx$$

If we let $u = 1+x^4$, then $du = 4x^3dx$, and we have

$$\begin{aligned}\frac{2}{3}\pi \cdot \frac{1}{4}\int \sqrt{u}\,du &= \frac{\pi}{6}\cdot\frac{2}{3}u^{3/2}\\ &= \frac{\pi}{9}(1+x^4)^{3/2}\Big|_0^2 = \frac{\pi}{9}(17^{3/2} - 1^{3/2})\\ &\approx 7.68\pi\end{aligned}$$

The surface area is about 7.68π units^2.

At present, our ability to solve many of these problems is limited by our lack of ability in integration techniques. We will be using arc length and surface areas as we improve our skills at integration.

Exercise Set 5.4

In Exercises 1–6, find the lengths of the curves over the indicated interval.

1. $y = \frac{1}{3}(x^2+2)^{3/2}$, over $[0, 3]$
2. $y = x^{3/2}$, over $[0, 8]$
3. $9x^2 = 4y^3$, from the point $(0, 0)$ to the point $(2\sqrt{3}, 3)$
4. $y = \dfrac{x^4}{4} + \dfrac{1}{8x^2}$, over $[1, 2]$
5. $y = \dfrac{x^3}{6} + \dfrac{1}{2x}$, over $[1, 3]$
6. $x = \dfrac{y^3}{3} + \dfrac{1}{4y}$, over $[1, 3]$

In Exercises 7–12, find the surface area of the solid generated by rotating the given curves around the indicated axis.

7. $y = 4x$ from $x = 0$ to $x = 2$ about the x-axis
8. $y = \frac{12}{5}x$ from $x = 0$ to $x = 8$ about the x-axis
9. $y = x$ from $(0, 0)$ to $(2, 2)$ about the x-axis
10. $y^2 = 4x$ from $x = 0$ to $x = 8$ about the x-axis
11. $x = 4y$ from $y = 1$ to $y = 3$ about the x-axis
12. $x = \sqrt{y}$ from $y = 0$ to $y = 6$ about the y-axis

Solve Exercises 13 and 14.

13. *Construction* A suspension bridge has its roadbed supported by vertical cables that are connected to a cable that is strung between two towers and from each tower to the ground. Each cable from the top of the tower to the ground fits the equation $y = \frac{2}{75}x^{3/2}$. Find the length of one of thc cables, if it meets the ground 225 m from the base of the tower.
14. *Interior design* A lamp shade is formed by rotating the graph of $y = (4-x)^3$ over $[2, 4]$, around the x-axis, where x and y are in inches. What is the surface area of the lamp shade?

In Your Words

15. Explain how to find the length of a graph.

16. Write an application in your technology area of interest that requires you to find either the arc length of a (nonlinear) curve or the surface area of an irregular shape. Give your problem to a classmate and see if he or she understands and can solve your problem. Rewrite the problem as necessary to remove any difficulties encountered by your classmate.

5.5 CENTROIDS

Thus far we have used integration to study moving objects as if they were particles that had mass but no size. Objects, however, are made up of many particles. If an object is moved, then its particles may have different kinds of behavior. This type of situation can be handled by studying what happens to one particular point of the object. This point is known as the center of mass or the centroid. In this section, we will study how to find the centroid for areas and for solids of revolution.

Suppose we had a metal bar 4 m long that we place along the x-axis with the center of the bar at the origin, one end at 2, and the other end at -2. Next, suppose we put a 10-kg mass at $x = 2$ and a 4-kg mass at $x = -2$. We would like to find the center of mass for this system. The **center of mass** of a system is the point where all the mass seems to be concentrated. If the system has a constant density, then the center of mass is called the **centroid**.

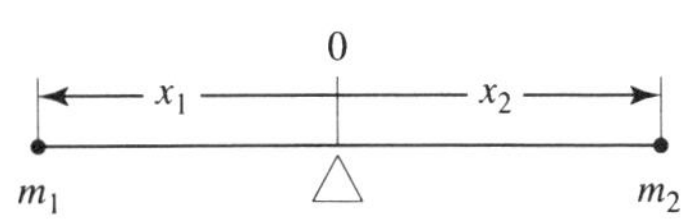

FIGURE 5.23

Before we solve this example, we will look at the idea behind the centroid. Suppose we had a system much like the one in Figure 5.23. The point 0 indicates the origin of the system. A mass m_1 is located x_1 units to the left of the origin. A second mass m_2 is located x_2 units to the right of the origin.

Associated with each mass is its moment with respect to a certain axis. In Figure 5.23, the axis is at the origin. Each **moment** is the product of the mass and its distance from the axis. For the system in Figure 5.23, the moment for mass 1 is m_1x_1 and for mass 2 the moment is m_2x_2. The **first moment** of this system is the sum of the individual moments. In this case, the first moment for the system is $m_1x_1 + m_2x_2$. We now select a new point $\bar{x}$ with the property that $(m_1+m_2)\bar{x}$ is equal to the first moment of the system. The location of $\bar{x}$ is the center of mass. Since $(m_1+m_2)\bar{x} = m_1x_1 + m_2x_2$, we have

$$\bar{x} = \frac{m_1x_1 + m_2x_2}{m_1 + m_2}$$

Continuing this process for more than two points gives the general formula for n points:

$$\bar{x} = \frac{\sum_{i=1}^{n} m_i x_i}{\sum_{i=1}^{n} m_i}$$

In this equation, the numerator is the first moment of the system and the denominator is the total mass of the system.

EXAMPLE 5.23

Four particles of masses 2, 4, 3, and 5 kg, respectively, are located on the x-axis at coordinates 3, −2, −3, and 6 m. Where is the center of mass of this system?

Solution Figure 5.24 indicates this by using stacks of 1-kg squares to represent the particles.

$$\bar{x} = \frac{m_1x_1 + m_2x_2 + m_3x_3 + m_4x_4}{m_1 + m_2 + m_3 + m_4}$$
$$= \frac{2(3) + 4(-2) + 3(-3) + 5(6)}{2 + 4 + 3 + 5} = \frac{19}{14}\text{ m}$$

The center of mass is at the coordinate $\frac{19}{14}$ m on the x-axis.

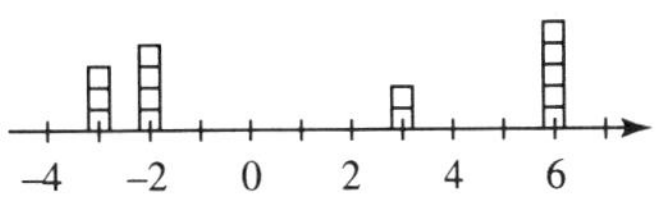

FIGURE 5.24

If we have several point masses located in a plane, then we locate the center of mass by separately finding the center of mass for the x-coordinates and for the y-coordinates. Thus, for the system in Figure 5.25, the center of mass would have coordinates

$$\bar{x} = \frac{m_1x_1 + m_2x_2 + m_3x_3}{m_1 + m_2 + m_3} \text{ and } \bar{y} = \frac{m_1y_1 + m_2y_2 + m_3y_3}{m_1 + m_2 + m_3}$$

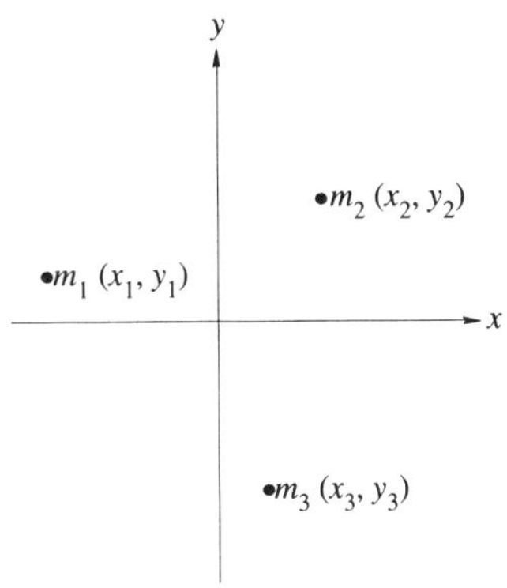

FIGURE 5.25

In general, we have the formulas in the box for the center of mass for n points.

Center of Mass for n Points

If the point masses $m_1, m_2, \ldots, m_n$ are located at $(x_1, y_1), (x_2, y_2), \ldots, (x_n, y_n)$, respectively, then the center of mass has the coordinates

$$\bar{x} = \frac{\sum_{i=1}^{n} m_i x_i}{m} \text{ and } \bar{y} = \frac{\sum_{i=1}^{n} m_i y_i}{m}$$

or

$$\bar{x} = \frac{M_y}{m} \text{ and } \bar{y} = \frac{M_x}{m}$$

where

$$M_y = \sum_{i=1}^{n} m_i x_i \text{ is the first moment around the } y\text{-axis,}$$
$$M_x = \sum_{i=1}^{n} m_i y_i \text{ is the first moment around the } x\text{-axis, and}$$
$$m = \sum_{i=1}^{n} m_i \text{ is the total mass of the system.}$$

EXAMPLE 5.24

Calculate the center of mass of the system with four particles of masses 10 g, 5 g, 3 g, and 8 g located respectively at the points $(2,-1)$, $(3,2)$, $(-4,1)$, and $(-3,4)$ (measured in cm).

Solution

$$\begin{aligned}\bar{x} &= \frac{10(2)+5(3)+3(-4)+8(-3)}{10+5+3+8}\\ &= \frac{-1}{26}\text{ cm}\\ \bar{y} &= \frac{10(-1)+5(2)+3(1)+8(4)}{10+5+3+8}\\ &= \frac{35}{26}\text{ cm}\end{aligned}$$

The center of mass is at $\left(-\frac{1}{26}, \frac{35}{26}\right)$.

EXAMPLE 5.25

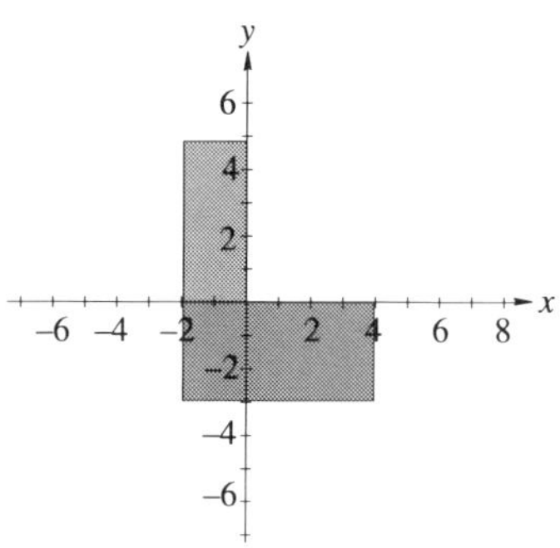

FIGURE 5.26

Find the center of mass of the area in Figure 5.26 if it has uniform density ρ.

Solution We will first divide the area into two rectangles. This division will be along the x-axis with region I above the x-axis and region II below. The geometric center of region I is $\left(-1, \frac{5}{2}\right)$ and of region II, $\left(1, -\frac{3}{2}\right)$. The mass of each region is the product of the density ρ (mass/unit area) and the area. The area of region I is 10 units and so its mass is 10ρ. The area of region II is 18 units and so its mass is 18ρ. The total mass of this figure is

$$m = 10\rho + 18\rho = 28\rho$$

The coordinates of the center of mass will be determined by adding the product of the coordinate of the centroid for each region by its mass. Thus, the x-coordinate is

$$\bar{x} = \frac{(-1)(10\rho)+(1)(18\rho)}{28\rho} = \frac{8\rho}{28\rho} = \frac{2}{7}$$

and the y-coordinate is

$$\bar{y} = \frac{\left(\frac{5}{2}\right)(10\rho)+\left(-\frac{3}{2}\right)(18\rho)}{28\rho} = \frac{-2\rho}{28\rho} = -\frac{1}{14}$$

The center of mass is at $\left(\frac{2}{7}, -\frac{1}{14}\right)$.

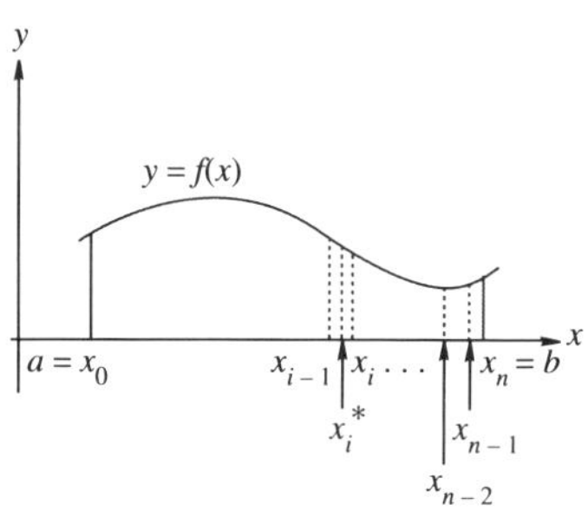

FIGURE 5.27

It is possible for the center of mass to be outside the system. You may have noticed that the density ρ was not used to locate the center of mass. This will happen as long as the object has uniform density.

Now let's apply this above technique to an arbitrary region. To begin, we will let the region be found by $y = f(x)$, $x = a$, $x = b$, and the x-axis, as shown in Figure 5.27. As we have done before, we will partition the interval $[a, b]$ into n subintervals as indicated in Figure 5.27.

Look at the shaded region in Figure 5.27. It is almost rectangular in shape and so its center is the midpoint of the interval $[x_{i-1}, x_i]$, which we will call $x_i{}^*$, and half the height of the rectangle, or $\frac{1}{2}f(x_i{}^*)$. The area of this rectangle is $f(x_i{}^*)(x_i - x_{i-1}) = f(x_i{}^*)\Delta x$, and so the mass of this rectangle is $\rho f(x_i{}^*)\Delta x$, where ρ is the density. The first moment of this shaded region is given by $\rho x_i{}^* f(x_i{}^*)\Delta x$. The first moment for all of these rectangular strips with respect to the y-axis is given by

$$\sum_{i=1}^{n} \rho x_i{}^* f(x_i{}^*)\Delta x$$

So, if the number of intervals increases, we get a first moment of

$$M_y = \lim_{\Delta x \to 0} \sum_{i=1}^{n} \rho x_i{}^* f(x_i{}^*)\Delta x$$

or $$M_y = \int_a^b \rho x f(x)dx$$

In a similar way, we can show that the first moment with respect to the x-axis is given by

$$M_x = \frac{1}{2}\int_a^b \rho[f(x)]^2 dx$$

The mass m of the region is

$$m = \int_a^b \rho f(x)dx$$

Combining these, we get the coordinate of the centroid as $(\bar{x}, \bar{y})$ where

$$\bar{x} = \frac{M_y}{m} \text{ and } \bar{y} = \frac{M_x}{m}$$

If $\rho = 1$, then the mass will be the same as the area.

EXAMPLE 5.26

Find the centroid of the plane region bounded by $y = 4 - x^2$ and $y = 0$.

Solution These two curves intersect at $x = -2$ and $x = 2$. Thus, we have

$$\begin{aligned} M_y &= \int_{-2}^{2} \rho x\left(4 - x^2\right)dx \\ &= \int_{-2}^{2} \rho\left(4x - x^3\right)dx \\ &= \rho\left(2x^2 - \frac{1}{4}x^4\right)\Bigg|_{-2}^{2} \\ &= \rho[(8-4) - (8-4)] = 0 \end{aligned}$$

$$M_x = \frac{1}{2}\int_{-2}^{2} \rho\left(4 - x^2\right)^2 dx$$

EXAMPLE 5.26 (Cont.)

$$\begin{aligned}
&= \frac{1}{2}\int_{-2}^{2} \rho\left(16-8x^2+x^4\right)dx \\
&= \frac{\rho}{2}\left(16x-\frac{8}{3}x^3+\frac{x^5}{5}\right)\Big|_{-2}^{2} \\
&= \frac{\rho}{2}\left(\frac{256}{15}+\frac{256}{15}\right) \\
&= \frac{256}{15}\rho
\end{aligned}$$

$$\begin{aligned}
m &= \int_{-2}^{2} \rho(4-x^2)dx \\
&= \rho\left(4x-\frac{x^3}{3}\right)\Big|_{-2}^{2} \\
&= \rho\left(\frac{16}{3}+\frac{16}{3}\right) = \frac{32}{3}\rho
\end{aligned}$$

Thus we have the following coordinates:

$$\begin{aligned}
\bar{x} &= \frac{M_y}{m} = \frac{0}{32\rho/3} = 0 \\
\bar{y} &= \frac{M_x}{m} = \frac{256\rho/15}{32\rho/3} = \frac{8}{5}
\end{aligned}$$

The centroid of this region is $\left(0, \frac{8}{5}\right)$.

If the region is between two curves, $f(x)$ and $g(x)$, with $f(x) \geq g(x)$, we have the following, more general, formulas. Notice that if $g(x) = 0$, we have the previous equations.

Moments, Mass, and Centroid of a Plane Region

If f and g are continuous functions with $f(x) \geq g(x)$ on the interval $[a, b]$, then the moments, mass, and centroid of the region of uniform density ρ bounded by the graphs of $y = f(x)$ and $y = g(x)$, on $[a, b]$, are

$$\begin{aligned}
M_x &= \frac{1}{2}\int_a^b \rho([f(x)]^2-[g(x)]^2)dx \\
M_y &= \int_a^b \rho x[f(x)-g(x)]dx \\
m &= \int_a^b \rho[f(x)-g(x)]dx \\
\bar{x} &= \frac{M_y}{m} \text{ and } \bar{y} = \frac{M_x}{m}
\end{aligned}$$

EXAMPLE 5.27

Find the centroid of the region bounded by $y = x^3$ and $y = \sqrt{x}$.

Solution These two curves intersect at $x = 0$ and $x = 1$. As we can see in Figure 5.28, $\sqrt{x} \geq x^3$ in this interval. Using the new formulas, we get

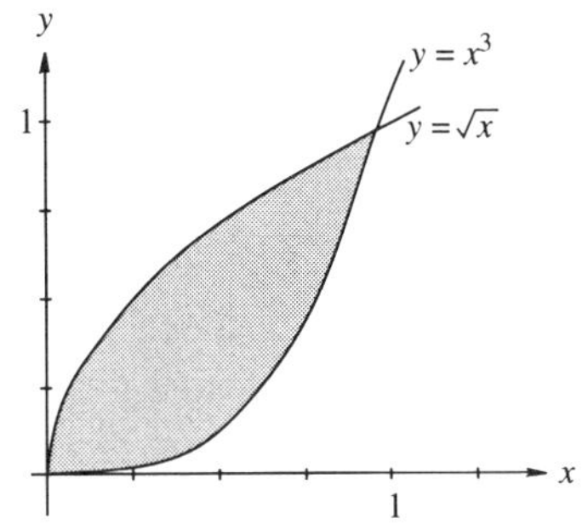

FIGURE 5.28

$$\begin{aligned}
M_y &= \int_0^1 \rho x \left(\sqrt{x} - x^3\right) dx \\
&= \int_0^1 \rho \left(x^{3/2} - x^4\right) dx \\
&= \rho \left[\frac{2}{5} x^{5/2} - \frac{1}{5} x^5\right]_0^1 = \frac{1}{5}\rho \\
M_x &= \frac{1}{2} \int_0^1 \rho \left[\left(\sqrt{x}\right)^2 - \left(x^3\right)^2\right] dx \\
&= \frac{1}{2} \int_0^1 \rho \left(x - x^6\right) dx \\
&= \rho \frac{1}{2} \left(\frac{1}{2} x^2 - \frac{1}{7} x^7\right)\Bigg|_0^1 = \frac{5}{28}\rho \\
m &= \int_0^1 \rho \left(\sqrt{x} - x^3\right) dx \\
&= \rho \left(\frac{2}{3} x^{3/2} - \frac{1}{4} x^4\right)\Bigg|_0^1 = \frac{5}{12}\rho
\end{aligned}$$

Thus, we have

$$\begin{aligned}
\bar{x} &= \frac{M_y}{m} = \frac{1/5\rho}{5/12\rho} = \frac{12}{25} \\
\bar{y} &= \frac{M_x}{m} = \frac{5/28\rho}{5/12\rho} = \frac{12}{28} = \frac{3}{7}
\end{aligned}$$

The centroid is at $\left(\frac{12}{25}, \frac{3}{7}\right)$. ▪

EXAMPLE 5.28

A thin plate covers a region bounded by the x-axis, $y = x^2 + 1$, and $x = 2$. The plate's density at the point (x, y) is $\rho = 5x$. Find the total mass and the center of mass of the plate.

Solution

$$\begin{aligned}
M_y &= \int_0^2 \rho x \left(x^2 + 1\right) dx \\
&= \int_0^2 (5x) x \left(x^2 + 1\right) dx \\
&= \int_0^2 \left(5x^4 + 5x^2\right) dx = x^5 + \frac{5}{3} x^3\Bigg]_0^2 = \frac{136}{3}
\end{aligned}$$

EXAMPLE 5.28 (Cont.)

$$\begin{aligned}
M_x &= \frac{1}{2}\int_0^2 \rho\left(x^2+1\right)^2 dx \\
&= \frac{1}{2}\int_0^2 5x\left(x^4+2x^2+1\right) dx \\
&= \frac{1}{2}\int_0^2 \left(5x^5+10x^3+5x\right) dx \\
&= \frac{1}{2}\left(\frac{5}{6}x^6+\frac{5}{2}x^4+\frac{5}{2}x^2\right)\Bigg]_0^2 = \frac{153}{3} \\
m &= \int_0^2 \rho\left(x^2+1\right) dx \\
&= \int_0^2 \left(5x^3+5x\right) dx = \frac{5}{4}x^4+\frac{5}{2}x^2\Bigg]_0^2 = 30 \\
\overline{x} &= \frac{M_y}{m} = \frac{136/3}{30} = \frac{136}{90} = \frac{68}{45} \\
\overline{y} &= \frac{M_x}{m} = \frac{155/3}{30} = \frac{31}{18}
\end{aligned}$$

Thus, the total mass is 30 units and the center of mass is at $\left(\frac{68}{45}, \frac{31}{18}\right)$.

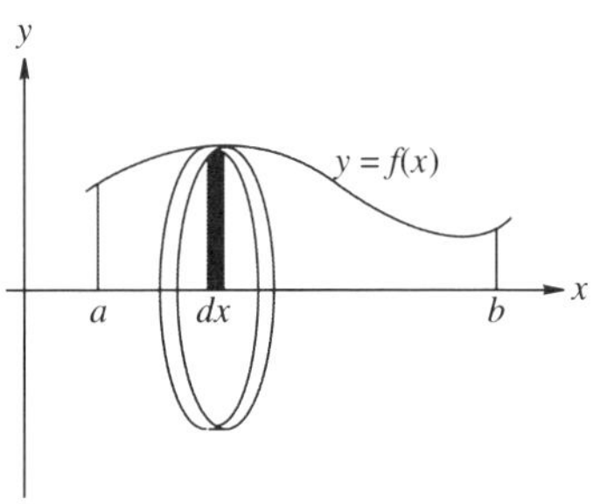

FIGURE 5.29

Centroid of a Solid of Revolution

This method can be extended to solids of revolution. As in the past, we will revolve the region bounded by the curve $y = f(x)$, $x = a$, $x = b$, and the x-axis around the x-axis, as shown in Figure 5.29. Because of the symmetry of the rotation, we know that the centroid is on the axis of rotation. In this case $\bar{y} = 0$. We need only locate $\bar{x}$.

The value of $\bar{x}$ will be $\dfrac{M_y}{m}$, where M_y is the first moment of the solid with respect to the y-axis and m is the mass. We already know that

$$m = \rho V = \pi \int_a^b \rho[f(x)]^2\, dx$$

where V is the volume.

The moment of a typical disk, as shown in Figure 5.29, is the product of x, ρ, and the volume of the disk, $\pi\,[f(x)]^2\, dx$. If we sum these for all the disks, we obtain

$$M_y = \pi \int_a^b \rho x[f(x)]^2 dx$$

Putting these together produces the following result.

Centroid of a Solid of Revolution

If f is a continuous function on the interval $[a,b]$, then the centroid of the solid generated by revolving the region bounded by $y=f(x)$, $x=a$, $x=b$, and the x-axis about the x-axis is at $(\bar{x},0)$, where

$$\bar{x}=\frac{M_y}{m}$$

when $M_y=\pi\int_a^b \rho x\,[f(x)]^2\,dx$ and $m=\rho V=\pi\int_a^b \rho[f(x)]^2\,dx$.

EXAMPLE 5.29

Find the centroid of the solid formed by revolving the region bounded by $y=x^2$, $y=0$, and $x=2$ about the x-axis, as shown in Figure 5.30.

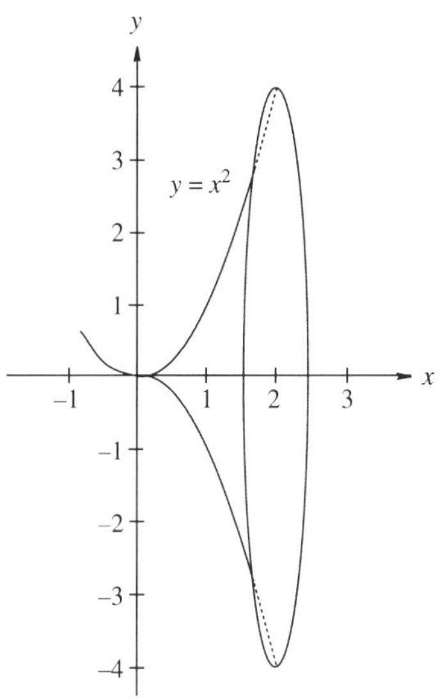

FIGURE 5.30

Solution Because this is revolved around the x-axis, $\bar{y}=0$. To determine $\bar{x}$, we need m and M_y. We get m from V.

$$\begin{aligned}
V &= \pi\int_0^2 [x^2]^2\,dx \\
&= \pi\int_0^2 x^4\,dx = \pi\frac{x^5}{5}\bigg|_0^2 = \frac{32}{5}\pi \\
m &= \rho V = \frac{32}{5}\rho\pi \\
M_y &= \pi\int_0^2 \rho x\,\left[x^2\right]^2 dx \\
&= \pi\int_0^2 \rho x^5\,dx = \rho\pi\frac{x^6}{6}\bigg|_0^2 = \frac{32}{3}\rho\pi \\
\bar{x} &= \frac{M_y}{m} = \frac{32/3\rho\pi}{32/5\rho\pi} = \frac{5}{3}
\end{aligned}$$

The centroid is at $\left(\frac{5}{3},0\right)$. ▪

Note If a solid is revolved around the y-axis, we would follow very similar procedures, only the centroid would be at $(0,\bar{y})$.

EXAMPLE 5.30

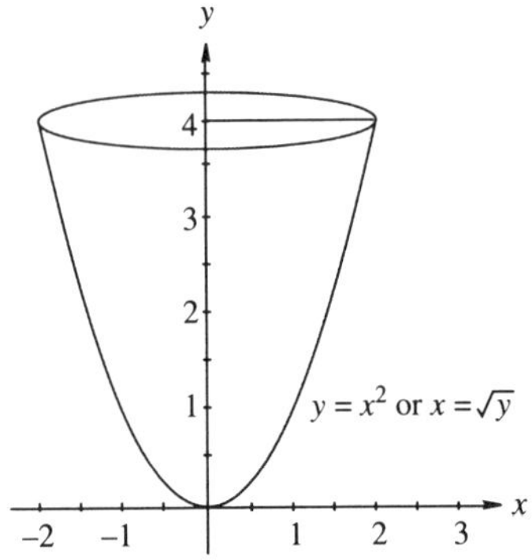

FIGURE 5.31

Find the centroid of the solid formed by revolving the region bounded by $y = x^2$, $y = 4$, and the y-axis around the y-axis as in Figure 5.31.

Solution Because it is revolved around the y-axis, $\bar{x} = 0$. To find $\bar{y}$, we need M_x and m. Again, we get m from V.

$$V = \pi \int_0^4 [\sqrt{y}]^2\, dy$$
$$= \pi \int_0^4 y\, dy = \pi \frac{y^2}{2}\bigg|_0^4 = 8\pi$$
$$m = \rho V = 8\rho\pi$$
$$M_x = \pi \int_0^4 \rho y\, [\sqrt{y}]^2\, dy$$
$$= \pi \int_0^4 \rho y^2\, dy = \pi\rho \frac{y^3}{3}\bigg]_0^4 = \frac{64}{3}\rho\pi$$
$$\bar{y} = \frac{M_x}{m} = \frac{64/3\rho\pi}{8\rho\pi} = \frac{8}{3}$$

The centroid is at $\left(0, \frac{8}{3}\right)$.

Note If a region is bounded by two functions $f(x)$ and $g(x)$ or $h(y)$ and $k(y)$, then the washer method should be used.

EXAMPLE 5.31

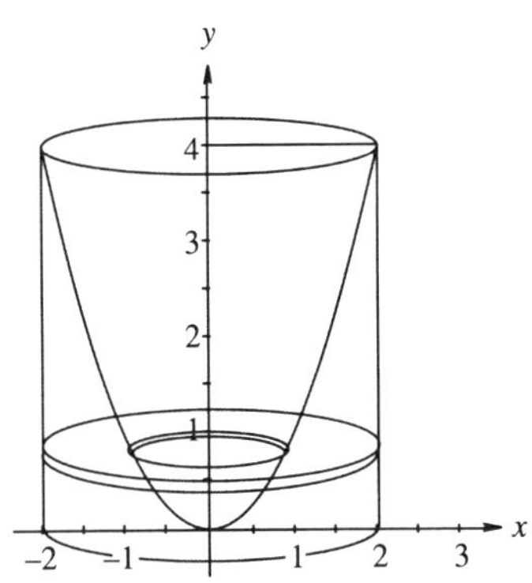

FIGURE 5.32

Find the centroid of the solid formed by revolving the region in Example 5.29 around the y-axis.

Solution Because it is revolved around the y-axis, we know that $\bar{x} = 0$.

We will use the washer method, as shown in Figure 5.32. To determine $\bar{y}$, we need M_x and m.

$$V = \pi \int_0^4 \left[2^2 - (\sqrt{y})^2\right] dy = \pi \int_0^4 (4 - y) dy$$
$$= \pi \left(4y - \frac{y^2}{2}\right)\bigg|_0^4 = 8\pi$$
$$m = \rho V = 8\pi\rho$$
$$M_x = \pi \int_0^4 \rho y \left[2^2 - (\sqrt{y})^2\right] dy = \pi \int_0^4 \rho(4y - y^2) dy$$
$$= \rho\pi \left(2y^2 - \frac{y^3}{3}\right)\bigg|_0^4 = \frac{32}{3}\pi\rho$$
$$\bar{y} = \frac{M_x}{m} = \frac{32/3\rho\pi}{8\pi\rho} = \frac{4}{3}$$

The centroid is at $\left(0, \frac{4}{3}\right)$.

Exercise Set 5.5

In Exercises 1–4, find the centroid of the system with masses located at the indicated points. Each mass is in grams and the units are in centimeters.

1. $m_1 = 4$ at $(5, 2)$ and $m_2 = 6$ at $(-3, 7)$
2. $m_1 = 3$ at $(-1, 4)$, $m_2 = 4$ at $(-2, -5)$, and $m_3 = 13$ at $(1, 2)$
3. $m_1 = 2$ at $(1, 5)$, $m_2 = 3$ at $(-1, 4)$, and $m_3 = 5$ at $(6, -4)$
4. $m_1 = 1$ at $(1, 1)$, $m_2 = 2$ at $(2, -2)$, $m_3 = 3$ at $(-3, 3)$, and $m_4 = 4$ at $(-4, -4)$

In Exercises 5–8, find the coordinates of the centroid for the indicated figures.

5.

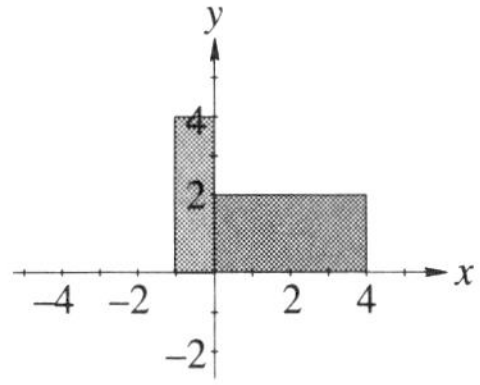

6.

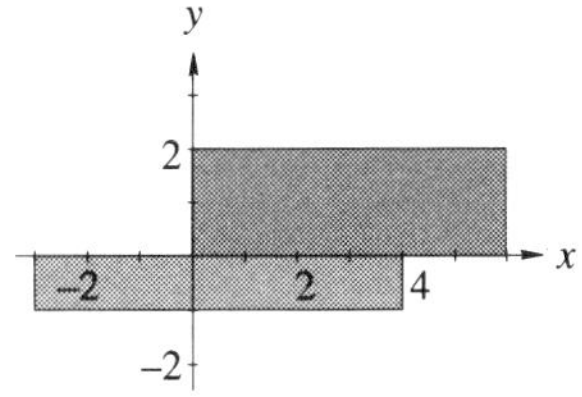

7.

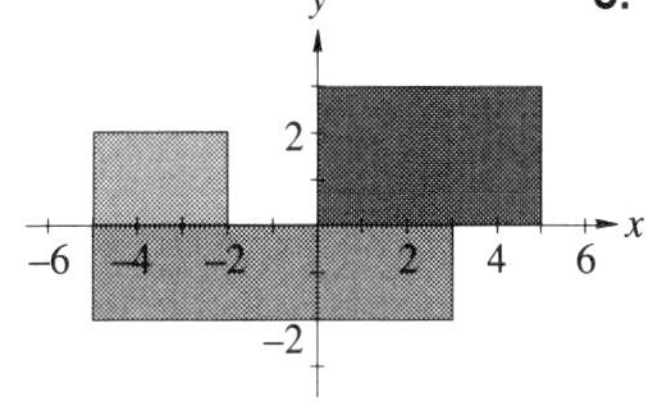

8.

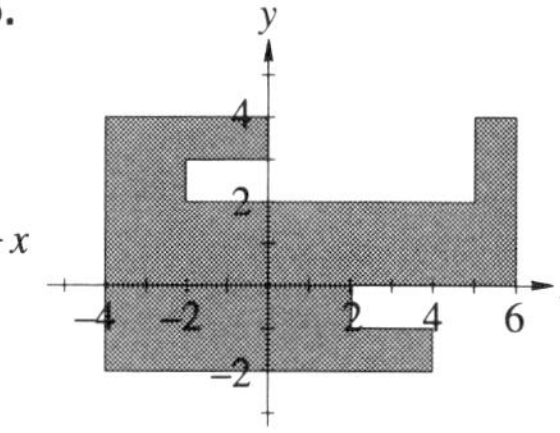

In Exercises 9–14, calculate the centroid of the plane region bounded by the given curve and the *x*-axis over the indicated interval.

9. $y = 2x + 3$, $[0, 3]$
10. $y = x^3$, $[0, 2]$
11. $y = x^{1/3}$, $[0, 8]$
12. $y = x^4$, $[-1, 2]$
13. $y = \sqrt{x + 4}$, $[0, 5]$
14. $y = x^2 + 16$, $[0, 4]$

In Exercises 15–22, find the centroids of the plane regions bounded by the given curves and lines.

15. $y = x^2$, $y = 4x$, $x = 0$, $x = 2$
16. $y = x^3$, $y = 2x$, $x = 0$, $x = 1$
17. $y = 2 - x$, $y = x^2$
18. $y = x^{3/2}$, $y = x$
19. $y = x^2$, $y = 18 - x^2$
20. $y = \sqrt{x}$, $y = x$
21. $y = x$, $y = 12 - x^2$
22. $y = x^2$, $y = x^3$

In Exercises 23–34, find the centroids of the solid of revolution formed by rotating the enclosed region around the indicated axis.

23. $y = x^3$, $y = 0$, $x = 2$, about the x-axis
24. $y = x^3$, $x = 0$, $y = 8$, about the y-axis
25. $y = x^4$, $y = 1$, about the y-axis
26. $y = (x - 1)^2$, $y = 1$, about the y-axis
27. $x + y = 6$, $x = 3$, $y = 0$, $x = 0$, about the x-axis
28. $x + y = 8$, $x = 4$, $y = 0$, $x = 8$, about the x-axis
29. $x + y = 8$, $x = 4$, $y = 0$, about the y-axis
30. $x^2 + y^2 = 16$, $x = 0$, $y = 0$, about the x-axis
31. $x^2 - y^2 = 1$, $x = 3$, about the x-axis
32. $x^2 - y^2 = 1$, $x = 3$, $y = 0$, about the y-axis
33. $y = x^2$, $y = \sqrt{x}$, about the x-axis
34. $y = x^4$, $y = x^2$, about the y-axis

Solve Exercises 35–38.

35. *Construction* The ends of two thin steel rods of equal lengths are welded together to make a right-angle frame, as shown in Figure 5.33.

FIGURE 5.33

(a) Locate the frame's centroid, if the rods are each 1 m long and 1 cm wide.
(b) Locate the frame's centroid, if the rods are each h units long and w units wide.

36. *Product design* A plastic pylon used at work sites is in the shape of a frustum of a cone on a square base, as shown in Figure 5.34. The radii of the bases of the cone are $2\frac{1}{2}$ in. and 7 in. and the height between the bases is 28 in.

(a) Determine the centroid of the portion of the pylon that is in the shape of a frustum of a cone.
(b) If the square base is 1 in. thick and 18 in. on each side, determine the centroid of the entire pylon.

37. Determine the mass and the center of mass of a rod of length 40 cm, where the density ρ of the rod at a point x cm from one end is $5x + 1$ g/cm.

38. Determine the mass and the center of mass of a rod of length 8 m, where the density ρ of the rod at a point x m from one end is $x + 5$ kg/m.

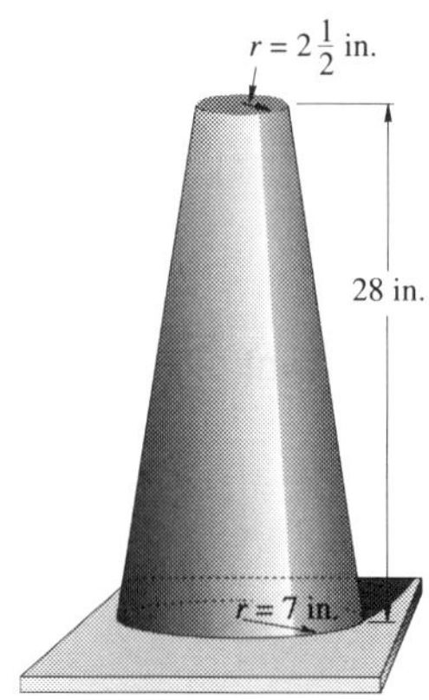

FIGURE 5.34

In Your Words

39. Explain the meaning of the moments, mass, and centroid of a plane region.

40. Describe how to determine the centroid of an irregular-shaped plane region.

≡ 5.6 MOMENTS OF INERTIA

In Section 5.5, the first moment around the y-axis (the line $x = 0$) of n masses was defined as

$$\begin{aligned} M_y &= \sum_{i=1}^{n} m_i x_i \\ &= m_1x_1 + m_2x_2 + m_3x_3 + \cdots + m_nx_n \end{aligned}$$

For other problems, other kinds of moments are more useful. In this section, we will study the moments of inertia. The **moment of inertia** is the measure of the tendency of an object to resist a change in motion. The moment of inertia is also called the **second moment** and for a system of n masses around the y-axis it is defined as

$$\begin{aligned} I_y &= \sum_{i=1}^{n} m_i {x_i}^2 \\ &= m_1 {x_1}^2 + m_2 {x_2}^2 + m_3 {x_3}^2 + \cdots + m_n {x_n}^2 \end{aligned}$$

If all of these masses were at the same distance r from the axis of rotation, we would have

$$\begin{aligned} I_y &= m_1 r^2 + m_2 r^2 + m_3 r^2 + \cdots + m_n r^2 \\ &= (m_1 + m_2 + m_3 + \cdots + m_n) r^2 \\ &= m r^2 \end{aligned}$$

where m is the total mass of the system. The number r is called the **radius of gyration**.

EXAMPLE 5.32

Find the moment of inertia and the radius of gyration about the x- and y-axes for three masses of 5 kg, 2 kg, and 4 kg located, respectively, at $(2,1)$, $(4,0)$, and $(-3,-5)$, where the units are in meters.

Solution

$$I_y = \sum_{i=1}^{3} m_i x_i^2 = 5(2)^2 + 2(4)^2 + 4(-3)^2 = 88\,\text{kg}\cdot\text{m}^2$$

Since $I_y = m{r_y}^2$, we know that ${r_y}^2 = \frac{I_y}{m}$ or $r_y = \sqrt{\frac{I_y}{m}}$. The total mass of the system is $m = 5 + 2 + 4 = 11$ kg. Thus, $r_y = \sqrt{\frac{88}{11}} = \sqrt{8} = 2\sqrt{2} \approx 2.83$ m.

With respect to the x-axis, we get the results

$$I_x = 5(1)^2 + 2(0)^2 + 4(-5)^2 = 105\,\text{kg}\cdot\text{m}^2$$

and $$r_x = \sqrt{\frac{105}{11}} \approx 3.09\,\text{m}$$

These answers mean that a mass of 11 kg placed on the line $x = 2.83$ (or $x = -2.83$) will have the same rotational inertia about the y-axis as the three objects. Similarly, placing an 11-kg mass on the line $y = 3.09$ (or $y = -3.09$) will produce the same rotational inertia about the x-axis as the three objects. ▪

It is often of interest to determine the radius of gyration for a system that is rotated about the origin, r_0. In this case, $r_0 = \sqrt{{r_x}^2 + {r_y}^2}$. In Example 5.32, the radius of gyration with respect to the origin is $r_0 = \sqrt{8 + \frac{105}{11}} \approx 4.19$. Thus, placing an 11-kg mass at 4.19 m from the origin will produce the same rotational inertia about the origin as the three objects in Example 5.32.

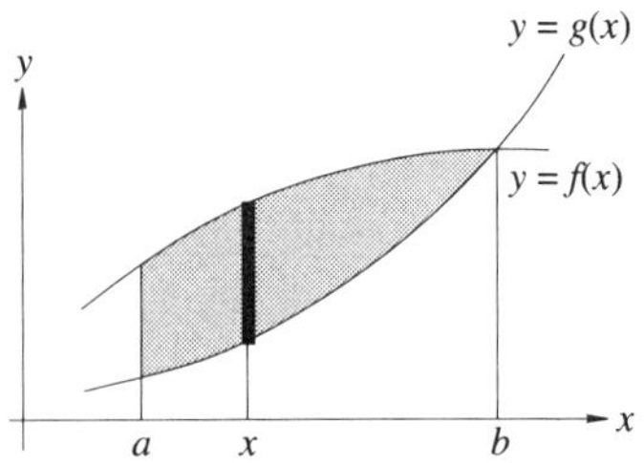

FIGURE 5.35

Moments of Inertia for a Region

Suppose an area is bounded by the curves of two functions such as $y = f(x)$ and $y = g(x)$ and the lines $x = a$ and $x = b$ with $f(x) \geq g(x)$, as shown in Figure 5.35. The moment of inertia for this region with respect to the y-axis is the sum of the moments of inertia for the individual elements of the region. The mass of each element is $\rho[f(x) - g(x)]\,dx$, where ρ is the density (mass per unit area) and $[f(x) - g(x)]\,dx$ is the area of the element. If this element is x-units from the y-axis, then the moment of inertia for the element is $\rho x^2[f(x) - g(x)]\,dx$. If we did this for all the elements, then their sum would give us the moment of inertia.

Moment of Inertia

If a region is bounded by the curves of two functions such as $y = f(x)$ and $y = g(x)$ and by the lines $x = a$ and $x = b$ with $f(x) \geq g(x)$, then the **moment of inertia** of this area with respect to the y-axis, I_y, is

$$I_y = \rho \int_a^b x^2[f(x) - g(x)]\,dx$$

where ρ is the density of the region.

If we have the moment of inertia with respect to the y-axis and the mass of the region, then we can determine the radius of gyration of the region as shown in the following box.

Radius of Gyration

The radius of gyration r_y with respect to the y-axis is

$$r_y = \sqrt{\frac{I_y}{m}}$$

where m is the mass of the region. ($m = \rho A$ where A is the area of the region.)

EXAMPLE 5.33

Find the moment of inertia and the radius of gyration for the region bounded by $y = \sqrt[3]{x}$, $x = 8$, and the x-axis with respect to the y-axis, if the region has a constant density of 3 kg/m^2. Assume all dimensions are meters.

EXAMPLE 5.33 (Cont.)

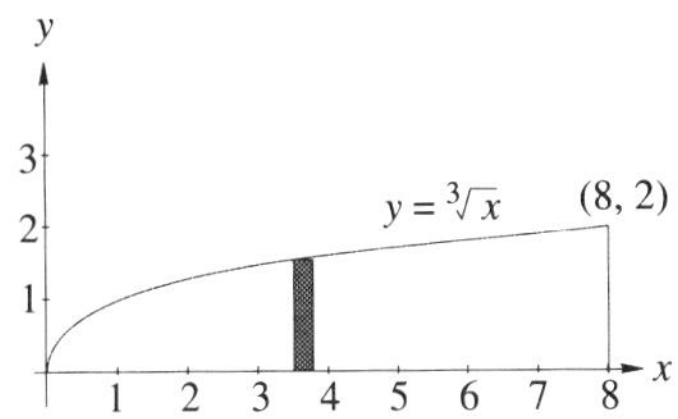

FIGURE 5.36

Solution The region is sketched in Figure 5.36. Here

$$\rho = 3\,\text{kg/m}^2,\ f(x) = \sqrt[3]{x},\ \text{and } g(x) = 0$$

$$\begin{aligned} I_y &= 3\int_0^8 x^2(\sqrt[3]{x} - 0)dx \\ &= 3\int_0^8 x^{7/3}\,dx \\ &= 3\left(\frac{3}{10}\right)x^{10/3}\Big|_0^8 = 921.6\,\text{kg}\cdot\text{m}^2 \\ A &= \int_0^8 \sqrt[3]{x}\,dx \\ &= \left(\frac{3}{4}\right)x^{4/3}\Big|_0^8 = 12\,\text{m}^2 \end{aligned}$$

Since $m = \rho A$, we have $m = 3A = 36$ kg.

$$\begin{aligned} r_y &= \sqrt{\frac{I_y}{m}} \\ &= \sqrt{\frac{921.6}{36}} = \sqrt{25.6} \approx 5.06\,\text{m} \end{aligned}$$

EXAMPLE 5.34

Find the moment of inertia and the radius of gyration of the region in Example 5.33 with respect to the x-axis.

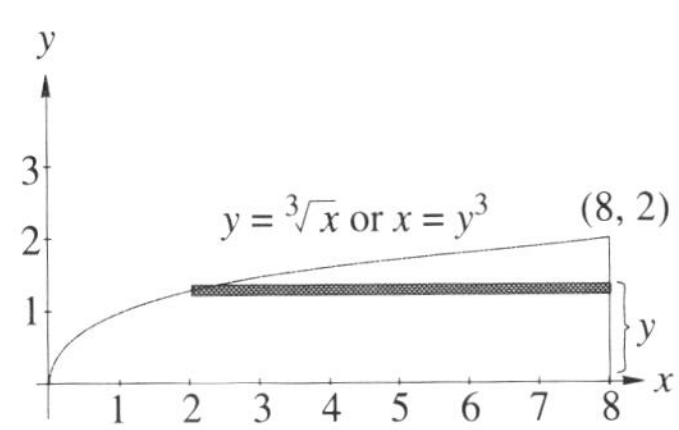

FIGURE 5.37

Solution The region is the same as before, only this time a typical element is parallel to the x-axis, as shown in Figure 5.37. Each element is y units from the x-axis and has the area $(8 - y^3)dy$. Thus, we have

$$\begin{aligned} I_x &= 3\int_0^2 y^2(8 - y^3)dy = 3\int_0^2 (8y^2 - y^5)dy \\ &= 3\left(\frac{8}{3}y^3 - \frac{1}{6}y^6\right)\Big|_0^2 \\ &= 3\left(\frac{64}{3} - \frac{64}{6}\right) = 32\,\text{kg}\cdot\text{m}^2 \\ A &= \int_0^2 (8 - y^3)dy \\ &= \left(8y - \frac{1}{4}y^4\right)\Big|_0^2 = (16 - 4) = 12\,\text{kg} \\ m &= \rho A = 3A = 36\,\text{kg} \\ r_x &= \sqrt{\frac{I_x}{m}} \\ &= \sqrt{\frac{32}{36}} = \frac{\sqrt{8}}{3} \approx 0.94\,\text{m} \end{aligned}$$

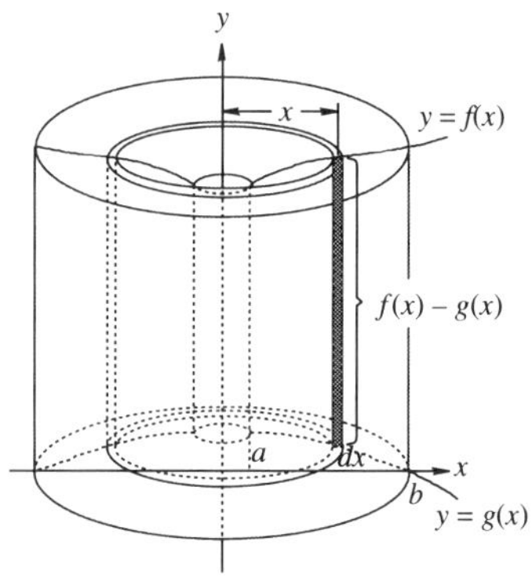

FIGURE 5.38

Moments of Inertia of a Solid of Revolution

We next move to moments of inertia and radii of gyration for solids of revolution. For this we will use the shell method. Consider a region bounded by the curve of the function $y = f(x)$, $x = a$, $x = b$, and $y = g(x)$. The volume of a typical element, such as the one in Figure 5.38, is $2\pi xy\,dx$, where x represents the radius of the cylindrical shell, y, or $f(x) - g(x)$ [assuming $f(x) \geq g(x)$], is its height, and dx is its thickness. The second moment of this shell is $\rho(2\pi xy\,dx)x^2$. Expressing the sums of the second moments as an integral, we get the moment of inertia with respect to the y-axis, I_y, as shown in the following box.

Moment of Inertia; Radius of Gyration

The **moment of inertia** of a solid of revolution formed by generating a region around the y-axis is

$$I_y = 2\pi\rho \int_a^b x^3 y\,dx = 2\pi\rho \int_a^b x^3[f(x) - g(x)]\,dx$$

and the **radius of gyration** is

$$r_y = \sqrt{\frac{I_y}{m}}$$

where m is the mass of the volume of the solid of revolution.

A similar formula exists for the moment of inertia with respect to the x-axis. This second moment, I_x, is

$$I_x = 2\pi\rho \int_c^d y^3 x\,dy = 2\pi\rho \int_c^d y^3[h(y) - k(y)]\,dy$$

where the region is bounded by $x = h(y)$, $x = k(y)$, $y = c$, and $y = d$, and where $h(y) \geq k(y)$ on $[c, d]$.

EXAMPLE 5.35

Find the moment of inertia and the radius of gyration of the solid formed by rotating the region bounded by $y = \sqrt[3]{x}$, $x = 8\,\text{cm}$, and $y = 0$ around the y-axis. This solid will be made from aluminum, which has a density of $2.70\,\text{g/cm}^3$.

Solution For this problem $a = 0$, $b = 8$, $f(x) = \sqrt[3]{x}$, $g(x) = 0$, and $\rho = 2.70$.

$$\begin{aligned} I_y &= 2\pi\rho \int_0^8 x^3\left(\sqrt[3]{x} - 0\right)dx \\ &= 5.40\pi \int_0^8 x^{10/3}\,dx \end{aligned}$$

EXAMPLE 5.35 (Cont.)

$$= 5.40\pi\left(\frac{3}{13}\right)x^{13/3}\Big|_0^8$$
$$= \frac{16.2\pi}{13}(8192) \approx 32070.924\,\text{g}\cdot\text{cm}^2$$

The mass is

$$m = \rho V = 2\pi\rho\int_0^8 x\left(\sqrt[3]{x}-0\right)dx$$
$$= 5.40\pi\int_0^8 x^{4/3}dx$$
$$= 5.40\pi\left(\frac{3}{7}\right)x^{7/3}\Big|_0^8$$
$$= \frac{16.2\pi}{7}(128) \approx 930.630\,\text{g}$$

The radius of gyration is

$$r_y = \sqrt{\frac{I_y}{m}}$$
$$= \sqrt{\frac{32070.924}{930.630}} \approx 5.87\,\text{cm}$$

EXAMPLE 5.36

Find the moment of inertia and the radius of gyration of the solid formed by rotating the region bounded by $y = \sqrt[3]{x}$, $x = 8$ cm, and $y = 0$ around the x-axis. This solid will be made from nylon, which has a density of 1.14 g/cm^3.

Solution This is rotated around the x-axis, so we will use functions of y. Here $h(y) = 8$, $k(y) = y^3$, $c = 0$, $d = 2$, and $\rho = 1.14$. We will not use the value of ρ until we have finished.

$$I_x = 2\pi\rho\int_0^2 y^3\left(8-y^3\right)dy$$
$$= 2\pi\rho\int_0^2 (8y^3 - y^6)dy$$
$$= 2\pi\rho\left(2y^4 - \frac{1}{7}y^7\right)\Big|_0^2$$
$$\approx 27.43\pi\rho \approx 86.17\rho\,\text{g}\cdot\text{cm}^2$$

Since $\rho = 1.14$, $I_x = 98.23\,\text{g}\cdot\text{cm}^2$. The mass is ρV and so

$$m = 2\pi\rho\int_0^2 y(8-y^3)dy$$

EXAMPLE 5.36 (Cont.)

$$= 2\pi\rho\int_0^2 (8y - y^4)dy$$

$$= 2\pi\rho\left(4y^2 - \frac{1}{5}y^5\right)\Big|_0^2 = 19.2\pi\rho \approx 60.32\rho g$$

The radius of gyration is

$$r_x = \sqrt{\frac{I_x}{m}}$$

$$= \sqrt{\frac{27.43\pi\rho}{19.2\pi\rho}} = \sqrt{\frac{27.43}{19.2}} \approx 1.20\,\text{cm}$$

Exercise Set 5.6

In Exercises 1–4, find the moment of inertia and the radius of gyration with respect to the origin, r_0, of the given masses at the given points.

1. 3 g at $(4, 0)$ and 5 g at $(-3, 0)$ cm
2. 6 g at $(0, 2)$, 3 g at $(0, -5)$, and 1 g at $(0, 4)$
3. 4 kg at $(2, 1)$ and 3 kg at $(-1, 4)$ m
4. 3 g at $(2, 3)$, 4 g at $(-2, -4)$, and 3 g at $(-4, 5)$

In Exercises 5–14, find the moment of inertia and radius of gyration about the given axis. Unless specified, let ρ represent the density.

5. The region bounded by $y = x^2$, the y-axis, $y = 2$ about y-axis
6. The region of Exercise 5 about x-axis
7. The region bounded by $x = 0$, $y = 0$, $x = 3$, $y = 5$ about the x-axis
8. The region in Exercise 7 about the y-axis
9. The region bounded by $y = x^3$ and $y = x^2$ about the x-axis
10. The region in Exercise 9 about the y-axis
11. The region bounded by $y = 4x - x^2$ and the x-axis, $\rho = 5\text{ g/cm}^2$ about y-axis
12. The region bounded by $y^3 = x^2$, $y = 4$ and the y-axis, $\rho = 4\text{ g/cm}^2$ about the x-axis
13. The region bounded by $y = \frac{1}{x^2}$, $y = x^2$, $x = 2$, $x = 1$, $\rho = 8\text{ g/cm}^2$ about the y-axis
14. The region in Exercise 13 about the x-axis

In Exercises 15–24, find the moment of inertia and the radius of gyration for the solid formed by rotating the region about the given axis. Unless specified, let ρ represent the density.

15. The region bonded by $y = x^2$, the y-axis, and $y = 2$ about the y-axis
16. The region in Exercise 15, but rotated about the x-axis
17. The region bounded by $y = x^2$ and $y = 4x$ about the y-axis
18. The region in Exercise 17 about the x-axis
19. The region bounded by $x = 0$, $y = 0$, $x = 4$ and $y = 6$ about the x-axis
20. The region in Exercise 19, but rotated about the y-axis
21. The region bounded by $y = 4x - x^2$ and $x + y = 4$, $\rho = 3\text{ g/cm}^3$ about y-axis
22. The region bounded by $y = \sqrt{4 - x}$ and $x + 2y = 4$, $\rho = 5\text{ g/cm}^3$, about the x-axis.

23. The region bounded by $y = \frac{1}{x}$, $y = x^3$, $x = 1$, $x = 2$, $\rho = 5\text{ g/cm}^3$, about the y-axis

24. The region in Exercise 23 about the x-axis

In Your Words

25. Without looking in the text, explain the concept of moment of inertia.

26. What is meant by a radius of gyration?

5.7 WORK AND FLUID PRESSURE

The last applications of the integral that we will consider in this chapter deal with the topic of work. **Work** is defined as the product of the force exerted on an object and the distance the object is moved by that force. For example, if a force of 10 lb is lifted a distance of 5 ft, the work performed is (10 lb) × (5 ft) = 50 ft·lb. In the same way, if a force of 20 dynes is lifted 9 cm the worked performed is (20 dynes) × (9 cm) = 180 dyne·*cm* = 180 ergs. Just as an erg is a dyne·cm, the joule is the equivalent of the Newton-meter (N · m).

Most work problems are not as simple as these two. If they were, we certainly would not be discussing them in a chapter about applications of integral calculus.

Hooke's Law

Consider the work done to stretch a spring. When you begin to stretch the spring, it takes very little force. The further the spring is stretched, more force is required. Hooke's law states that the force F required to stretch a spring a distance x is proportional to x. Thus, if k is the constant of proportionality, $F = kx$ and so the force is a function of x or $F(x)$.

We want to determine the amount of work that is done to stretch the spring from point a to point b. As usual, we partition the interval $[a, b]$ into n subintervals of lengths Δx. If $x_i{}^*$ is the midpoint of one of the intervals, then the work done to move the spring through that interval is approximately $\Delta W = F(x_i{}^*)\Delta x$. Adding the work over all of the intervals and taking the limits as $\Delta x \to 0$, we get the following formula.

Work

The **work** W done by a continuous force directed along the x-axis from $x = a$ to $x = b$ is

$$W = \int_a^b F(x)dx$$

EXAMPLE 5.37

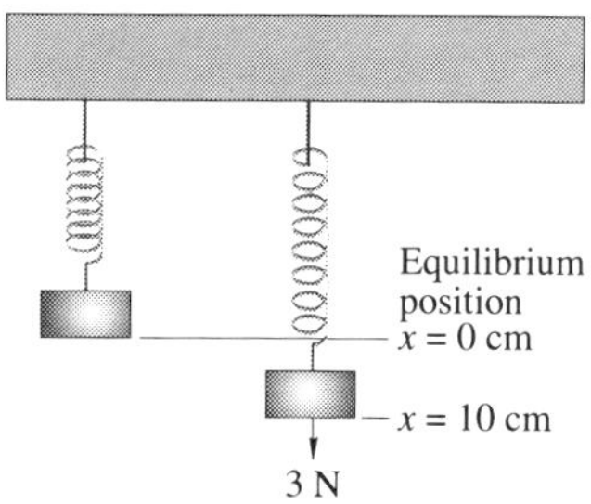

FIGURE 5.39

Application

A spring is stretched 10 cm by a force of 3 N, as shown in Figure 5.39. How much work is needed to stretch the spring 40 cm?

Solution According to Hooke's law, $F = kx$. We are given $F = 3$ N when $x = 10\,\text{cm} = 0.1$ m. Substituting these into the formula for Hooke's law produces

$$F = kx$$
$$3 = k(0.1)$$
$$30 = k$$

Thus, for this particular spring, $F = 30x$ or the function $F(x) = 30x$. The amount of work needed to stretch this spring $40\,\text{cm} = 0.4\,\text{m}$ is

$$W = \int_0^{0.4} F(x)\,dx$$
$$= \int_0^{0.4} 30x\,dx = 15x^2\Big|_0^{0.4} = 2.4\,\text{N}\cdot\text{m}$$

The work needed to stretch the spring is $2.4\,\text{N}\cdot\text{m} = 2.4$ J. (The symbol "J" is used for joule or joules.)

EXAMPLE 5.38

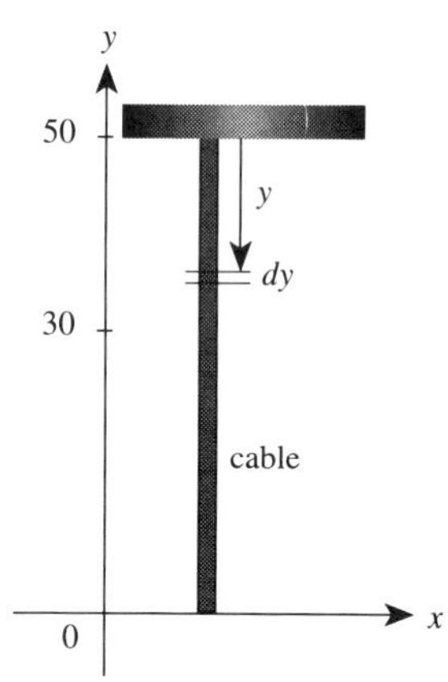

FIGURE 5.40

Application

Find the work done in winding 30 m of a 50-m cable, if the cable has a density of 4.5 kg/m.

Solution First we let y denote the length of the cable that has been wound up at any time. The amount of cable that has yet to be wound is $50 - y$. As usual, we will subdivide the interval $[0, 30]$ into n subdivisions. A typical subdivision is shown in Figure 5.40. This subdivision has a length dy and is y units from the top. The mass of a subdivision is $\rho\,dy$ and its weight is $\rho g\,dy$, where ρ is the density and g the gravitational constant. In this example, ρ was given as 4.5 kg/m and, since this problem is in the metric system, $g = 9.8\,\text{m/s}^2$. Thus, the force needed at any point is

$$(4.5)(9.8)(\text{amount of cable left to be wound}) = (4.5)(9.8)(50 - y)$$
$$= 44.1(50 - y)$$

The work needed to move this element from its initial position to the top is $44.1(50 - y)dy$. Since only 30 m are going to be wound, the work is

$$W = \int_0^{30} 44.1(50 - y)dy$$
$$= 44.1\left(50y - \frac{1}{2}y^2\right)\Big|_0^{30}$$
$$= 44.1(1\,050)\,\text{J}$$
$$= 46\,305\,\text{J}$$

Electrical Charges

According to Coulomb's law, the electric force between two charges, q_1 and q_2, in a vacuum is directly proportional to the product of these two changes and inversely proportional to the square of the distance r between them. Thus, the force is a function of r and we have

$$F(r) = k\frac{q_1 q_2}{r^2}$$

where $k = 8.988 \times 10^9\,\mathrm{N{\cdot}m^2/C^2}$. Using the formula for work

$$W = \int F(r)\,dr$$

we can determine the work done when two charges move toward or away from each other.

Application

EXAMPLE 5.39

Find the work done when two protons, $q = 1.6 \times 10^{-19}$ C each, move until they are 100 nm apart, if they were originally separated by 1 m.

Solution We have $q_1 = q_2 = 1.6 \times 10^{-19}$. The particles move from 1 m apart to $100\,\mathrm{nm} = 100 \times 10^{-9}\,\mathrm{m} = 10^{-7}$ m apart.

$$\begin{aligned} W &= \int_1^{10^{-7}} \frac{(8.988 \times 10^9)(1.6 \times 10^{-19})(1.6 \times 10^{-19})}{r^2}\,dr \\ &= 2.301 \times 10^{-28} \int_1^{10^{-7}} \frac{dr}{r^2} \\ &= 2.301 \times 10^{-28} \left(\frac{-1}{r}\right)\Bigg|_1^{10^{-7}} \\ &= -2.301 \times 10^{-21}\,\mathrm{J} \end{aligned}$$

The negative sign means that the force must be done *on* the system to move the particles together. Thus, it takes 2.301×10^{-21} J acting on these protons to move them from 1 m apart to 100 nm apart.

Pumping Liquids

As a third example of a problem where the force is variable, consider the work needed to pump a liquid out of, or into, a tank. In these examples, we will use the basic idea that work = (weight of object) × (distance moved).

Application

EXAMPLE 5.40

An oil tank is in the shape of a right-circular cylinder with a height of 40 ft, a radius of 10 ft, and is filled with oil to a depth of 30 ft. How much work is required to pump the oil out of the top of the tank, if oil weighs 54.8 lb/ft^3?

Solution We will position the tank as shown in Figure 5.41. The position of the top of the tank is at $x = 0$ and the bottom at $x = 40$. The work needed to pump a

EXAMPLE 5.40 (Cont.)

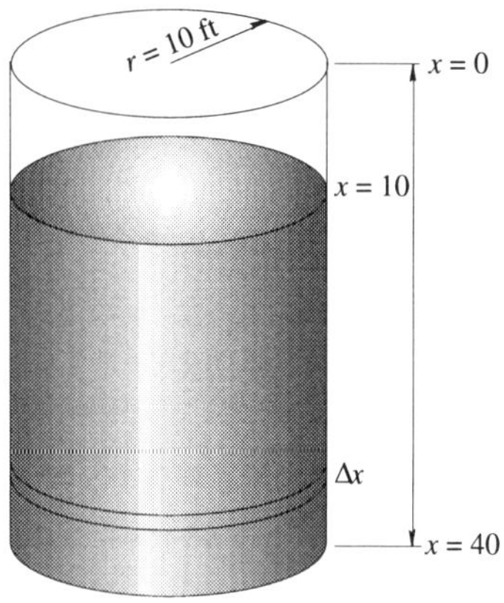

FIGURE 5.41

certain volume of oil over the top depends on the weight of this volume of oil and its distance from the top.

The oil fills the tank from $x = 10$ to $x = 40$. We will partition the interval $[10, 40]$ into subintervals and let Δx be the length of one interval. The volume at this interval is $\pi r^2 h$, and since $r = 10$ and $h = \Delta x$, we have the volume as $100\pi\,\Delta x$. We are told that oil weighs $54.8\,\text{lb/ft}^3$, so the weight of this layer is $(54.8)(100\pi\,\Delta x)$. If the interval is located at $x_i{}^*$, then the work done to lift this layer is $x_i{}^*(54.8)(100\pi\,\Delta x)$.

The layers of oil occur from 10 ft to 40 ft, thus the work required is given by

$$\begin{aligned}\int_{10}^{40}(54.8)(100\pi)x\,dx &= 5480\pi\left(\frac{1}{2}x^2\right)\Bigg|_{10}^{40}\\ &= 4{,}110{,}000\pi\,\text{ft}\cdot\text{lb}\\ &\approx 12{,}911{,}946\,\text{ft}\cdot\text{lb}\end{aligned}$$

It takes a total of 12,911,946 ft·lb to pump the oil out of the top of the tank. ■

The procedure used in Example 5.40 follows the following guidelines for finding the work needed to pump all the liquid out of a tank.

Guidelines for Finding Work Done During Pumping

1. Draw a picture with a coordinate system.
2. Determine the mass of a thin horizontal slab of the liquid.
3. Find an expression for the work needed to lift this slab to its destination.
4. Integrate the work expression from step 3 from the bottom of the liquid to the top.

Application

EXAMPLE 5.41

Suppose a hemispherical tank of radius 10 ft is on a platform so that the bottom is 30 ft off the ground. How much work is done filling the tank from a source of water at ground level, if the water is pumped through a hole in the bottom?

Solution We will put the origin of the vertical axis at ground level. Then the bottom of the tank is at 30 ft and the top is at 40 ft, as shown in Figure 5.42a.

Divide the interval $[30, 40]$ into n subintervals. Suppose a typical interval is x_i units from the top of the tank, as shown in Figure 5.42b. The slice of water taken at this point has a radius y_i. Because the tank is a hemisphere, we know that the radius r of the tank is the hypotenuse of the right triangle formed by x_i and y_i. This means that $x^2 + y^2 = r^2$. The volume of this slice of water is $\Delta V = \pi y^2 \Delta x = \pi(r^2 - x^2)\Delta x$. The force required to lift this slice is

$$\Delta w = 62.4\pi(r^2 - x^2)\Delta x$$

EXAMPLE 5.41 (Cont.)

where the specific weight of water is $w = 62.4\,\text{lb/ft}^3$. Now all we need is the distance this slice was lifted. Since it was lifted from $x = 0$ to a point x_i units below $x = 40$, it was lifted a total of $40 - x_i$ units. Putting this together with the fact that $r = 10$, we get

$$\begin{aligned} W &= \int_0^{10} 62.4\pi(10^2 - x^2)(40 - x)dx \\ &= 62.4\pi \int_0^{10} (4{,}000 - 100x - 40x^2 + x^3)dx \\ &= 62.4\pi \left(4{,}000x - 50x^2 - \frac{40}{3}x^3 + \frac{1}{4}x^4\right)\Bigg|_0^{10} \\ &\approx 4{,}737{,}521.7\,\text{ft}\cdot\text{lb} \end{aligned}$$

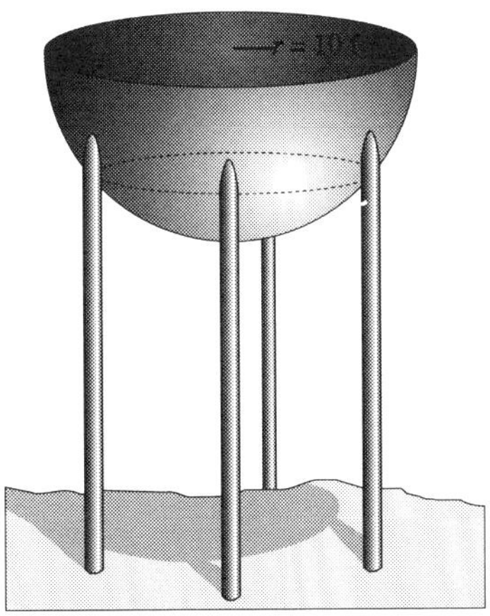

FIGURE 5.42a

Fluid Pressure

When a container is built to hold a fluid, it is important to know the force F caused by the liquid pressure on the sides of the container. The pressure P exerted by a liquid of mass density ρ at depth h units below the surface of the liquid is defined as

$$P = \rho g h$$

where g is the acceleration of gravity, $32\,\text{ft/s}^2 \approx 9.8\,\text{m/s}^2$.

Note In the English system we are given the weight density w as lb/ft^3. Here $w = \rho g$.

If a flat plate is submerged in a fluid and placed in a horizontal position, the pressure on the plate is the same at all points. The total force on one side of the plate is the product of the pressure P and the area A of the side $F = PA = \rho g h A$.

Application

EXAMPLE 5.42

A tank 3 m long and 2 m wide is filled with 1.5 m of water. What is the force exerted on the bottom of the tank?

Solution The density of water is about $\rho = 1\,000\,\text{kg/m}^3$. The area of the bottom of the tank is $A = (3\,\text{m})(2\,\text{m}) = 6\,\text{m}^2$. The pressure is $P = (1\,000)(9.8)(1.5) = 14\,700\,\text{N/m}^2$. $F = PA = (14\,700)(6) = 88\,200\,\text{N}$.

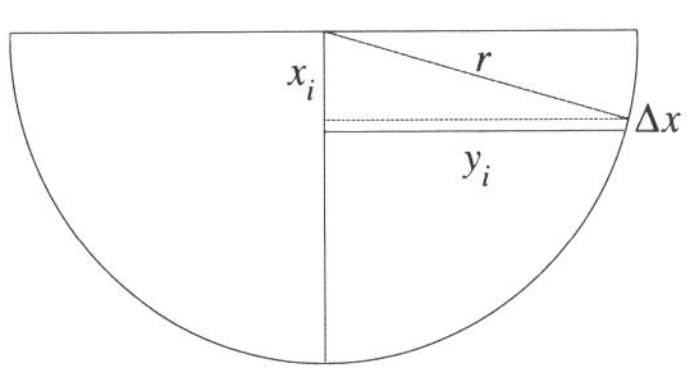

FIGURE 5.42b

If a flat plate submerged in a fluid of density ρ is placed in a vertical position, then the pressure varies as the depth changes. If the plate is subdivided into n horizontal strips, as shown by the typical strip in Figure 5.43, then the pressure is almost constant on each strip.

As shown in Figure 5.43, each horizontal strip has a width dy and length $L(y)$. So, the area is $L(y)\,dy$. We introduce a coordinate system, as shown in the figure. If the plate extends over the interval $[c, d]$ on the y-axis, then for each y in this interval, the depth of each strip is $h(y)$. If dy is small, the pressure at any point in the horizontal

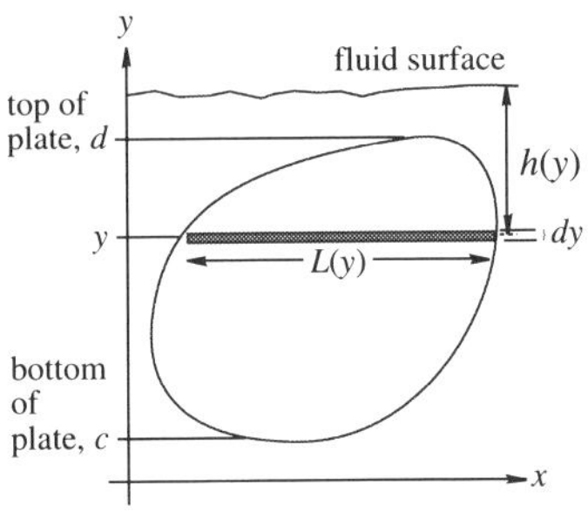

FIGURE 5.43

strip is $\rho h(y)$. Thus, the force on each rectangular strip is given by the product of the pressure on any point in the strip and the area of the strip, or $\rho h(y)L(y)dy$. Integrating this over the depth of the plate, we get the following formula.

Fluid Force

The total force F exerted by a fluid of constant density ρ on a submerged vertical plate from depth $y = c$ to $y = d$ is given by

$$F = \rho \int_c^d h(y)L(y)dy$$

where $h(y)$ is the depth of the fluid at y and $L(y)$ is the horizontal length of the plate at y.

Application

EXAMPLE 5.43

A tank has a cross-section in the shape of a trapezoid with a lower base of 4 ft, and upper base 8 ft. It is filled with 2.5 ft of water. What is the total pressure due to the water on one end of the tank?

Solution A sketch of the tank is shown in Figure 5.44. An xy-coordinate system has been placed on the figure to help in the calculations. The y-axis is partitioned into n subintervals. A typical strip formed by this partition is shaded in the figure. The force against this ith strip is $P_i A_i = 62.4(2.5 - y_i)A_i$, where $h(y_i) = 2.5 - y_i$ is the depth of the strip and A_i is its area.

The length of the strip is approximately $2x_i$. But the edge of the tank, line BC, passes through $(2, 0)$ and $(4, 2.5)$. This line has the equation $y = \frac{2.5}{2}x - 2.5$ and so $x = \frac{2}{2.5}y + 2 = \frac{4}{5}y + 2$. Thus the length of the ith strip is $2\left(\frac{4}{5}y_i + 2\right)$ and its area $A_i = 2\left(\frac{4}{5}y_i + 2\right)\Delta y = \left(\frac{8}{5}y_i + 4\right)\Delta y_i$.

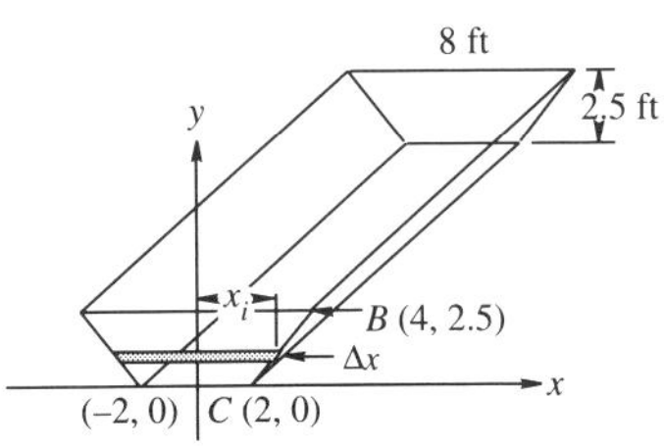

FIGURE 5.44

We know that the force F_i on this ith strip is

$$F_i = 62.4\,(2.5 - y_i)\left(\frac{8}{5}y_i + 4\right)\Delta y_i$$

Integrating, we get the total force

$$\begin{aligned} F &= \int_0^{2.5} 62.4\,(2.5 - y)\left(\frac{8}{5}y + 4\right)dy \\ &= 62.4\int_0^{2.5}\left(10 - \frac{8}{5}y^2\right)dy \\ &= 62.4\left(10y - \frac{8}{15}y^3\right)\Bigg|_0^{2.5} \\ &= 62.4(16.67) \approx 1{,}040\,\text{lb} \end{aligned}$$

∎

Application

EXAMPLE 5.44

A tank truck, like the milk truck in the photograph in Figure 5.45a, is used to carry liquids. If the tank on this truck is in the shape of a circular cylinder with a radius of 3 ft and length 35 ft, determine the force on one end of this truck when it is full of milk, if milk has a density of 64.5 lb/ft^3.

Courtesy of H.P. Hood Inc.

FIGURE 5.45a

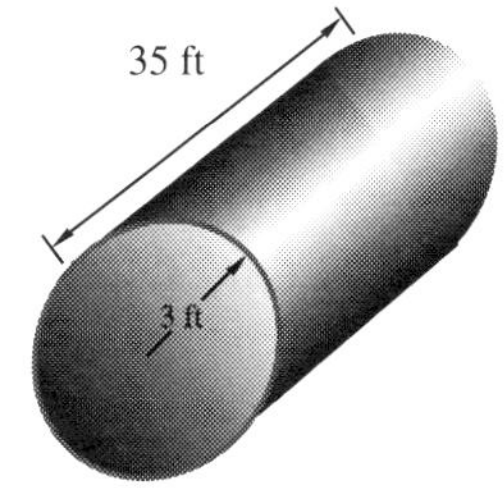

FIGURE 5.45b

Solution A sketch of the tank is shown in Figure 5.45b. A cross-section of an end of the tank is shown in Figure 5.45c. A coordinate system and a typical horizontal strip have been drawn in the figure. Since this is a circular cylinder and the origin of the coordinate system is placed at the center of the base, the circle is described by the equation $x^2 + y^2 = 3^2$, so the length of the strip is $x = 2\sqrt{9 - y^2}$. Since the width of the strip is dy, its area is $A_i = 2\sqrt{9 - y^2}\,dy$. As can be seen in Figure 5.45c, the depth of this strip is $3 - y$. Thus, the force on this strip is $F_i = \rho(3 - y)2\sqrt{9 - y^2}\,dy$ and the total force on the end of the tank is

$$
\begin{aligned}
F &= 2\rho \int_{-3}^{3} (3 - y)\sqrt{9 - y^2}\,dy \\
&= 6\rho \int_{-3}^{3} \sqrt{9 - y^2}\,dy + 2\rho \int_{-3}^{3} (-y)\sqrt{9 - y^2}\,dy
\end{aligned}
$$

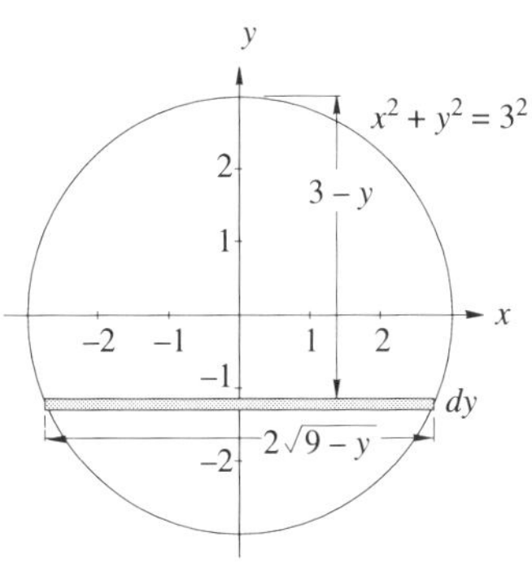

FIGURE 5.45c

While we do not yet know how to integrate the left-hand integral, we do know that its integral will be the area of one-half a circle of radius 3. From geometry, we know that the area of a complete circle of radius 3 is $\pi r^2 = \pi 3^2 = 9\pi$ and so the desired integral will be one-half that amount, or $\frac{9}{2}\pi$. Substituting this in the previous equation, we have

$$
\begin{aligned}
F &= 6\rho\frac{9}{2}\pi + 2\rho \int_{-3}^{3} (-y)(9 - y^2)^{1/2}\,dy \\
&= 27\pi\rho + \frac{2}{3}\,\rho(9 - y^2)^{3/2}\Big|_{-3}^{3} = 27\pi\rho
\end{aligned}
$$

Since milk has a density of $\rho = 64.5$ lb/ft^3, the total force on the end of the tank is $1{,}741.5\pi \approx 5{,}471.08$ lb/ft^3. Notice that the length of the tank had no effect on the force exerted on the ends.

Exercise Set 5.7

Solve Exercises 1–36.

1. *Physics* A force of 3 lb stretches a spring 4 in. How much work is done in stretching the spring 10 in.?
2. *Physics* If the spring in Exercise 1 is stretched from 1 ft to 2 ft, how much work is done?
3. *Physics* A force of 6 N stretches a spring 20 cm. How much work is done to stretch the spring 1 m?
4. *Physics* How much work is done to stretch the spring in Exercise 3 from 10 cm to 20 cm?
5. *Physics* If the work done in stretching a spring from rest at 0.50 m to a length of 0.60 m is 0.20 J, find the force needed to keep the spring at this length.
6. *Physics* A spring has a natural (or unstretched) length of 10 in. A force of 5 lb compresses the spring to a length of 8 in. Find the work needed to compress it from 9 in. to 6 in.
7. *Physics* If 105 ft·lb of work was needed to stretch the spring in Exercise 1 from its equilibrium position, how far was it stretched?
8. *Mechanical engineering* A cable 50 ft long and weighting 2 lb/ft is hanging from a winch. How much work is needed to wind 20 ft of it up?
9. *Mechanical engineering* A cable 30 m long weighing 2.5 N/m is hanging from a winch. How much work is needed to wind all of the cable onto the winch?
10. *Mechanical engineering* A chain with a mass of 4.7 kg/m has a length of 15 m. If there is a load of 87 kg at the end of the chain, what work is done in winding the chain and load up 10 m?
11. *Mechanical engineering* A chain 12 ft long is lying on the floor. If the chain weighs 0.75 lb/ft, how much work is needed to raise one end to a height of 20 ft?
12. *Mechanical engineering* A chain 5 m long with a density of 0.4 kg/m is lying on the floor. How much work is needed to raise one end to a height of 10 m?
13. *Mechanical engineering* A tank with a mass of 45 kg is filled with 1 000 L of water. The water is used to spray chemical cleaner off the side of a building. If the water is used at the rate of 20 L/min and the tank is raised at 0.50 m/min, how much work is done in raising the tank 20 m up the wall?
14. *Nuclear physics* Find the work needed to move a charge of $+5 \times 10^{-7}$ C and a charge of -2×10^{-7} C until they are 1 cm apart, if they were originally 3 mm apart. (Note: Change all distances to meters.)
15. *Nuclear physics* A charge of $+4 \times 10^{-19}$ C is 5 cm from a charge of $+5 \times 10^{-8}$ C. How much work is needed to move them until they are 10 nm apart? ($1\text{ nm} = 10^{-9}$ m.)
16. *Nuclear physics* A proton has a charge of $+1.6 \times 10^{-18}$ C and an electron has a charge of -1.6×10^{-19} C. How much work is needed to move them from 530 nm apart to 1 mm apart?
17. *Physics* According to Newton's law of gravitation, the gravitational force of attraction between two objects that are r units apart is given by $F(r) = \dfrac{k}{r^2}$. If two objects are 1 cm apart, how much work is needed to separate them by a distance of 1 m? (Leave answer in terms of k.)
18. *Physics* If two objects are 1 ft apart, how much work is needed to separate them until they are 20 ft apart? (Leave answer in terms of k.)
19. *Space technology* How much work is required to lift a 1,000-lb satellite from the surface of the Earth to an orbit of 1,000 mi above the Earth's surface? (See Exercise 17 with $k = 16 \times 10^9$ mi^2·lb and the radius of the Earth as 4,000 mi.)
20. *Petroleum engineering* A cylindrical tank 4 ft in radius and 8 ft high is filled with petroleum. How much work is needed to pump the oil out over the top? (The density of oil is about 54.8 lb/ft^3.)
21. *Petroleum engineering* How much work is required to pump all the oil over the top of a full cylindrical tank that is 4 m in diameter and 8 m high? (The mass density of petroleum varies. For this, and subsequent exercises, let the mass density $\rho = 880$ kg/m^3.)

22. *Recreation* A swimming pool is 2 m deep, 10 m long, and 7 m wide. If it is filled with water, how much work is needed to empty the pool, if the water is pumped over the top?

23. *Recreation* How much work is needed to empty the pool in Exercise 22, if the water was only 1.5 m deep?

24. *Petroleum engineering* The oil in a cylindrical tank that is 10 ft high and has a diameter of 6 ft is to be pumped to a height 5 ft above the top of the tank. How much work is required?

25. *Petroleum engineering* An upright conical tank is filled with oil from a reservoir that is 15.0 m below the bottom of the tank. If the top of the tank has a radius of 2.40 m and the height of the tank is 3.60 m, what work is done in filling the tank when the density of the oil is 847 kg/m^3?

26. *Wastewater technology* A water tank is in the shape of a hemisphere with a radius of 4 m. If the tank is already filled with water, how much work is needed to pump all the water over the top of the tank? (The density of water is 1000 kg/m^3.)

27. *Wastewater technology* How much work is required to pump the top 2 m of water out the top of the tank in Exercise 26?

28. *Wastewater technology* If the tank in Exercise 26 is placed on a platform so that the bottom of the tank is 10 m above the ground, how much work is required to fill the tank with water to the depth of 3 m from ground level?

29. *Recreation* **(a)** What is the pressure of the water on the bottom of the tank in Exercise 22? **(b)** What is the force of the water on the bottom of this tank?

30. *Petroleum engineering* **(a)** What is the pressure of the petroleum on the bottom of the tank in Exercise 24? **(b)** What is the force of the petroleum on the bottom of this tank?

31. *Recreation* If the pool in Exercise 22 is full of water, what is the total force due to water pressure on one short side of the pool?

32. *Recreation* For the pool in Exercise 22, what is the total force due to water pressure on a long side of the pool?

33. *Civil engineering* A vertical flood gate is in the shape of an isosceles triangle which has a base of 3 m and an altitude of 2 m. If the base of the floodgate is on the surface of the water, what is the force on its face?

34. *Recreation* A rectangular swimming pool is 40.0 m long, 1.00 m deep at the shallow end, and 3.00 m deep at the deep end. Assuming that all of the sides are vertical, find the total force on the long side of the swimming pool when it is full.

35. *Transportation* A tanker truck is in the shape of a right-circular cylinder. The radius of the cylinder is 1.6 m and the density of gasoline is about $\rho = 680$ kg/m^3. If the tanker is half filled, what is the total force due to the pressure of the gasoline on one end of the cylinder?

36. *Transportation* A tanker truck is in the shape of a right-circular cylinder with a radius of 4 ft. If the tank is half full of gasoline, and the weight of gasoline is about 42.5 lb/ft^3, what is the total force due to the pressure of the gasoline on one end of the tanker?

In Your Words

37. Explain what is meant by the concept of work as used in this text.

38. Without looking in the text, describe the four guidelines for finding the work done during pumping.

CHAPTER 5 REVIEW

Important Terms and Concepts

Arc length
Average value
Centroid
Fluid pressure
Hooke's law
Mass
Moment of inertia
Moments
First
Second
Radius of gyration
Root mean square
Surface area
Volumes of revolution
Disk method
Shell method
Washer method
Work

Review Exercises

In Exercises 1–4, find the average value and the root mean square of the function over the given interval.

1. $f(x) = 4x$, over $[1, 6]$
2. $g(x) = x^3$, over $[0, 2]$
3. $h(x) = x^2 - 4$, over $[2, 4]$
4. $k(t) = 4.9t^2 - 2.8t - 4$, over $[0, 2.5]$

In Exercises 5–12, find (a) the volume and (b) the centroid of the solid of revolution generated by revolving the enclosed region about the indicated axis.

5. $y = 6x$, $x = 1$, $x = 5$, $y = 0$, about the x-axis
6. $y = -x$, $x = 1$, $x = 4$, $y = 0$, about the y-axis
7. $y = x^4$, $x = 1$, $x = 2$, $y = 0$, about the x-axis
8. $y = \sqrt[3]{x^2}$, $x = 0$, $x = 1$, $y = 0$, about the x-axis
9. $y = x^3$, $y = \sqrt[3]{x}$, about the y-axis
10. $y = x^3$, $y = \sqrt[3]{x}$, about the x-axis
11. $y = x^2$, $y = 9 - x^2$, about the y-axis
12. $y = x^3$, $y = 4x$, about the line $x = 5$

In Exercises 13–14, find the lengths of the curves over the indicated interval.

13. $x = \sqrt{y^3}$ from $y = 0$ to $y = 4$
14. $y = \sqrt{8x^3}$ from $x = 0$ to $x = 2$

In Exercises 15–16, find (a) the lengths of the curves over the indicated interval and (b) the surface area of the solid generated by rotating the curve around the indicated axis.

15. $y = \dfrac{x^4}{4} + \dfrac{1}{8x^2}$ from $x = 1$ to $x = 3$, about the x-axis
16. $x = \dfrac{y^6}{6} + \dfrac{1}{16y^4}$ from $y = 1$ to $y = 2$, about the y-axis

Solve Exercises 17–8.

17. *Physics* An unstretched spring 0.6 m long is stretched to a length of 0.8 m by a force of 5 N. How much work is needed to stretch the spring to a length of 1.2 m?
18. *Electricity* The voltage across a 60-μF capacitor is zero. What is the voltage after 0.001 s, if a current of 0.40 A charges the capacitor?
19. *Electricity* The electric current, as a function of time, for a certain circuit is $I = 4t - t^3$. What is the average value of the current, with respect to time, for the first 2 s?
20. *Physics* If a ball is dropped from the top of the Washington Monument, which is 555 ft high, what is its average height and average velocity between the time it is dropped and the time it strikes the ground?

21. *Construction* A steel beam weighing 1,000 lb hangs from a 50-ft cable. The cable weighs 5 lb/ft. How much work is done in winding 30 ft of cable with a winch?

22. *Civil engineering* A dam is shaped like a trapezoid with a height of 100 ft. It is 400 ft long at the top and 200 ft long at the bottom. When the water level behind the dam is level with the top, what is the total force that the water exerts on the dam?

23. *Petroleum engineering* A tank is in the shape of a right-circular cylinder with a radius of 6 m and a height of 16 m. If it is filled with petroleum, how much work is needed to pump it out the top? (Let $\rho = 880\text{ kg/m}^3$.)

24. *Waste technology* A tank is in the shape of an inverted cone with a radius of 6 ft and a height of 16 ft. If it is filled with water, how much work is needed to pump it out the top?

CHAPTER 5 TEST

1. Find the average value of $f(x) = 3x^2 + 1$ over the interval $[1, 5]$.
2. Determine the root mean square of $g(x) = x^3 - 1$ over $[1, 3]$.
3. Find the volume of the solid generated by revolving the region bounded by $y = \sqrt{9x+1}$, $x = 0$, $x = 7$, and the x-axis about the x-axis.
4. Find the centroid of the solid of revolution generated by revolving the region bounded by $y = x^2 + 1$, $x = 0$, $x = 4$, and $y = 0$, about the x-axis.
5. Find the length of the curve $y = 1 + x^{3/2}$ from $x = 1$ to $x = 6$.
6. Find the surface area of the solid generated by rotating the curve $y = \frac{1}{6}x^3$ from $x = 1$ to $x = 3$ around the x-axis.
7. How much work does it take to pump the water from a full upright circular-cylindrical tank of radius 5 m and height 10 m to a point 4 m above the top of the tank?
8. A vertical gate is in the shape of a right triangle with its base on top (and the right angle at the bottom), as shown in Figure 5.46. If the base is 5 ft below the surface of the water and the altitude of the gate is 3 ft, what is the force against the gate?

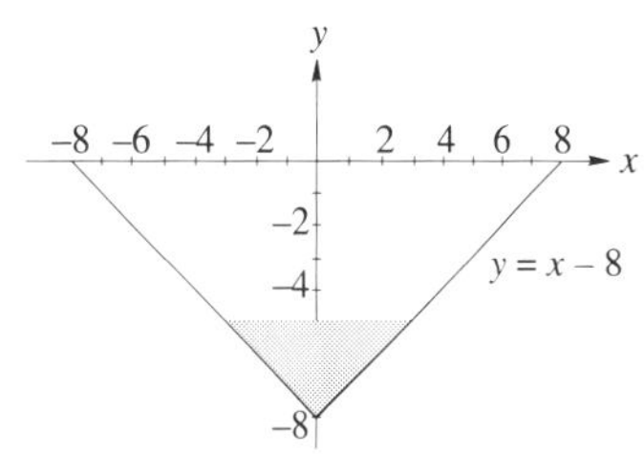

FIGURE 5.46

Appendix

A

Table of Integrals

Basic Forms

1. $\int u\,dv = uv - \int v\,du$

2. $\int u^n\,du = \frac{1}{n+1}u^{n+1} + C,\ n \neq -1$

3. $\int \frac{du}{u} = \ln|u| + C$

4. $\int e^u\,du = e^u + C$

5. $\int a^u\,du = \frac{1}{\ln a}a^u + C$

6. $\int \sin u\,du = -\cos u + C$

7. $\int \cos u\,du = \sin u + C$

8. $\int \sec^2 u\,du = \tan u + C$

9. $\int \csc^2 u\,du = -\cot u + C$

10. $\int \sec u \tan u\,du = \sec u + C$

11. $\int \csc u \cot u\,du = -\csc u + C$

12. $\int \tan u\,du = \ln|\sec u| + C$

$= -\ln|\cos u| + C$

13. $\int \cot u\,du = \ln|\sin u| + C$

14. $\int \sec u\,du = \ln|\sec u + \tan u| + C$

15. $\int \csc u\,du = \ln|\csc u - \cot u| + C$

16. $\int \frac{du}{\sqrt{a^2-u^2}} = \sin^{-1}\frac{u}{a} + C$

17. $\int \frac{du}{a^2+u^2} = \frac{1}{a}\tan^{-1}\frac{u}{a} + C$

18. $\int \frac{du}{u\sqrt{u^2-a^2}} = \frac{1}{a}\sec^{-1}\frac{u}{a} + C$

19. $\int \frac{du}{a^2-u^2} = \frac{1}{2a}\ln\left|\frac{u+a}{u-a}\right| + C$

20. $\int \frac{du}{u^2-a^2} = \frac{1}{2a}\ln\left|\frac{u-a}{u+a}\right| + C$

Forms Involving $\sqrt{a^2 - u^2}$

21. $\int \sqrt{a^2-u^2}\,du = \frac{u}{2}\sqrt{a^2-u^2} + \frac{a^2}{2}\sin^{-1}\frac{u}{a} + C$

22. $\int u^2\sqrt{a^2-u^2}\,du =$

$\frac{u}{8}(2u^2-a^2)\sqrt{a^2-u^2} + \frac{a^4}{8}\sin^{-1}\frac{u}{a} + C$

23. $\int \frac{\sqrt{a^2-u^2}}{u}\,du =$

$\sqrt{a^2-u^2} - a\ln\left|\frac{a+\sqrt{a^2-u^2}}{u}\right| + C$

24. $\int \frac{\sqrt{a^2-u^2}}{u^2}\,du = -\frac{1}{u}\sqrt{a^2-u^2} - \sin^{-1}\frac{u}{a} + C$

25. $\int \frac{u^2\,du}{\sqrt{a^2-u^2}} = -\frac{u}{2}\sqrt{a^2-u^2} + \frac{a^2}{2}\sin^{-1}\frac{u}{a} + C$

26. $\int \frac{du}{u\sqrt{a^2-u^2}} = -\frac{1}{a}\ln\left|\frac{a+\sqrt{a^2-u^2}}{u}\right| + C$

27. $\int \frac{du}{u^2\sqrt{a^2-u^2}} = -\frac{1}{a^2u}\sqrt{a^2-u^2} + C$

28. $\int (a^2-u^2)^{3/2}\,du =$
$-\frac{u}{8}(2u^2-5a^2)\sqrt{a^2-u^2} + \frac{3a^4}{8}\sin^{-1}\frac{u}{a} + C$

29. $\int \frac{du}{(a^2-u^2)^{3/2}} = \frac{u}{a^2\sqrt{a^2-u^2}} + C$

Forms Involving $\sqrt{u^2 \pm a^2}$

30. $\int \sqrt{u^2\pm a^2}\,du = \frac{u}{2}\sqrt{u^2\pm a^2} \pm$
$\frac{a^2}{2}\ln\left|u+\sqrt{u^2\pm a^2}\right| + C$

31. $\int u^2\sqrt{u^2\pm a^2}\,du = \frac{u}{8}\left(2u^2\pm a^2\right)\sqrt{u^2\pm a^2} -$
$\frac{a^4}{8}\ln\left|u+\sqrt{u^2\pm a^2}\right| + C$

32. $\int \frac{\sqrt{u^2+a^2}}{u}\,du =$
$\sqrt{u^2+a^2} - a\ln\left|\frac{a+\sqrt{u^2+a^2}}{u}\right| + C$

33. $\int \frac{\sqrt{u^2-a^2}}{u}\,du = \sqrt{u^2-a^2} - a\cos^{-1}\frac{a}{u} + C$

34. $\int \frac{\sqrt{u^2\pm a^2}}{u^2}\,du =$
$-\frac{\sqrt{u^2\pm a^2}}{u} + \ln\left|u+\sqrt{u^2\pm a^2}\right| + C$

35. $\int \frac{du}{\sqrt{u^2\pm a^2}} = \ln\left|u+\sqrt{u^2\pm a^2}\right| + C$

36. $\int \frac{du}{u\sqrt{u^2+a^2}} = -\frac{1}{a}\ln\left|\frac{a+\sqrt{u^2+a^2}}{u}\right| + C$

37. $\int \frac{du}{u\sqrt{u^2-a^2}} = \frac{1}{a}\cos^{-1}\left|\frac{a}{u}\right| + C$
$= \frac{1}{a}\sec^{-1}\left|\frac{u}{a}\right|$

38. $\int \frac{u^2\,du}{\sqrt{u^2\pm a^2}} = \frac{u}{2}\sqrt{u^2\pm a^2} \mp$
$\frac{a^2}{2}\ln\left|u+\sqrt{u^2\pm a^2}\right| + C$

39. $\int \frac{du}{u^2\sqrt{u^2\pm a^2}} = \mp\frac{\sqrt{u^2\pm a^2}}{a^2u} + C$

40. $\int \left(u^2\pm a^2\right)^{3/2}du = \frac{u}{4}\left(u^2\pm a^2\right)^{3/2} \pm$
$\frac{3a^2u}{8}\sqrt{u^2\pm a^2} + \frac{3a^4}{8}\ln\left|u+\sqrt{u^2\pm a^2}\right| + C$

41. $\int \frac{\left(u^2+a^2\right)^{3/2}}{u}\,du = \frac{1}{3}\left(u^2+a^2\right)^{3/2} +$
$a^2\sqrt{u^2+a^2} - a^3\ln\left|\frac{a+\sqrt{u^2+a^2}}{u}\right| + C$

42. $\int \frac{\left(u^2-a^2\right)^{3/2}}{u}\,du =$
$\frac{1}{3}\left(u^2-a^2\right)^{3/2} - a^2\sqrt{u^2-a^2} - a^3\sec^{-1}\frac{u}{a} + C$

43. $\int \frac{du}{\left(u^2\pm a^2\right)^{3/2}} = \frac{\pm u}{a^2\sqrt{u^2\pm a^2}} + C$

44. $\int \frac{du}{u\left(u^2+a^2\right)^{3/2}} =$
$\frac{1}{a^2\sqrt{u^2+a^2}} - \frac{1}{a^3}\ln\left|\frac{a+\sqrt{u^2+a^2}}{u}\right| + C$

45. $\int \frac{du}{u\left(u^2-a^2\right)^{3/2}} = \frac{-1}{a^2\sqrt{u^2-a^2}} - \frac{1}{|a^3|}\sec^{-1}\frac{u}{a} + C$

Forms Involving $a + bu$

46. $\int \frac{u\,du}{a+bu} = \frac{1}{b^2}(a+bu-a\ln|a+bu|) + C$

47. $\int \frac{u^2\,du}{a+bu} =$
$\frac{1}{2b^3}\left[(a+bu)^2 - 4a(a+bu) + 2a^2\ln|a+bu|\right] + C$

48. $\int \frac{du}{u(a+bu)} = \frac{1}{a}\ln\left|\frac{u}{a+bu}\right| + C$

49. $\displaystyle\int \frac{u\,du}{(a+bu)^2} = \frac{1}{b^2}\left(\frac{a}{a+bu} + \ln|a+bu|\right) + C$

50. $\displaystyle\int \frac{du}{u(a+bu)^2} = \frac{1}{a(a+bu)} - \frac{1}{a^2}\ln\left|\frac{a+bu}{u}\right| + C$

51. $\displaystyle\int u\sqrt{a+bu}\,du = \frac{2}{15b^2}(3bu-2a)(a+bu)^{3/2} + C$

52. $\displaystyle\int \frac{u\,du}{\sqrt{a+bu}} = \frac{1}{3b^2}(bu-2a)\sqrt{a+bu} + C$

53. $\displaystyle\int \frac{u^2\,du}{\sqrt{a+bu}} =$

$\displaystyle\frac{2}{15b^3}(8a^2+3b^2u^2-4abu)\sqrt{a+bu} + C$

54. $\displaystyle\int \frac{du}{u\sqrt{a+bu}}$

$$= \begin{cases} \dfrac{1}{\sqrt{a}}\ln\left|\dfrac{\sqrt{a+bu}-\sqrt{a}}{\sqrt{a+bu}+\sqrt{a}}\right| + C & \text{if } a > 0 \\ \dfrac{2}{\sqrt{-a}}\tan^{-1}\sqrt{\dfrac{a+bu}{-a}} + C, & \text{if } a < 0 \end{cases}$$

55. $\displaystyle\int \frac{\sqrt{a+bu}}{u}\,du = 2\sqrt{a+bu} + a\int \frac{du}{u\sqrt{a+bu}}$

Answers to Odd-Numbered Exercises

ANSWERS FOR CHAPTER 1

Exercise Set 1.1

1. 9 **3.** 7 **5.** 28 **7.** $-1/2$ **9.** 1

11. $\dfrac{x_1^2+x_1-2}{x_1-1}=x_1+2$ **13.** $2x_1+h$

15. **(a)** 29 liters/min; **(b)** 39 L/m; **(c)** 34 L/m **17.** **(a)** 30, **(b)** 27, **(c)** 25.5, **(d)** 24.75, **(e)** 24.3, **(f)** 24.15 **19.** See *Computer Programs* **21.** **(a)** 36.5 mi, **(b)** 43.8 mph

Exercise Set 1.2

1. 84 **3.** 11.9375 **5.** 181 **7.** 72.28125
9. 175.75 **11.** 30.85 **13.** See *Computer Programs*
15. \$2,480 **17.** 1.206 Ω

Exercise Set 1.3

1.

x	0.9	0.99	0.999	0.9999	1.0001	1.001	1.01	1.1
$f(x)=3x$	2.7	2.97	2.997	2.9997	3.0003	3.003	3.03	3.3

$\lim\limits_{x\to 1} 3x = 3$

3.

x	−1.1	−1.01	−1.001	−1.0001	−0.9999	−0.999	−0.99	−0.9
$h(x)=x^2+2$	3.21	3.0201	3.002	3.0002	2.9999	2.998	2.9801	2.81

$\lim\limits_{x\to -1}(x^2+2)=3$

5.

x	−0.1	−0.01	−0.001	−0.0001	0.0001	0.001	0.01	0.1
$f(x)=\dfrac{\tan x}{x}$	1.0033467	1.0000333	1.0000003	1	1	1.0000003	1.0000333	1.0033467

$\lim\limits_{x\to 0}\dfrac{\tan x}{x}=1$

7.

x	0.9	0.99	0.999	0.9999	1.0001	1.001	1.01	1.1
$h(x)=\dfrac{x}{x-1}$	−9	−99	−999	−9999	10001	1001	101	11

$\lim\limits_{x\to 1}\dfrac{x}{x-1}$ Does not exist

9. **(a)** 0, both sides of graph near $x=-1$ converge on 0; **(b)** Does not exist, graph goes toward -2 when $x<2$ and goes toward 1 for $x>2$; **(c)** 4, both sides of graph near $x=3$ converge on 4.

11.

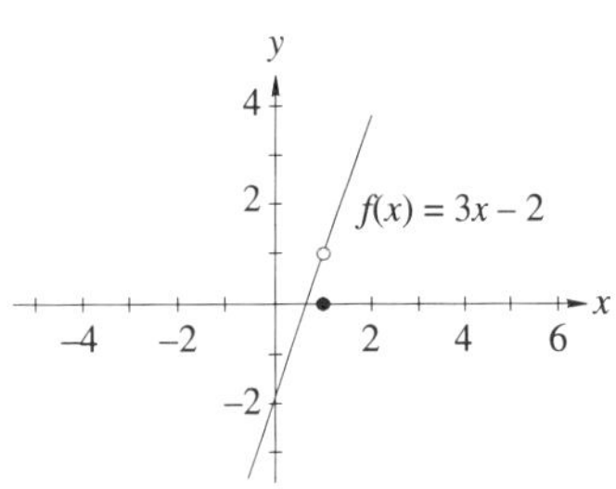

$\lim_{x\to 1} f(x)$ Does not exist

13.

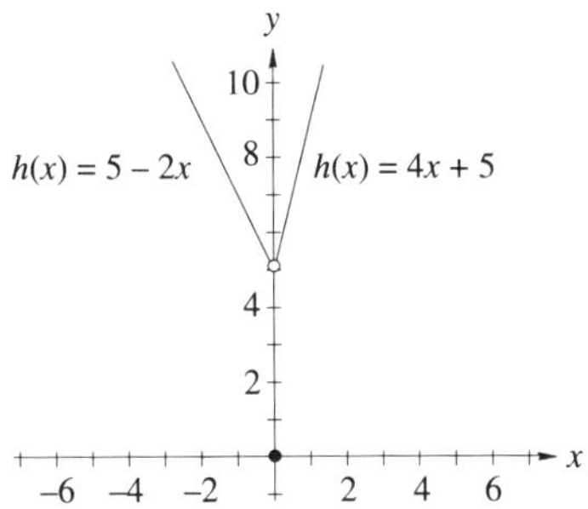

$\lim_{x\to 1} h(x)$ Does not exist

15.

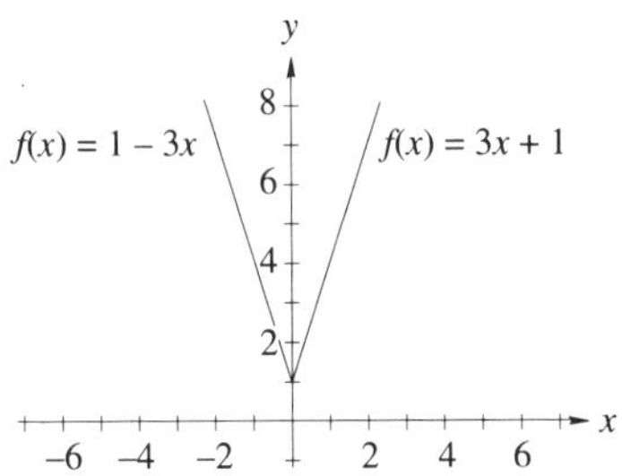

$\lim_{x\to 1} g(x) = 1$

17.

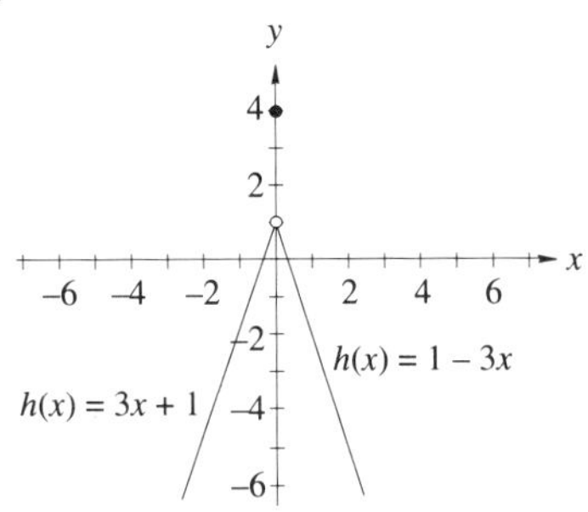

$\lim_{x\to 1} h(x)$ Does not exist

19. -15 **21.** -4 **23.** 8 **25.** 0 **27.** 4

Exercise Set 1.4

1. **(a)** -2; **(b)** 1; **(c)** 4; **(d)** 4 **3.** 3 **5.** ∞ **7.** -4
9. -1 **11.** -1 **13.** 0 **15.** 0 **17.** 0 **19.** 2
21. 0 **23.** $-\infty$ **25.** **(a)** 0.02381, **(b)** 0.0556, **(c)** 0
27. **(a)** \$4.75, **(b)** \$4.75, **(c)** \$4.75

Exercise Set 1.5

1. 8 **3.** 5 **5.** 1 **7.** 6 **9.** 9 **11.** 5
13. 8 **15.** 27 **17.** 1 **19.** $\sqrt{6}$ **21.** 2 **23.** -2
25. 2 **27.** $\frac{13}{9}$ **29.** 8 **31.** $\frac{2}{5}$ **33.** 0

Exercise Set 1.6

1. Continuous **3.** Not continuous, $\lim_{x\to 1-} h(x) = 4 \neq 3 = h(1)$ **5.** Not continuous, $k(2)$ is not defined **7.** Not continuous, $\lim_{x\to -2-} g(x) = 4 \neq g(-2)$
9. Not continuous, $\lim_{x\to 2-} j(x) = 0 = j(2)$ **11.** None
13. -1 **15.** $-3, 2$ **17.** None **19.** 2
21. $(-\infty,-3), (-3,\infty)$ **23.** $(-\infty,-2), (-2,2), (2,\infty)$
25. $[-3,\infty)$ **27.** $(-\infty,-3], [3,\infty)$ **29.** $(-\infty,0)$, $(0,\infty)$ **31.** **(a)** yes
(b)

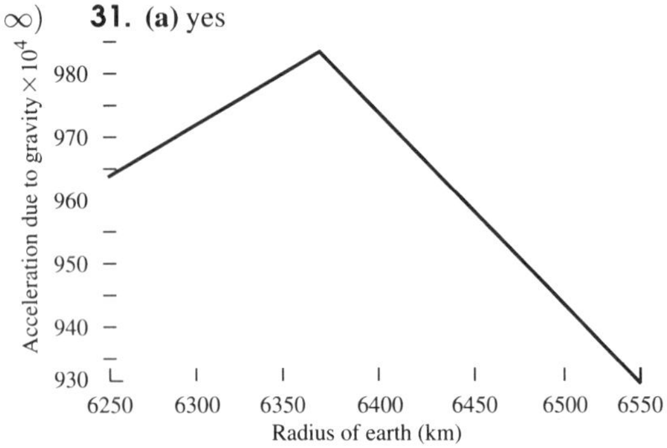

33. **(a)** \$3,585, **(b)** \$3,585, **(c)** \$3,585, **(d)** Yes, **(e)** \$14,155, **(f)** \$14,155, **(g)** \$14,155, **(h)** Yes

Review Exercises

1. 2 **3.** -10 **5.** $b+1$ **7.** 21 **9.** 65.96875
11. 0 **13.** 6 **15.** 3 **17.** 0 **19.** 0 **21.** 5
23. -3 **25.** $(-\infty,-7)(-7,\infty)$ **27.** $(-\infty,-5), (5,\infty)$

Chapter 1 Test

1. 2 **3.** 13 **5.** 2 **7.** $x = 2$ and $x = 3$

ANSWERS FOR CHAPTER 2

Exercise Set 2.1

1. 3 **3.** -3 **5.** $6x$ **7.** $4x$ **9.** $-8x$
11. $-12x^2$ **13.** $1-9t^2$ **15.** $\dfrac{-1}{(x+1)^2}$ **17.** $\dfrac{8x}{(1-x^2)^2}$
19. $32t-6$ **21.** $\dfrac{x}{\sqrt{x^2+4}}$ **23.** $\dfrac{1-x}{\sqrt{2x-x^2}}$ **25.** 3; 3
27. 98, 66, 2 **29.** 0, 3 **31.** $\dfrac{-10}{21^2} = \frac{-10}{441}, \frac{10}{441}$ **33.** $\frac{3}{4}$, $-\frac{3}{4}$ **35.** 0, $\frac{15}{64}$, $-\frac{20}{27}$

Exercise Set 2.2

1. 0 **3.** 7 **5.** $18x$ **7.** $5x^{14}$ **9.** $-10x^{-3}$ **11.** $-x^{1/2}$ **13.** $\frac{2\sqrt{3}}{\sqrt{x}}$ **15.** $-x^{-4/3} = \frac{-1}{x\sqrt[3]{x}}$ **17.** $18x+3$ **19.** $x^2+x-5-\frac{4}{3}x^{-3}$ **21.** $4x^4-\frac{9}{2}x^2-x^{-3}+2x^{-4}$ **23.** $12\sqrt{3}x^3+\frac{3\sqrt{5}}{2\sqrt{x}}+4x$ **25.** $32t-32$ **27.** $-8.0t+t^{-1/2}$ **29.** $\frac{\sqrt{6}}{2}t^{-1/2}-6t^{1/2}+\frac{2}{3}t^{-1/3}$ **31.** $\frac{35}{3}x^{4/3}$ **33.** $\frac{-8}{x^3}$ **35.** $-\frac{2}{3}x^{-7/6}$ **37.** $\frac{9}{2}x^{1/2}-7-3x^{-5/2}$ **39.** $f'(x)=3x^2+6x$, $f'(x)=0$ when $x=0$ or $x=-2$ **41.** $g'(x)=6x^2+2x-4$, $g'(x)=0$ when $x=-1$ or $x=2/3$ **43.** $j(x)=12t^2-\frac{1}{t^2}$, $j(t)=0$ when $t=\frac{1}{\sqrt[4]{12}}\approx 0.537285$ **45.** **(a)** 146,382 people, **(b)** 63.52 increase in population per month **47.** **(a)** 172 ft, **(b)** 71 ft/s **49.** **(a)** 133,912 people, **(b)** 1,493 people per year, **(c)** 139,886 people, **(d)** 1,494 people per year **51.** 120 W/A **53.** **(a)** $P(n)=R(n)-C(n)=(78n-0.025n^2)-(9{,}800+22.5n)=55.5n-0.025n^2-9{,}800$, **(b)** $P'(n)=55.5-0.05n$, **(c)** \$53/telephone **55.** **(a)** 605 ft above sea level, **(b)** $t=11$ s, **(c)** $h(2)=621$ ft, $v(2)=48$ ft/s, **(d)** $h(4)=861$ ft, $v(4)=192$ ft/s, **(e)** $h(6)=1{,}325$ ft, $v(6)=240$ ft/s, **(f)** $h(10)=1{,}005$ ft, $v(10)=-720$ ft/s, **(g)** $v(11)=-1{,}320$ ft/s

Exercise Set 2.3

1. $12x-19$ **3.** $18x^2-14x-8$ **5.** $-54x^2+12x+36$ **7.** $16t^3-72t^2+72t$ **9.** $(3w^3-4w^2+2w-5)(2w+w^{-2})+(9w^2-8w+2)(w^2-w^{-1})=15w^4-16w^3+6w^2-16w+4-5w^{-2}$ **11.** $\frac{(2x+3)(4)-(4x-1)(2)}{(2x+3)^2}=\frac{14}{(2x+3)^2}$ **13.** $\frac{(4x-1)(18x)-(9x^2+2)(4)}{(4x-1)^2}=\frac{36x^2-18x-4}{(4x-1)^2}$ **15.** $12s+\frac{1}{3s^3}$ **17.** $12t^2+\frac{4}{(t-2)^2}$ **19.** $2x+1$ **21.** $\frac{(3\phi^2-1)(2\phi)-\phi^2(6\phi)}{(3\phi^2-1)^2}$ **23.** $\frac{(x^3-3x-1)(9x^2-1)-(3x^3-x+1)(3x^2-3)}{(x^3-3x-1)^2}=\frac{-16x^3-12x^2+4}{(x^3-3x-1)^2}$ **25.** $\frac{t^{1/3}(6t-1)-(3t^2-t-1)\frac{1}{3}t^{-2/3}}{t^{2/3}}=\frac{5t^2-\frac{2}{3}t+1/3}{t^{4/3}}=\frac{15t^2-2t+1}{3t^{3/4}}$ **27.** $24w+\frac{2}{(w+1)^2}$ **29.** $\frac{2x}{(x^2+1)^2}$ **31.** $\frac{(2-x^2)(3x-x^3)(-3x^2)-(4-x^3)(5x^4-15x^2+6)}{(2-x^2)^2(3x-x^3)^2}=\frac{2x^7-20x^4-12x^3+60x^2-24}{(2-x^2)^2(3x-x^3)^2}$ **33.** $\frac{(s^2-1)(s-1)(6s^2)-2s^3[(s^2-1)+(s-1)(2s)]}{(s^2-1)^2(s-1)^2}=\frac{12s^5-10s^4-8s^3+6s^2}{(s^2-1)^2(s-1)^2}$ **35.** $\frac{(3x^3+1)(6x^2-22x+14)-(2x^3-11x^2+14x-3)(9x^2)}{(3x^3+1)^2}=\frac{33x^4-84x^3+33x^2-22x+14}{(3x^3+1)^2}$ **37.** $\frac{(t^2-1)}{(2t+1)}\left[\frac{(2t+1)-(t-1)(2)}{(2t+1)^2}\right]+\frac{(t-1)}{(2t+1)}\left[\frac{(2t+1)(2t)-(t^2-1)(2)}{(2t+1)^2}\right]=\frac{2t^3+3t^2-5}{(2t+1)^3}$ **39.** 16 **41.** Tangent: $y-2=-\frac{1}{2}(x-3)$ or $2y+x=7$ Normal: $y-2=2(x-3)$ or $y-2x-4$ **43.** Tangent: $y-3=-\frac{1}{3}(x-1)$ or $3y+x=10$ Normal: $y-3=3(x-1)$ or $y-3x=0$ **45.** Tangent: $y+2=\frac{1}{8}(x+2)$ or $8y-x=-14$ Normal: $y+2=-8(x+2)$ or $y+8x=-18$ **47.** 8.2 thousand people per year or 8,200 people per year **49.** $V'=\frac{6600}{(R+60)^2}$ **51.** **(a)** $N'(t)=\frac{9t^2+8t^{3/2}-6t-8\sqrt{t}-3}{2\left(3\sqrt{t}+2\right)^2\sqrt{t}}$, **(b)** 777 bacteria/hr

Exercise Set 2.4

1. $12(3x-6)^3$ **3.** $-25(5x-7)^{-6}$ **5.** $4(2x+3)(x^2+3x)^3$ **7.** $\frac{-4x}{(4x^2+7)^{3/2}}$ **9.** $3(t^4-t^3+2)^2(4t^3-3t^2)$ **11.** $3(u^3+2u^{-4})^2(3u^2-8u^{-5})$ **13.** $2[(x-2)(3x^2-x)]^2[(x-2)(6x-1)+(3x^2-x)]=2[(x-2)(3x^2-x)]^2(9x^2-14x+2)$ **15.** $6(v^2+1)^2(2v-5)^2+4v(v^2+1)(2v-5)^3$

17. $4\left[\left(x^3-6x^2\right)\left(5x-6x^2+x^3\right)\right]^3\left[\left(x^3-6x^2\right)\left(5-12x+3x^2\right)+\left(5x-6x^2+x^3\right)\left(3x^2-12x\right)\right]=4\left[\left(x^3-6x^2\right)\left(5x-6x^2+x^3\right)\right]^3\left(6x^5-60x^4+164x^3-90x^2\right)$

19. $\frac{(3x^2+2)^3(3x^2-7)}{\sqrt{x^3-7x}}+18x(3x^2+2)^2\sqrt{x^3-7x}=\frac{243x^8-945x^6-1{,}710x^4-732x^2-56}{\sqrt{x^3-7x}}$

21. $\dfrac{3(v^3-9)^2(v^2-4v)^2(2v-4)-(v^2-4v)(2)(3v^2)(v^3-9)}{(v^3-9)^4} = \dfrac{12v^7-96v^6+138v^5+540v^4-1728v^3+1728v^2}{(v^3-9)^3}$

23. $6(x^5+4)^5(5x^4)=30x^4(x^5+4)^5$
25. $4x(4x^2-5)^{-1/2}$ **27.** $(14x^2-6)(7x^3-9x)^{-1/3}$
29. $\dfrac{-6}{(x+3)^3}[\dfrac{1}{(x+3)^2}+1]^2$
31. $\frac{1}{3}[(x^3+4)^2-2(x^3+4)]^{-2/3}[2(x^3+4)-2](3x^2)=$
$2x^2\left[\left(x^3+4\right)^2-2\left(x^3+4\right)\right]^{-2/3}(x^3+3)$
33. $\left[16\left(3t^2-4t\right)^3-3\right](6t-4)=2{,}592t^7-12{,}096t^6+$
$20{,}736t^5-15{,}360t^4+4{,}096t^3-18t+12$
35. $6(9x^2+4x)^5(18x+4)$
37. $10(11x^5-2x+1)^9(55x^4-2)$ **39.** $\dfrac{-56(18x)}{(9x^2-4)^9}$
41. $\dfrac{3x}{\sqrt[4]{2x^2-5}}$ **43.** $\dfrac{-3072}{x^7}(\dfrac{128}{x^6}-1)$ **45.** $\dfrac{-480}{x^{11}}$
47. $y+8=192(x-2)$ or $192x-y=-394$
49. **(a)** $P=E'=\dfrac{dE}{dt}=144\left(1+4t^2\right)^2t$ W, **(b)** $P=E'=$
$\dfrac{dE}{dt}=\dfrac{144\left(1+4t^2\right)^2t}{1000}=0.144\left(1+4t^2\right)^2t$ kW,
(c) 243.36 kW **51.** **(a)** $P'(t)=\dfrac{250t\left(t^2+30\right)}{\left(t^2+15\right)^{3/2}}$,
(b) 271.8 ppm/day **53.** $R'=\dfrac{4n-4}{(n+1)^3}$
55. **(a)** $I'=-\dfrac{120X_C}{\left(900+X_C{}^2\right)^{3/2}}$, **(b)** -0.0504 A/Ω

Exercise Set 2.5

1. $-4/5$ **3.** $\dfrac{1}{2y}$ **5.** x/y **7.** $-9x/16y$
9. $\dfrac{4x}{2+y}$ **11.** $\dfrac{2x-y-5y^2}{x+10xy}$
13. $(2x+4y)/(1-4x-2y)$ **15.** $\dfrac{1+10x}{3+20y}$ **17.** $\dfrac{-8x}{3y^2}$
19. $\dfrac{y-3x^2}{18y^2-x}$ **21.** $\dfrac{1-2xy}{x^2}$ **23.** $-\dfrac{y^2}{x^2}$
25. $\dfrac{3(x+1)^{-2}-2xy+3(x+1)^{-2}}{x^2}=\dfrac{3-2xy(x+1)^2}{x^2(x+1)^2}$
27. $\dfrac{x^{-2}y-y+1}{x+x^{-1}}=\dfrac{y-x^2y+x^2}{x^3+x}$ **29.** $\dfrac{5+6xy^2-2x^3}{y-6x^2y}$
31. $\dfrac{2y\sqrt{x^2+y^2}+x^3}{2x\sqrt{x^2+y^2}-x^2y}$ **33.** $\dfrac{6x(x^2+y^2)^2}{1-6y(x^2+y^2)^2}$
35. $y+4=\frac{3}{4}(x-3)$ or $4y-3x+25=0$
37. $y-2=\frac{9}{16}(x+2)$ or $9x-16y+50=0$ **39.** tangent: 1; normal: -1 **41.** **(a)** $\dfrac{ds}{dt}=\dfrac{6\sqrt{st}-s}{8s\sqrt{st}+t}$, **(b)** 1.464 mi, **(c)** 0.399 mi/min = 23.94 mi/hr **43.** **(a)** 1.48. **(b)** The store should sell 1.48 television sets for each digital satellite that is sold.

Exercise Set 2.6

1. $24x-12$ **3.** $840x$ **5.** $2x^{-3}$ **7.** $2+\dfrac{2}{(t+1)^3}$
9. $-6u^{-4}+24u^{-5}=\dfrac{-6u+24}{u^5}$ **11.** $-\frac{1}{4}(x+1)^{-3/2}$
13. 48 **15.** $\frac{10}{9}x^{-8/3}$
17. $24x(x+1)^{-5}-24(x+1)^{-4}=-24(x+1)^{-5}$
19. $216w^2(3w^2+1)^{-4}-12(3w^2+1)^{-3}=$
$(180w^2-12)(3w^2+1)^{-4}$. **21.** $840x^3-480x$
23. $24x^{-5}$ **25.** $\dfrac{18x^3+18x^2+6x}{(3x-1)^3}$
27. $\frac{3}{4}x(x+1)^{-5/2}-(x+1)^{-3/2}=\dfrac{-x-4}{4(x+1)^{5/2}}$
29. $y'=\dfrac{2x-1}{8y}$, $y''=\dfrac{16y^2-(2x-1)^2}{64y^3}$
31. $y'=\dfrac{4}{2xy-1}$; $y''=\dfrac{-8(4x+2xy^2-y)}{(2xy-1)^2}$
33. $y'=\dfrac{-2y}{x}$; $y"=\dfrac{6y}{x^2}$ **35.** **(a)** $N'(t)=10t-0.2t^{3/2}$,
(b) $N''(t)=10-0.3t^{1/2}$ **37.** **(a)** $h'(t)=120-24t$,
(b) $h''(t)=-24$

Review Exercises

1. $4x-1$ **3.** $1-6x^2$ **5.** $\dfrac{1}{2\sqrt{x}}$ **7.** $6x+2$
9. $15x^4-\dfrac{12}{x^4}+\dfrac{3}{2\sqrt{x}}$ **11.** $\dfrac{3}{(x+3)^2}$
13. $(x-4)(3x^2-3)+(x^3-3x)=9x^2-30x+12$
15. $\dfrac{1}{2\sqrt{x+1}}$ **17.** $6x^5-10x^4+12x^3-18x^2-8x+8$
19. $\dfrac{4x^3+15x^2+7}{2\sqrt{(x^3+7)(x+5)}}$
21. $(x^2+5x)^2(-6x^2)(x^3+1)^{-3}+2(x^2+5x)(2x+5)(x^3+1)^{-2}=\dfrac{-2x^6-30x^5-100x^4+4x^3+30x^2+50x}{(x^3+1)^3}$
23. $3(x+1)^2(x-1)^2+2(x+1)(x-1)^3=$
$(x-1)^2(x+1)(5x+1)$ **25.** $-30x(x^2+1)^{-4}$
27. $2x(x^2-2)^{-2/3}-3x^3(x^2-2)^{-5/2}=$
$(x^2-2)^{-5/2}(-x^3-4x)$
29. $-4(u^2+\dfrac{1}{u}-\dfrac{4}{u^2})^{-5}(2u-u^{-2}+8u^{-3})$

31. $6x^5 + 12x^3 + 4x$ **33.** $-\frac{1}{2}$ **35.** $\dfrac{2y - 2x - y^2}{2xy - 2x}$

37. $\dfrac{y^2 - 2xy^5}{3x^2y^4 - 1}$ **39.** $42x^5 + 24x^2$ **41.** $24 - \dfrac{12}{x^4}$

43. $42 - 420x^{-6}$ **45.** $30 + \frac{3}{8}x^{-5/2}$

47. $\dfrac{6x^2y - 8x + 2y}{(x^2 - 1)^2}$ **49.** $y' = \dfrac{2x - y}{x - 2y}$; $y'' = \dfrac{2x^3 - 8x^2y + 4x^2 + 8xy^2 + 2xy - 2y^2}{(x - 2y)^3}$ **51.** 18 2/9

53. Tangent $\frac{17}{6}$; Normal $-\frac{6}{17}$ **55.** 0, −2

57. $y - 2 = \frac{5}{4}(x + 2)$ or $4y - 5x - 18 = 0$ **59.** $\pm\dfrac{\sqrt{3}}{3}$

Chapter 2 Test

1. $15x^2$ **3.** $2x^5(x^6 - 2)$ **5.** $\dfrac{2 - 4x}{(4x^2 + 1)^{1/2}}$

7. $y' = \dfrac{3x^2 - 2y - y^3}{2x + 3y^2x}$ **9.** $v(t) = 9t + 27$

ANSWERS FOR CHAPTER 3

Exercise Set 3.1

1. $v(t) = 6t - 12$; $a(t) = 6$

t	0	1	2	3	4	5
$s(t)$	5	−4	−7	−4	5	20

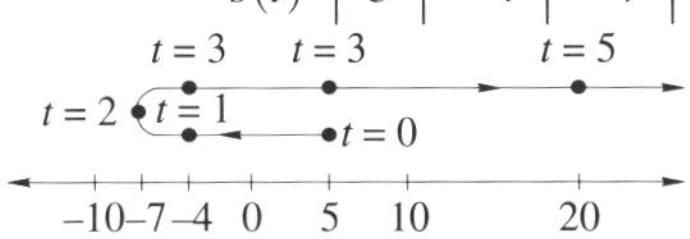

3. $v(t) = 3t^3 - 12$; $a(t) = 6t$

t	−4	−3	−2	−1	0	1	2	3	4
$s(t)$	−14	11	18	13	2	−9	−14	−7	18
$v(t)$	36	15	0	−9	−12	−9	0	15	36
$a(t)$	−24	−18	−12	−6	0	6	12	18	24

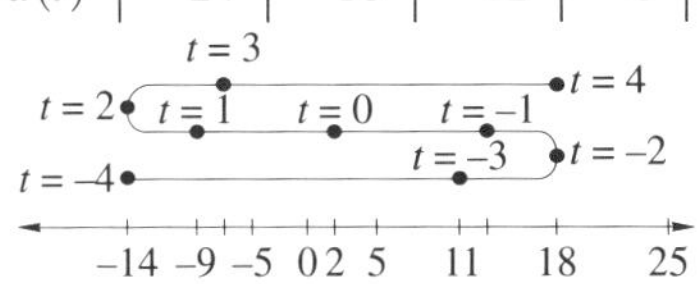

5. $v(t) = 1 - 4/t^2$; $a(t) = 8/t^3$

t	1	2	3	4
$s(t)$	5	4	17/3	5
$v(t)$	−3	0	0.5555	0.75
$a(t)$	8	1	0.2963	0.125

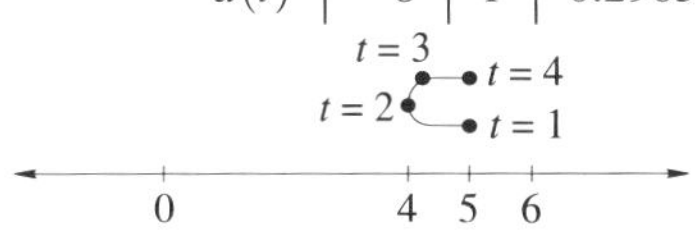

7. $v(t) = t^{-1/2} - \frac{1}{2}t^{-3/2}$; $a(t) = -\frac{1}{2}(t^{-3/2} - \frac{3}{4}t^{-5/2})$

t	1	2	3	4
$s(t)$	3	3.5355	4.0415	4.5
$v(t)$	0.5	0.5303	0.4811	0.4375
$a(t)$	0.25	−0.0442	−0.0481	−0.0391

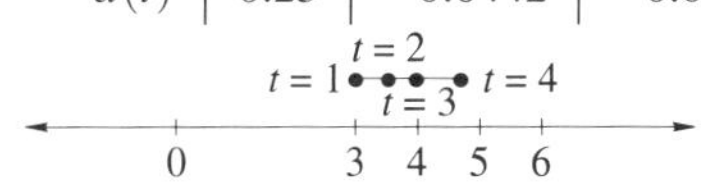

9. **(a)** 9 s; **(b)** $v = 144$ ft/s downward; $a = 32$ ft/s² downward; **(c)** 0 ft (ground level) **11.** **(a)** 6 s; **(b)** $v - 29.4$ m/s; $a = -9.8$ m/s²; **(c)** 0 m **13.** **(a)** $\dfrac{dF}{ds} = F(s) = \dfrac{-100}{s^3}$ dynes/cm; **(b)** $\frac{-100}{27}$ dynes/cm **15.** **(a)** $\frac{9}{5}$; **(b)** $\frac{5}{9}$; **(c)** $\frac{5}{9}$ **17.** **(a)** $\dfrac{dQ}{dt} = -50 + t$ gallons/min; **(b)** −46 gal/min; **(c)** −42 gal/min **19.** **(a)** 4.0 m/s; **(b)** 8.0 m; **(c)** 4.0 s **21.** **(a)** $v = -N\dfrac{d\phi}{dt} = -\dfrac{60}{t^{1/2}} + 40t$; **(b)** −20 V at $t = 1.0$ s and 340 V at $t = 9.0$ s **23.** 46.24π m²/s **25.** **(a)** $P = \dfrac{6,41}{V}$; **(b)** $\dfrac{dP}{dV} = -\dfrac{6.41}{V^2}$ **27.** **(a)** $v(t) = s'(t) = 6t - 0.05t^2$, **(b)** $a(t) = v'(t) = s''(t) = 6 - 0.1t$, **(c)** $|v(t)| = |6t - 0.05t^2|$

Exercise Set 3.2

1. Critical values: $x = 4$; increasing on $(4, \infty)$; decreasing on $(-\infty, 4)$; Maximum: none; Minimum: $f(4) = -16$
3. Critical values: $x = 0$; increasing on $(-\infty, 0)$; decreasing on $(0, \infty)$; Maximum: $m(0) = 16$; Minimum: none
5. Critical values: $x = 2$; increasing on $(-\infty, 2)$; decreasing on $(2, \infty)$; Maximum: $j(2) = 31$; Minimum: none
7. Critical values: $x = 2$; increasing on $(2, \infty)$; decreasing on $(-\infty, 2)$; Maximum: none; Minimum: $f(2) = -15$
9. Critical values: $t = \pm 1$; increasing on $(-\infty, -1)$ and $(1, \infty)$; decreasing on $(-1, 1)$; Maximum: $q(-1) = 2$; Minimum: $q(1) = -2$ **11.** Critical values: $x = 0, \pm 6$;

increasing on $(-6,0)$ and $(6,\infty)$; decreasing on $(-\infty,-6)$ and $(0,6)$; Maximum: $y(0)=16$; Minimum: $y(-6)=y(6)=-308$ **13.** Critical values: $z=-4$; increasing on $(-\infty,\infty)$ decreasing nowhere; Max: none; Minimum: none. **15.** Critical values: $v=-1,0,2$; increasing on $(-1,0)$ and $(2,\infty)$; decreasing on $(-\infty,-1)$ and $(0,2)$; Maximum: $g(0)=12$; Minimum: $g(-1)=7$ and $g(2)=-20$. **17.** Critical values: none; increasing nowhere; decreasing on $(-\infty,-1)$ and $(-1,\infty)$; Maximum: none; Minimum: none **19.** Critical values: $x=-1$; increasing on $(-1,0)$ and $(0,\infty)$; decreasing on $(-\infty,-1)$; Maximum: none; Minimum: $g(-1)=3$ **21.** Critical values: $x=0$; increasing on $(0,1)$ and $(1,2)$; decreasing on $(-\infty,0)$ and $(2,\infty)$; Maximum: $f(2)=-4$; Minimum: $f(0)=0$ **23.** Critical values: $t=\pm1,\pm\frac{\sqrt{2}}{2}$; increasing on $\left(-\frac{\sqrt{2}}{2},\frac{\sqrt{2}}{2}\right)$; decreasing on $\left(-1,\frac{-\sqrt{2}}{2}\right)$ and $\left(\frac{\sqrt{2}}{2},1\right)$; Maximum: $f\left(\frac{\sqrt{2}}{2}\right)=0.5$; Minimum: $f\left(-\frac{\sqrt{2}}{2}\right)=-0.5$
25. Critical values: $x=0.6,1$; increasing on $(-\infty,0.6)$ and $(1,\infty)$; decreasing on $(0.6,1)$; Maximum: $(0.6,0.326)$; Minimum: $(1,0)$ **27.** **(a)** 8 s; **(b)** 1274 ft
29. **(a)** $P=0.0$ W; **(b)** $t=2.0$ s **31.** **(a)** minimum when $R_2=0$; no maximum; **(b)** minimum $R=0$
33. **(a)** $P'(n)=-n+8.6$, **(b)** 8.6, **(c)** \$6,498
35. **(a)** $N'(t)=2t^{-1/2}+3t^{1/2}-2t^{3/2}$, **(b)** $t=2$ mo

Exercise Set 3.3

1. Maximum: $(2,12)$, Concave downward: $(-\infty,\infty)$

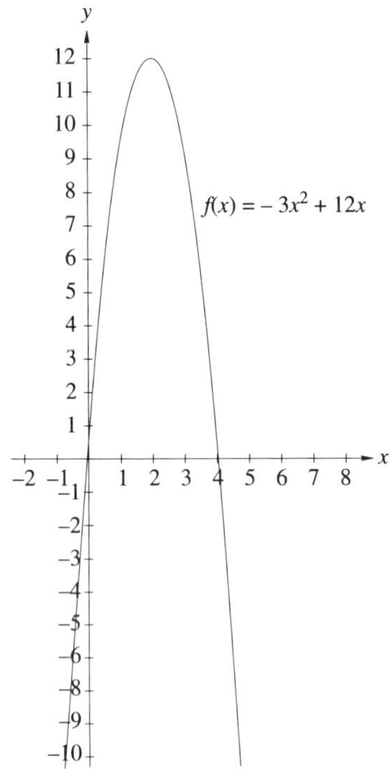

3. Maximum: $(-2,28)$; Minimum: $(2,-4)$; Concave upward: $(0,\infty)$; Concave downward: $(-\infty,0)$: Inflection point: $(0,12)$

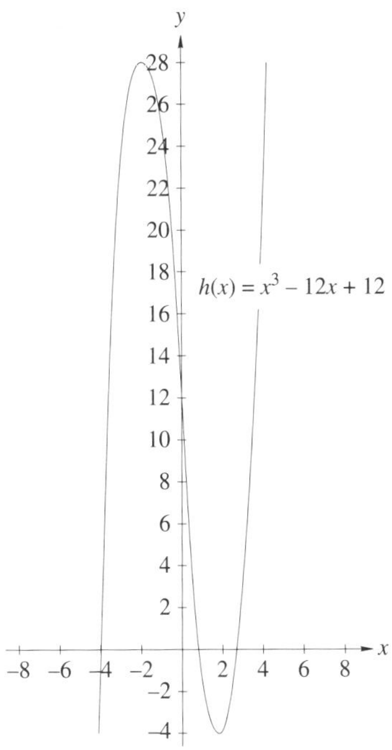

5. Maximum: none; Minimum: none; Concave upward: $(2, \infty)$, Concave downward: $(-\infty, 2)$: Inflection point: $(2, 4)$

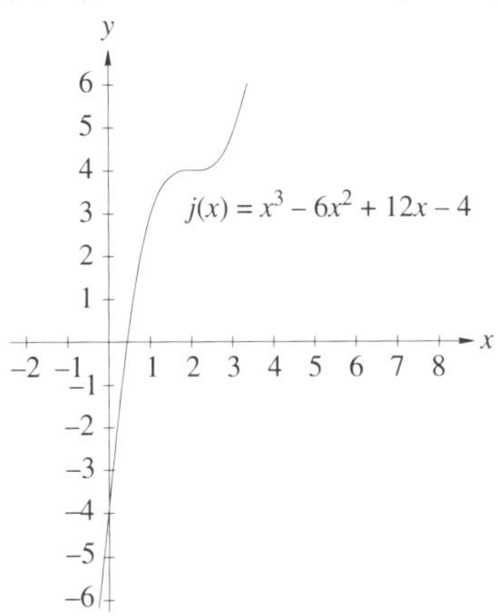

7. Maximum: none, Minimum: none; Concave upward: $\left(-\infty, \frac{1}{2}\right)$; Concave downward: $\left(\frac{1}{2}, \infty\right)$; Inflection point: $\left(\frac{1}{2}, -\frac{1}{2}\right)$

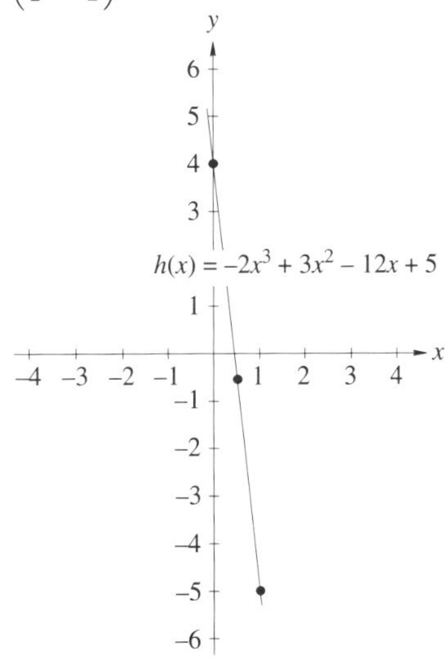

9. Maximum: $(-2, 10.67)$ and $(1, 1.67)$; Minimum: $(0, 0)$; Concave upward: $(-1.22, 0.55)$; Concave downward: $(-\infty, -1.22)$ and $(0.55, \infty)$; Inflection point: $(-1.22, 6.16)$ and $(0.55, 0.90)$

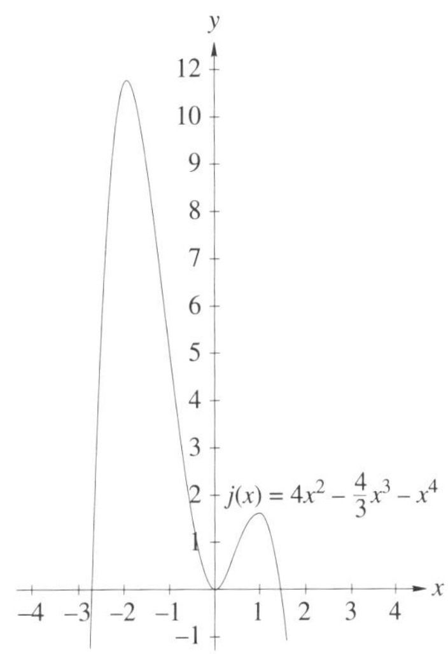

11. Maximum: none; Minimum: none; Concave upward: $(-\infty, 1)$; Concave downward: $(1, \infty)$; Inflection point: $(1, 0)$

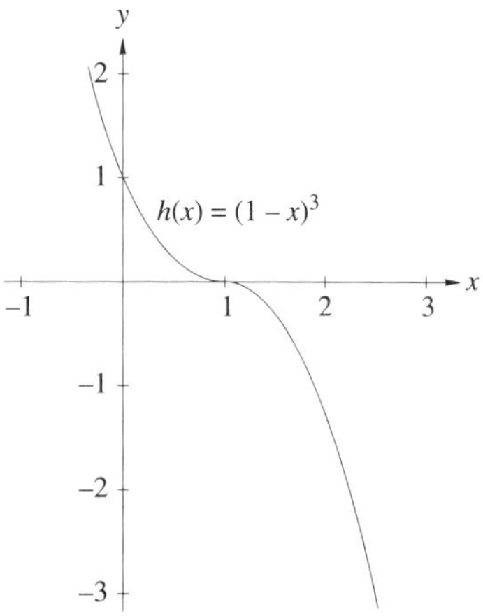

13. Maximum: none; Minimum: none; Concave upward: $(-\infty, 0)$; Concave downward: $(0, \infty)$; Inflection point: $(0, 2)$

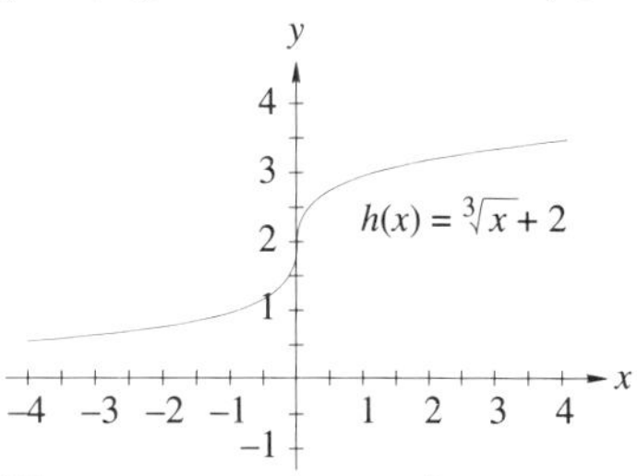

15. Maximum: $(-2, -4)$; Minimum: $(2, 4)$; Concave upward: $(0, \infty)$; Concave downward: $(-\infty, 0)$; Inflection point: none

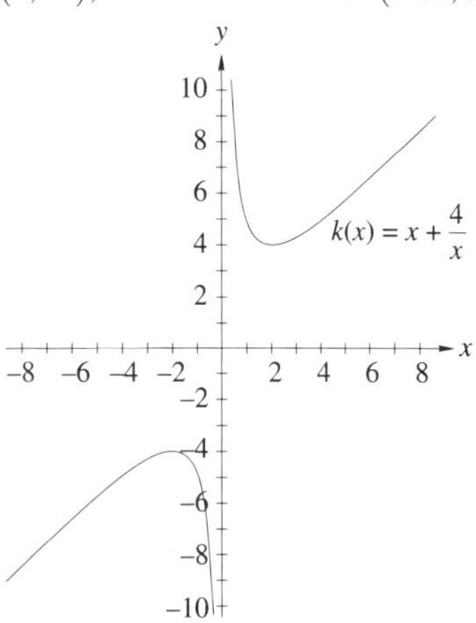

17. Maximum: none; Minimum: $(-1, 3)$; Concave upward: $(-\infty, 0)$ and $\left(\sqrt[3]{2}, \infty\right) \approx (1.26, \infty)$; Concave downward: $\left(0, \sqrt[3]{2}\right) \approx (0, 1.26)$; Inflection point: $\left(\sqrt[3]{2}, 0\right) \approx (1.26, 0)$

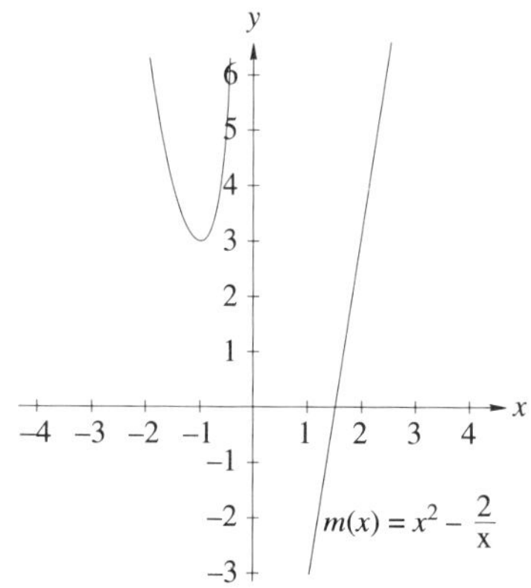

19. Maximum: $(-2, 7.333)$; Minimum: $(3, -13.5)$; Concave upward: $\left(\frac{1}{2}, \infty\right)$; Concave downward: $\left(-\infty, \frac{1}{2}\right)$; Inflection point: $(0.5, -3.083)$

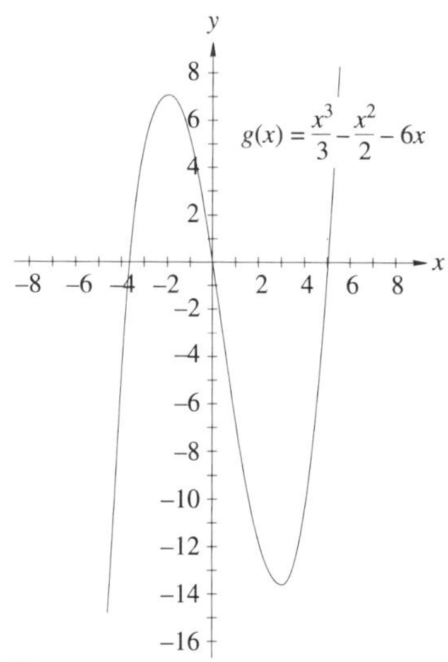

21. Maximum: $(1, 1)$; Minimum: $(0, 0)$; Concave upward: nowhere; Concave downward: $(0, \infty)$; Inflection point: none

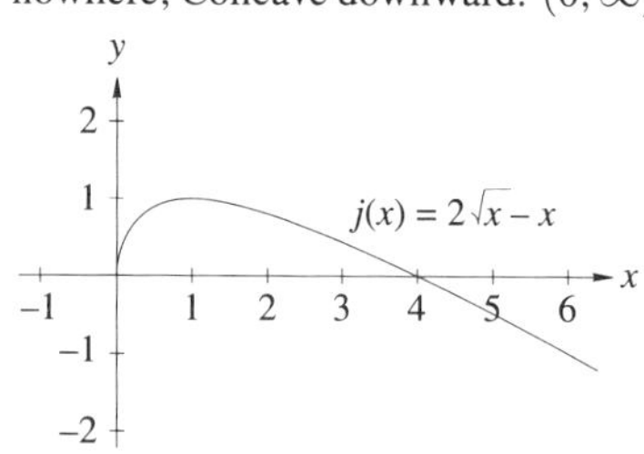

23. Maximum: none; Minimum: $(0, 2)$; Concave upward: nowhere, Concave downward: $(-\infty, \infty)$; Inflection point: none

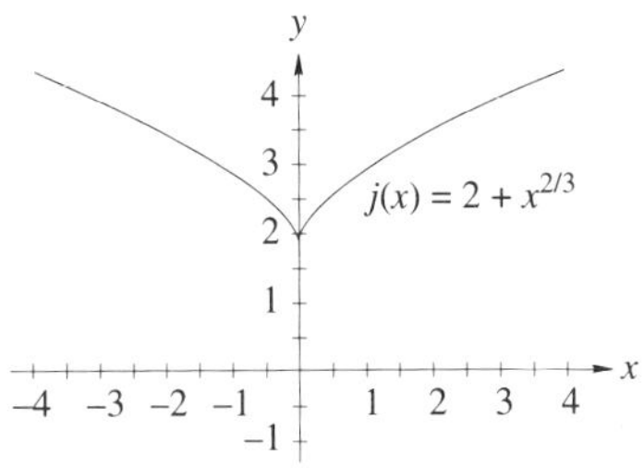

25. Maximum: $(0,0)$; Minimum: $(4.-15.119)$; Concave upward: $(-\infty,\infty)$; Concave downward: nowhere; Inflection point: none

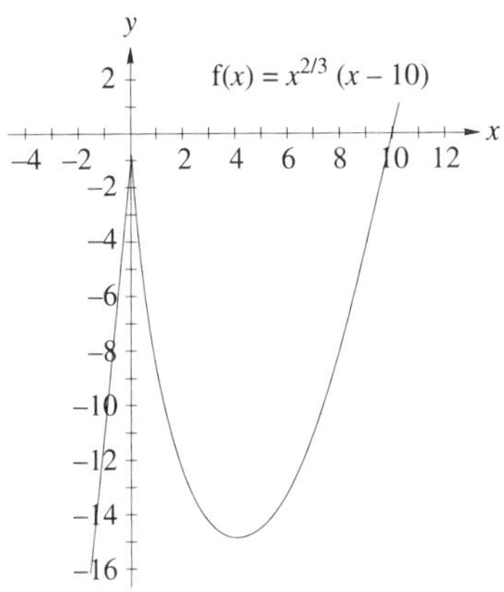

27. Maximum: $\left(1,\frac{1}{2}\right)$; Minimum: $(-1,-\frac{1}{2})$; Concave upward: $(-\sqrt{3},0)$, $(-\sqrt{3},0)$, $(\sqrt{3},\infty)$; Concave downward: $(-\infty,-\sqrt{3})$, $(0,\sqrt{3})$; Inflection points: $(-\sqrt{3},-\sqrt{3/4})$, $(\sqrt{3},\sqrt{3/4})$

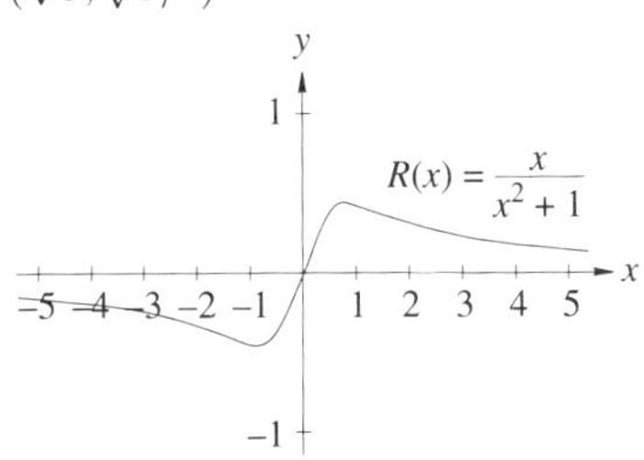

29. $t = 23$ **31.** **(a)** $N(9) = 116$, **(b)** $N'(4) = 8$ crimes/mo

Exercise Set 3.4

1. 2500 ft $\times$5000 ft; area $= 12{,}500{,}000\text{ ft}^2$ **3.** **(a)** 2000 ft along river $\times$ 3000 ft; **(b)** cost is \$12,000 **5.** 4 cm **7.** $2\sqrt{2} \approx 2.83$ units horizontally and $\frac{3}{2}\sqrt{2} \approx 2.12$ units vertically **9.** $4'' \times 4'' \times 2''$ **11.** radius $= 5.05$ cm, height $= 10.1$ cm **13.** 21.984 cm $\times$26.381 cm **15.** **(a)** $w = \sqrt{300} \approx 17.32$ in.; $d = \sqrt{600} \approx 24.49$ in. **(b)** $w = 2r\frac{\sqrt{3}}{3}$ in.; $d = 2r\frac{\sqrt{6}}{3}$ in. **17.** 80 mm $\times$160 mm **19.** 1.25 A **21.** $A = \sqrt{c/b}$ **23.** 2.24 m

25. **(a)** $t = 4.5\,s$; **(b)** $P = 0.0\,W$ **27.** about 3.1 m from the less intense light **29.** $0.578\,L$ **31.** $\frac{h}{d} = \frac{3\pi+2}{4}$ **33.** $12 - \frac{5}{3}\sqrt{3} \approx 9.11$ mi **35.** **(a)** 56.2×74.7 ft, **(b)** \$52,290

Exercise Set 3.5

1. $\frac{400}{3} = 133\frac{1}{3}\text{ cm}^2/\text{min}$ **3.** $\frac{4\sqrt{5}}{5} \approx 1.789$ ft/min **5.** 2.9104 km/min $=$ 174.6 km/h **7.** **(a)** $\frac{5}{8\pi} \approx 0.1989$ ft/min; **(b)** $\frac{10}{21\pi} \approx 0.1516$ ft/min; **(c)** $\frac{5}{12\pi} \approx 0.1326$ ft/min **9.** 12.4 mm/min **11.** $15\text{ cm}^2/\text{s}$ **13.** $11.25\text{ cm}^2/\text{s}$ **15.** $24\text{ mm}^2/\text{s}$ **17.** -8.3% of the volume per hour **19.** 387.2 m/s or 1394 km/h **21.** **(a)** -18.0 m/s^2; **(b)** -18.0 m/s^2 **23.** **(a)** Either 5.75 mi west and 5.56 mi north of A or 5.75 mi west and 5.56 mi south of A. **(b)** 14.86 mph **(c)** $\theta = 51.2°$ north of east or 38.8° east of north **25.** **(a)** $\frac{50}{7} \approx 7.143$ ft/s; **(b)** 7.143 ft/s; **(c)** 7.143 ft/s **27.** $0.0064\,\Omega/\text{s}$ **29.** $0.86\,\Omega/\text{min}$ **31.** decreasing $0.04725\,\Omega/\text{s}$ **33.** $-2.04\,\Omega/\text{min}$ **35.** **(a)** 218.47 Hz; **(b)** 20.17 Hz/s **37.** 0.06375 mm/min **39.** $-1.06\,\Omega/\text{s}$ **41.** $4.30\,\Omega/\text{s}$ **43.** $0.1\pi\text{ in.}^2/\text{yr} \approx 0.314\text{ in.}^2/\text{yr}$

Exercise Set 3.6

1. 1.2289 **3.** -1.2838; 1.5338 **5.** -3.3437; 3.3437 **7.** 2.5649 **9.** 1.1066 rad $\approx 63.40°$

Exercise Set 3.7

1. $dy = (4x^3 - 2x)dx$ **3.** $dy = (15x^2 - 2x + 1)dx$ **5.** $dy = \frac{-2}{3}(4-2x)^{-2/3}dx$ **7.** $dy = 1.0, \Delta y = 1.04$ **9.** $dy = 3.9, \Delta y = 3.9675$ **11.** $dy = -0.0074074, \Delta y = -0.0065195$ **13.** **(a)** $\frac{16}{3}\pi\text{ m}^3$; **(b)** $\frac{16}{300}\pi m^3 = \frac{4}{75}\pi\text{ m}^3$; **(c)** 1% **15.** **(a)** 0.96 in^2; **(b)** $0.00167 = 0.167\%$ **17.** $4\pi\text{ m}^2$ **19.** **(a)** $\pm$ 24.2 m^3, **(b)** $\pm 2.42\%$ **21.** $0.08\sigma T^4$ **23.** -0.03 A **25.** $\pm 2.18\,\Omega$ **27.** 1.0% **29.** $\pm 0.07\text{ mm}^3$ **31.** $450\pi \approx 1414\text{ mm}^3$ **33.** \$132.50 **35.** $19.2\pi \approx 60.3\text{ cm}^3$.

Exercise Set 3.8

1. $7x + C$ **3.** $2x - x^3 + C$ **5.** $x^4 - x^3 + x^2 + 9x + C$ **7.** $-\frac{1}{3}x^{-3} - x^{-2} + x^{-1} + 5x + C$ **9.** $\frac{2}{3}x^{3/2} + \frac{1}{2}x^2 + \frac{1}{x} + C$ **11.** $-5x^{-0.4} - 0.7x^{-5} - 4x^{-0.3} + C$ **13.** $\frac{1}{3}t^3 + t^2 + C$ **15.** $21t^2 - 5t + C$ **17.** **(a)** 6.12 s; **(b)** 195.96 ft/s **19.** **(a)** 5 s; **(b)** 525 ft **21.** **(a)** 17 s; **(b)** 1904 ft **23.** 160 m **25.** -2.5 m/s^2 **27.** **(a)** 3.9 rev; **(b)** 4.6 rev **29.** **(a)** $q = 2.2t^2 - 0.7t^3 + 5.0$; **(b)** 4.6 C **31.** **(a)** $\phi(t) = \frac{-1}{200}(t^2 - 3t^{4/3} - 4.0)$;

(b) $\phi(0.729) = 0.027\,\text{Wb}$ **33.** $f(x) = -2\sqrt{x} + 1$
35. $T = 2875(x+1)^{-2} + 25$

Review Exercises

1. Critical values: $0, \pm\frac{\sqrt{2}}{2}$; Maximum: $(0,0)$; Minimum: $\left(-\frac{\sqrt{2}}{2}, -0.25\right)$, $\left(\frac{\sqrt{2}}{2}, -0.457\right)$; Inflection point: $\left(\pm\frac{\sqrt{6}}{6}, -0.139\right)$; Concave upward: $(-\infty, -\frac{\sqrt{6}}{6})$, $(\frac{\sqrt{6}}{6}, \infty)$; Concave downward: $(-\frac{\sqrt{6}}{6}, \frac{\sqrt{6}}{6})$

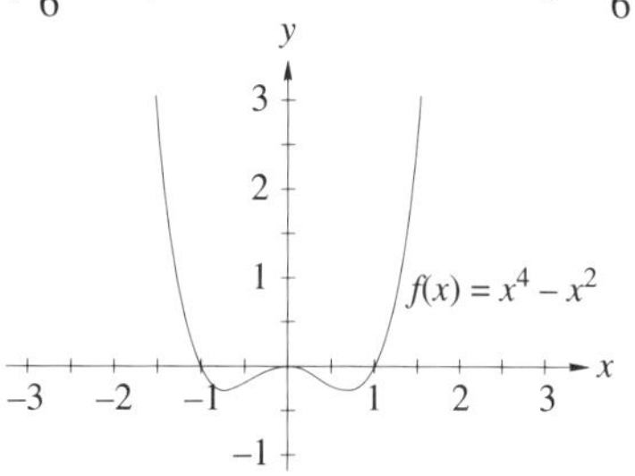

3. Critical values: $0, \pm 3$; Maximum: $(\pm 3, 81)$; Minimum: $(0,0)$; I.P. $(\pm\frac{\sqrt{3}}{3}, 5.889)$; Concave upward: $(-\frac{\sqrt{3}}{3}, \frac{\sqrt{3}}{3})$; Concave downward: $(-\infty, -\frac{\sqrt{3}}{3})$; $(\frac{\sqrt{3}}{3}, \infty)$

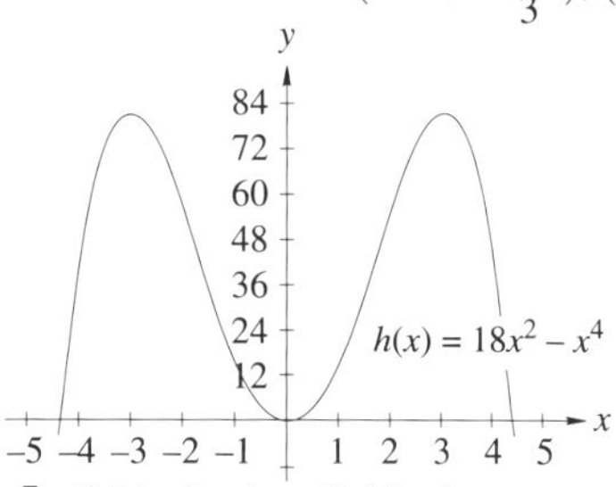

5. Critical value: 0; Maximum: none; Minimum: none; Inflection point: $(0,0)$; Concave upward: $(-\infty, 0)$; Concave downward: $(0, \infty)$

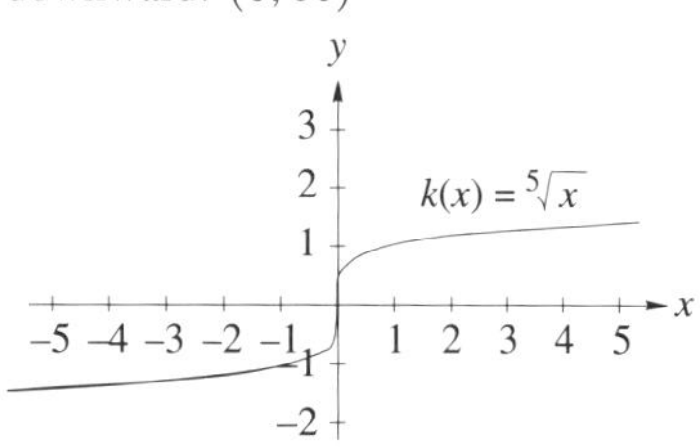

7. Critical value: 0; Maximum: $(0, -\frac{1}{4})$; Minimum: none; Inflection point: none; Asymptotes: $x = \pm 2$; $y = 1$; Concave upward: $(-\infty, -2)$, $(2, \infty)$; Concave downward: $(-2, 2)$

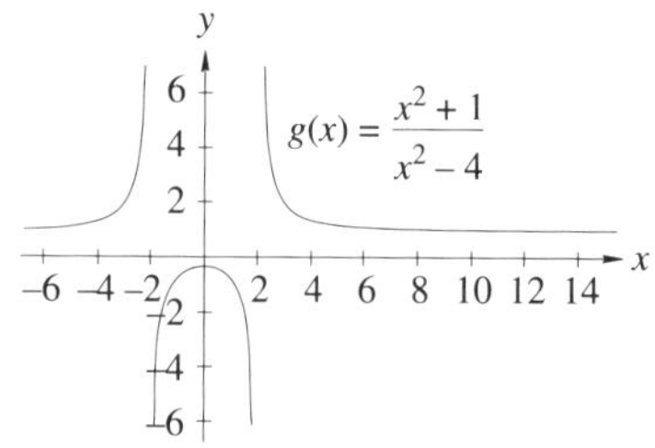

9. Critical values: none; Maximum: none; Minimum: none; Inflection point: $(-0.3275, 0.1475)$; Asymptotes: $x = -2$, $x = 1$, $y = 0$; Concave upward: $(-2, -0.3275)$, $(1, \infty)$; Concave downward: $(-\infty, -2)$, $(-0.3275, 1)$

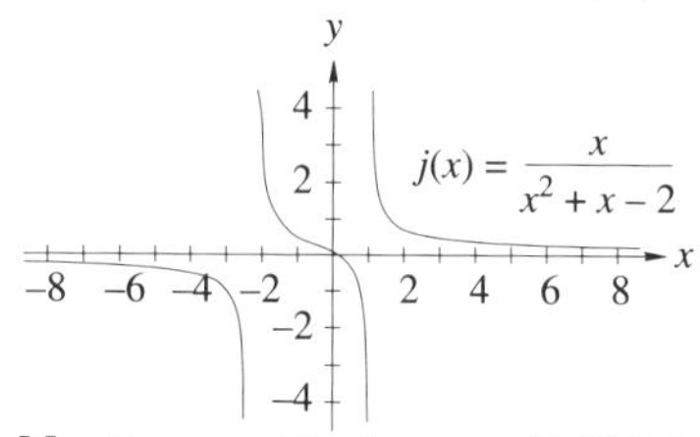

11. distance: Maximum at $(0.056, 0.028)$; Minimum at $(4, -76)$ and $(-4, -212)$; velocity: Maximum at $(4, -23)$ and $(-4, 121)$; Minimum at $(3, -26)$; acceleration: Minimum at $(-4, -42)$ and maximum at $(4, 6)$ **13.** $x^3 - 2x^2 + C$
15. $\frac{2}{3}x^{3/2} + \frac{1}{3}x^3 + \frac{3}{x} + C$ **17.** **(a)** $v(t) = 288 - 32t$; $a(t) = -32$; **(b)** height: 896 ft; velocity: 160 ft/s upward; acceleration: $-32\,\text{ft/s}^2$; **(c)** 1296 ft; **(d)** 9 s; **(e)** 18 s; **(f)** 288 ft/s downward **19.** 56.98 mph **21.** $300 \times 300 = 90000\,\text{m}^2$
23. **(a)** $0.09\pi \approx 0.2827\,\text{m}^3$; **(b)** $0.12\pi \approx 0.3770\,\text{m}^2$
25. 13.5 V **27.** $r = \sqrt[3]{\frac{10}{\pi}}\,\text{m} \approx 1.4710\,\text{m}$, $h = 2\sqrt[3]{\frac{10}{\pi}}\,\text{m} \approx 2.942\,\text{m}$ **29.** **(a)** 784 ft; **(b)** 244 ft/s

Chapter 3 Test

1. **(a)** $x = -2, 0$, and 2; **(b)** Maximum $(0,0)$, Minimum $(-2, -16)$ and $(2, -16)$, **(c)** Concave upward: $\left(-\infty, -\sqrt{\frac{4}{3}}\right)$ and $\left(\sqrt{\frac{4}{3}}, \infty\right)$, **(d)** Concave downward: $\left(-\sqrt{\frac{4}{3}}, \sqrt{\frac{4}{3}}\right)$

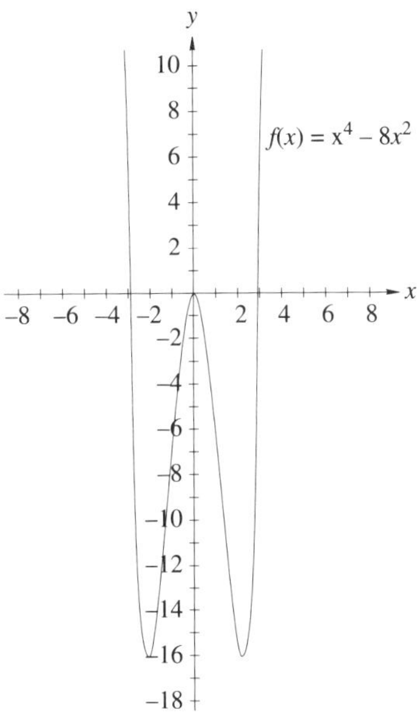

3. 2.646

5. **(a)** $v(t) = 3t^2 - 24t$, $a(t) = 6t - 24$; **(b)** Position: Maximum at $(0, 5)$, Minimum at $(4, -123)$, Velocity: Maximum at $(-2, 60)$, Minimum at $(4, -48)$; Acceleration: Maximum at $(4, 0)$, Minimum at $(-2, -36)$ **7.** 3.38 m

ANSWERS FOR CHAPTER 4

Exercise Set 4.1

1. 4 **3.** 28 **5.** 4 **7.** -6 **9.** $\frac{2}{3}$ **11.** $\frac{22}{3}$
13. 20 **15.** 72 **17.** $\frac{14}{3}$

Exercise Set 4.2

1. 30 **3.** 24 **5.** $33\frac{1}{3}$ **7.** 6 **9.** 18
11. -18 **13.** -16 **15.** 18 **17.** 20 **19.** -30
21. $\frac{37}{3}$ **23.** $-\frac{37}{3}$ **25.** 2 **27.** 60 **29.** $\frac{26}{3}$
31. $-\frac{7}{8}$ **33.** $\frac{1}{6}$ **35.** $\frac{4}{3}\sqrt{2} \approx 1.8856$ **37.** -2.9501
39. 10.7 m **41.** **(a)** 39.4 m/s; **(b)** 74.1 m; **(c)** 5.1258 s
43. **(a)** 1 000 000 m/s^2; **(b)** 5005 m **45.** 117 moose
47. **(a)** 2 ft, **(b)** 0.545 ft, **(c)** 3.92 ft

Exercise Set 4.3

1. $9x + C$ **3.** $2x^3 + C$ **5.** $\frac{2}{7}x^{7/2} + C$
7. $\frac{1}{4}t^4 + t + C$ **9.** $\frac{1}{3}y^3 + 2y^2 - 3y + C$
11. $\frac{3x\sqrt[3]{x}}{4} - 9\sqrt[3]{x} + C$ **13.** $\frac{1}{2}(x^2+3)^2 + C$
15. $-\frac{(4-2x^2)^2}{2} + C$ **17.** $-\frac{1}{4}(3-x^2)^4 + C$
19. $\frac{1}{3}(x^2+4)^{3/2} + C$ **21.** $\frac{1}{2}(\sqrt{x}-1)^4 + C$
23. $\frac{1}{45}(3x^3+1)^5 + C$ **25.** $\sqrt{x^2+3} + C$
27. $-\frac{1}{4}(x^2+3)^{-2} + C$ **29.** $\frac{9}{5}x^5 - 2x^3 + x + C$
31. $\frac{1}{2}x^2 - 3x + C$ **33.** $\frac{15}{4}(x^2-1)^{2/3} + C$
35. $\frac{1}{18}(1+3x^2)^3 + C$ **37.** $\frac{3}{20}(4x^3-5)^{5/3} + C$
39. $x + 2x^3 + \frac{9}{5}x^5 + C$ **41.** $\frac{1}{3}(4x^2+2x)^{3/2} + C$
43. $-\frac{1}{x-1} + C$ **45.** $\frac{16}{5}x^{10} - 16x^6 + 36x^2 + C$
47. $\frac{1}{5}(x^3-3x)^{5/3} + C$ **49.** $\frac{1}{12}(3x+1)^4 + C$
51. $\frac{2}{5}(x^4-2x)^{5/4} + C$ **53.** $\frac{2}{3}(t+7)^{3/2} + C$
55. $\frac{1}{3}y^3 + y^2 + 4y + C$ **57.** -8 **59.** $\frac{9}{2}$ **61.** $156\frac{1}{3}$
63. 139.5 **65.** 6,865,432.5
67. $L(x) = 0.3x + 0.001x^2 + 0.8$
69. $P(x) = 2x^2 + 35x - 150$
71. $T(x) = 2875(x+1)^{-2} + 25$

Exercise Set 4.4

1. 8

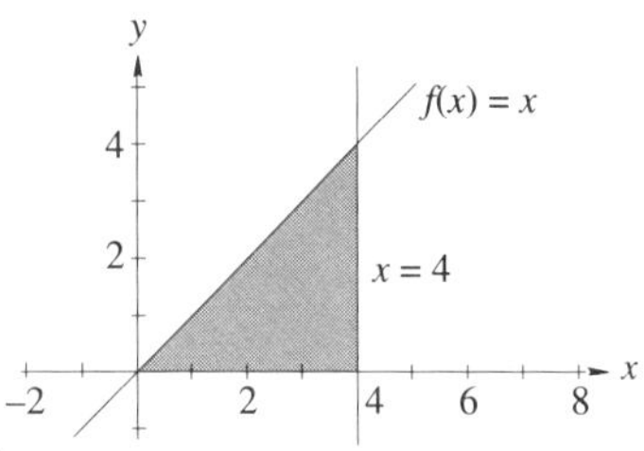

3. $\frac{64}{3}$

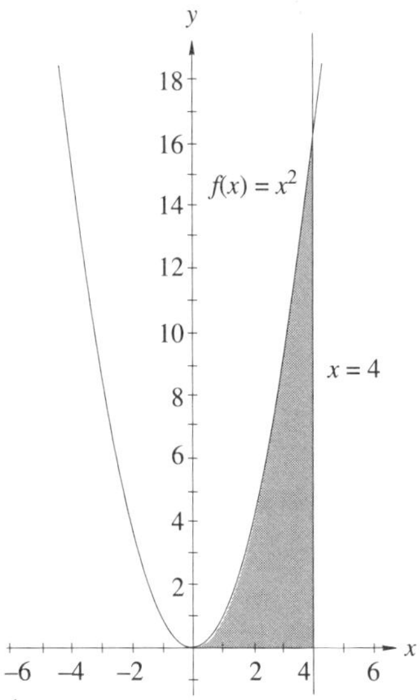

5. $160\frac{1}{6}$

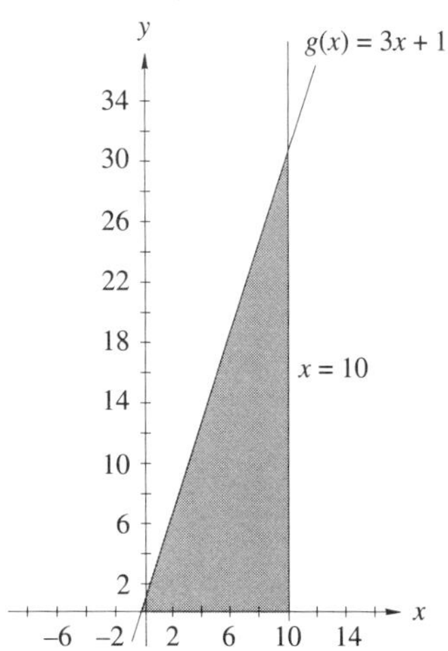

7. 12

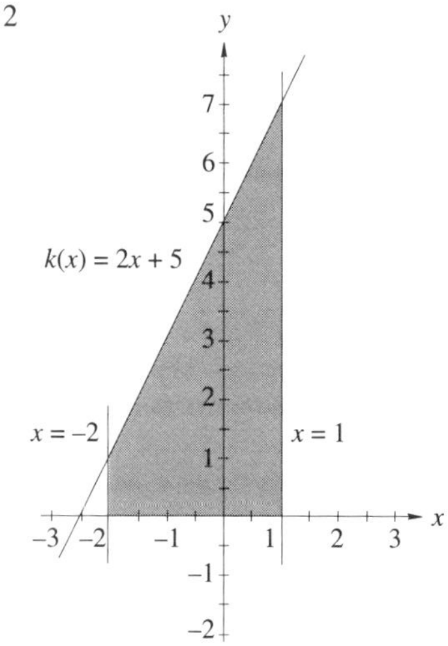

9. $\frac{22}{3}$

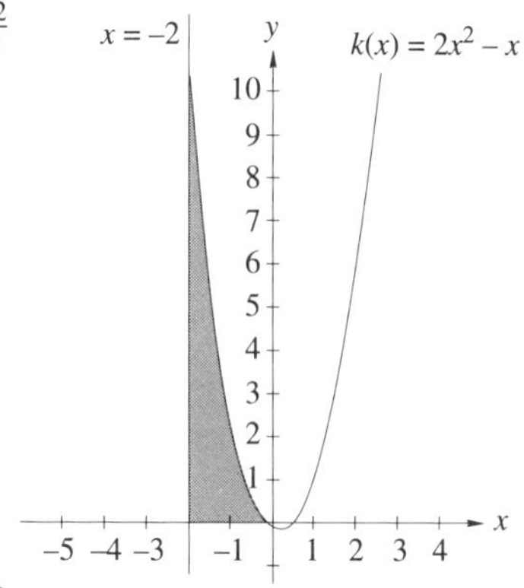

11. $\frac{32}{3}$

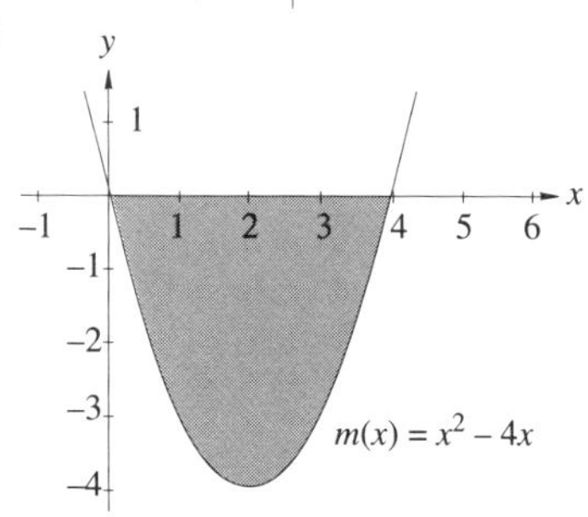

13. 36

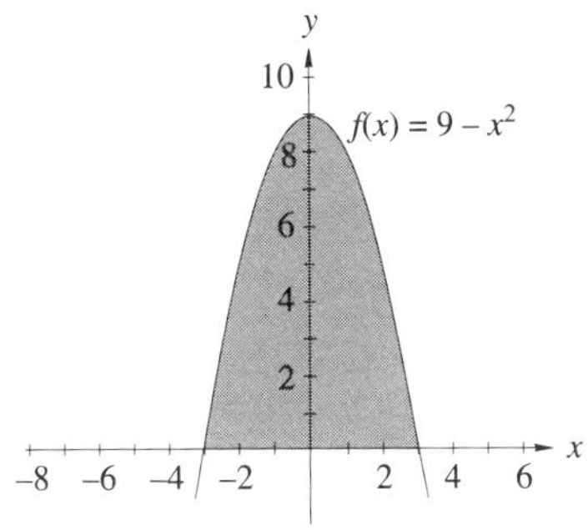

15. 16

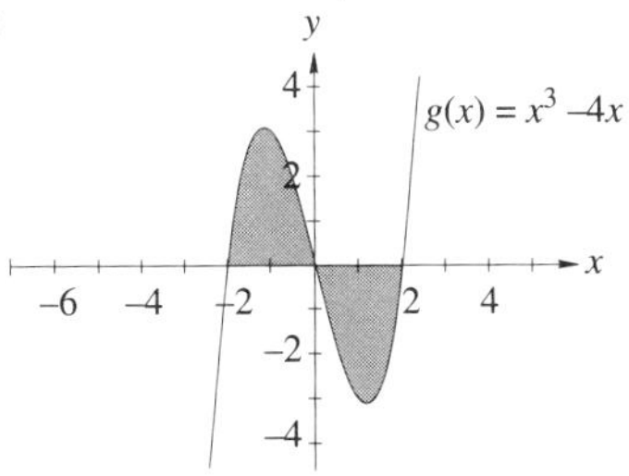

17. 18

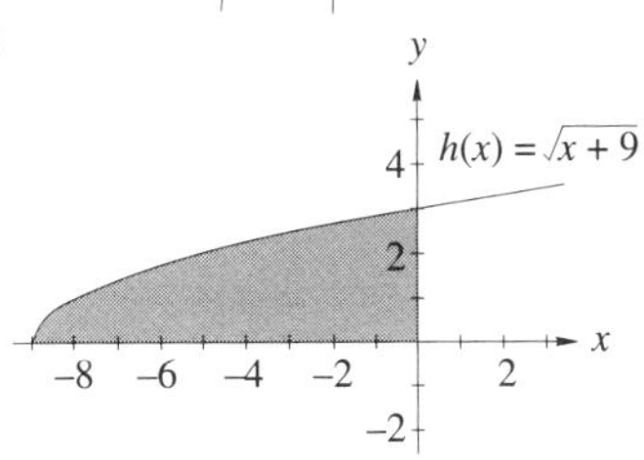

19. $\frac{4}{3}$

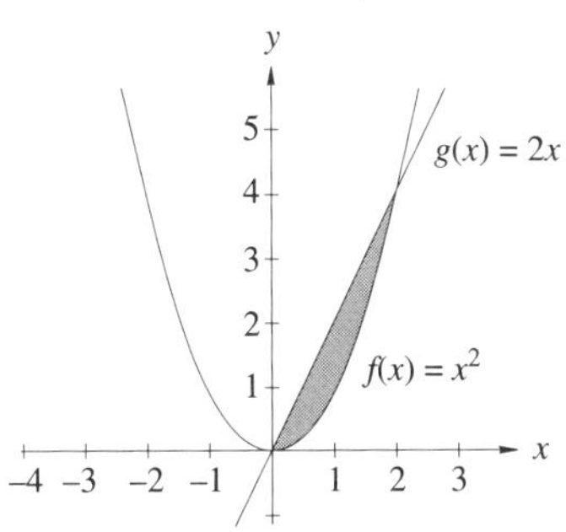

21. $8\sqrt{6} \approx 19.596$

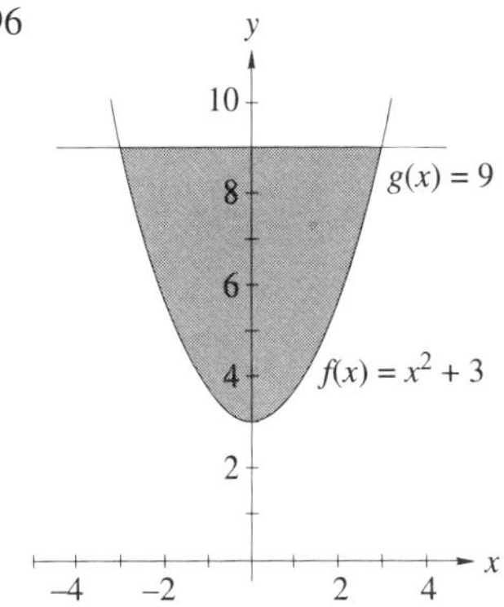

23. 21.333

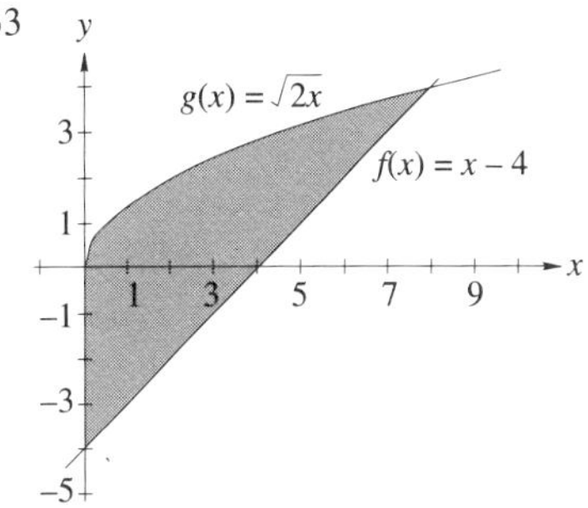

25. 20.8333

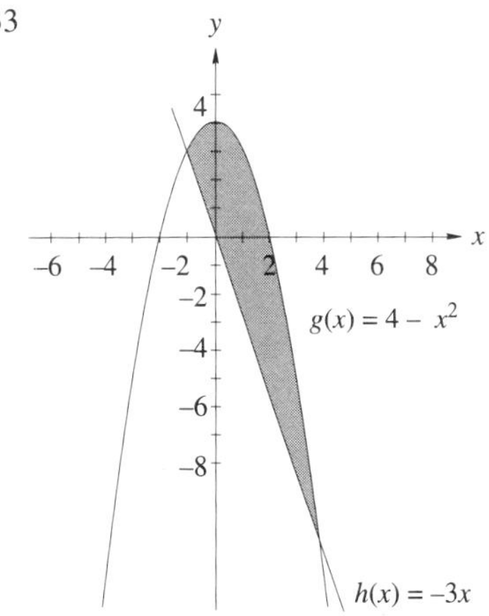

27. $\frac{125}{6} \approx 20.833$

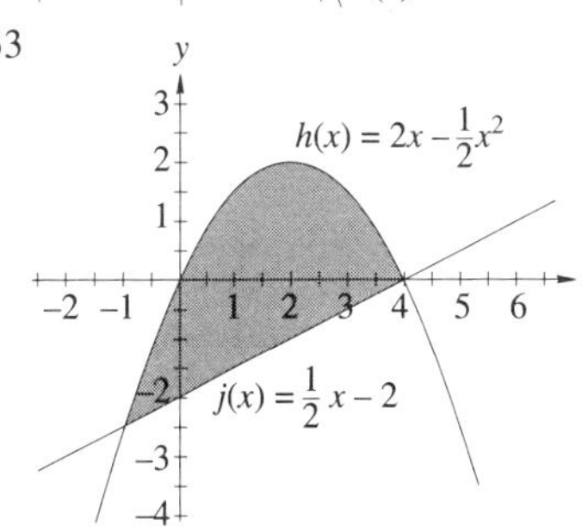

29. 18

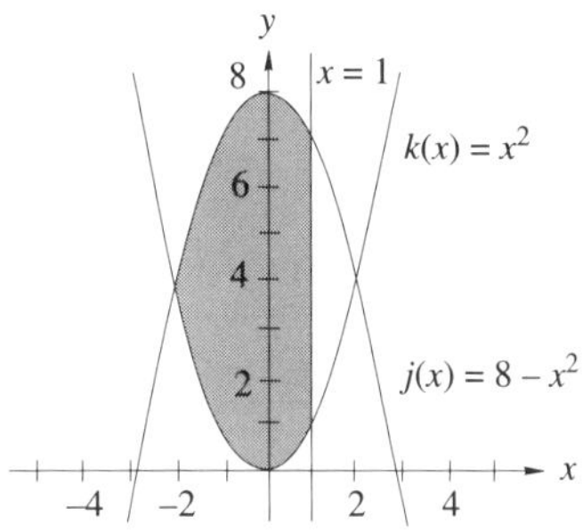

31. 8

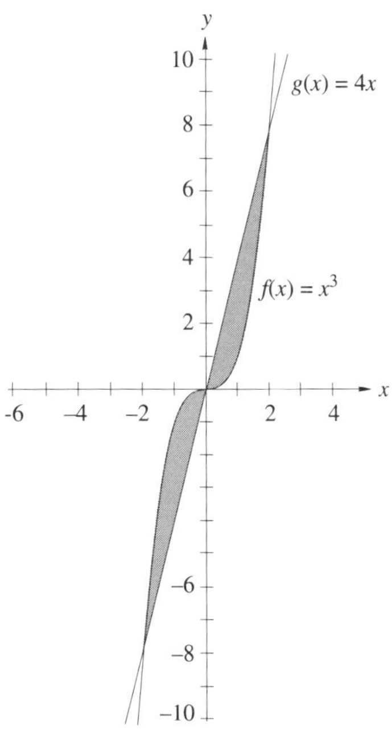

33. $41\frac{2}{3}$

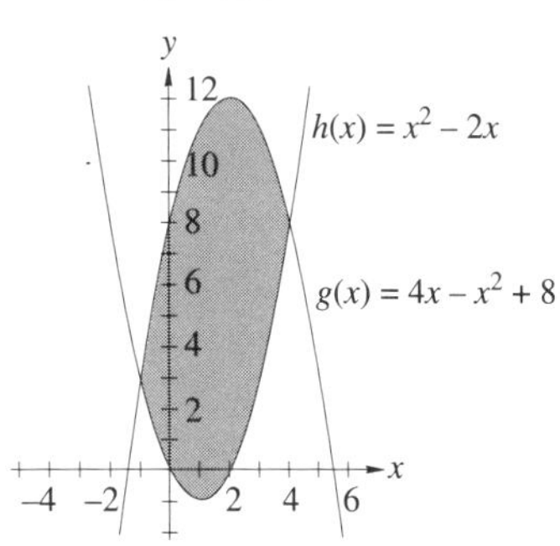

35. 2.25

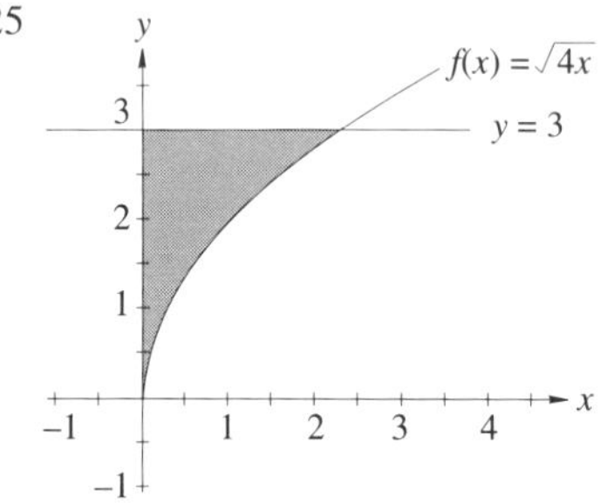

37. $\frac{64}{6} \approx 10.667$

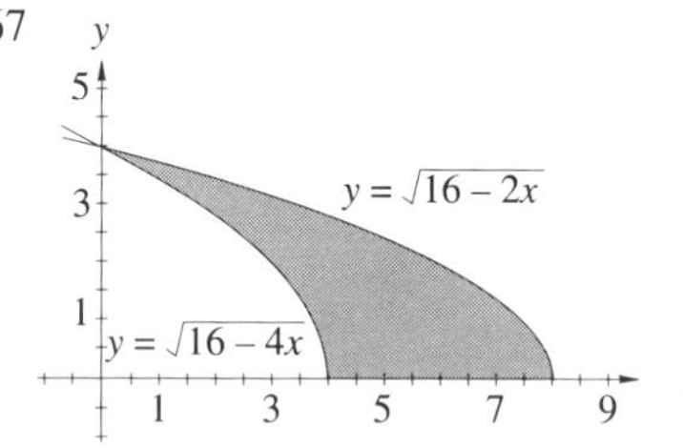

39. 0.375

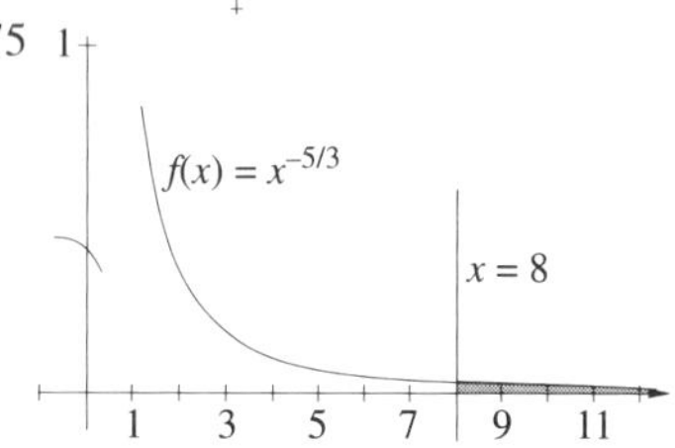

41. $\frac{1}{3}$

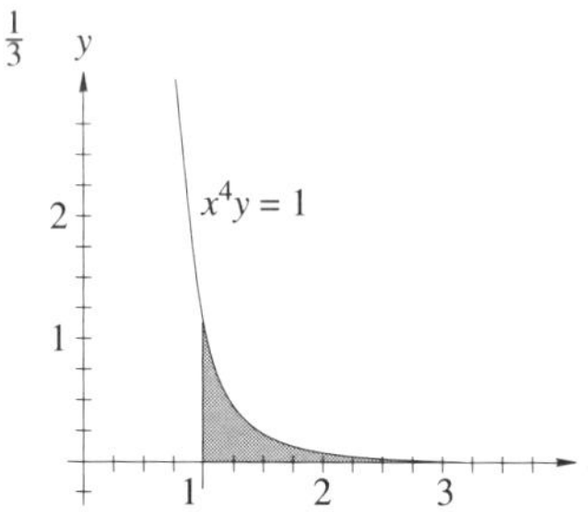

43. White. The blue area is $\frac{5}{12}$ and the white area is $\frac{7}{12}$. **45.** $33\,866\frac{2}{3} \approx 33\,866.67\text{ cm}^3$ **47.** 2630 Btu **49.** 139.7 rad/s

Exercise Set 4.5

1. **(a)** 6.264 789 1; **(b)** 6.200 102 86; actual value, 6.2 **3.** **(a)** 0.643 283; **(b)** 0.656 526 3; actual value $\frac{2}{3}$ **5.** **(a)** 1.628 968; **(b)** 1.610 847 **7.** **(a)** 1.103 210 7; **(b)** 1.098 726 **9.** **(a)** 3.139 926; **(b)** 3.141 592 6 **11.** **(a)** 3.251 744; **(b)** 3.241 238 **13.** **(a)** 7.909 233; **(b)** 7.912 321 **15.** **(a)** 2.141 030; **(b)** 2.170 342 **17.** **(a)** 30.85; **(b)** 30.833 33 **19.** **(a)** 24.646 25; **(b)** 24.629 167 **21.** **(a)** 26.76 μC; **(b)** 20.34 μC **23.** See *Computer Programs*

25. Trapezoidal: 11.264 acres, Simpson: 11.417 acres

Review Exercises

1. $\frac{1}{6}x^6+C$ **3.** 9 **5.** $\frac{9}{2}$ **7.** $-\frac{1}{3}(t+5)^{-3}+C$
9. $\frac{381}{7}\approx 54.429$ **11.** $\frac{1}{5}u^5+\frac{1}{3}u^{-3}+C$
13. $\frac{2}{3}(x^2-5)^{3/2}+C$ **15.** $-\frac{3}{8}(6-x^2)^{4/3}+C$ **17.** $-\frac{39}{2}$
19. $(y^3-5)^{1/3}+C$ **21.** $\frac{16}{35}$
23. $\frac{1}{3}(2x)^{3/2}-(2x)^{1/2}+C$ **25.** $-\frac{1}{4}$
27. $\frac{855}{2312}\approx 0.36981$ **29.** $\frac{1}{3}$ **31.** 53 **33.** $\frac{125}{6}$

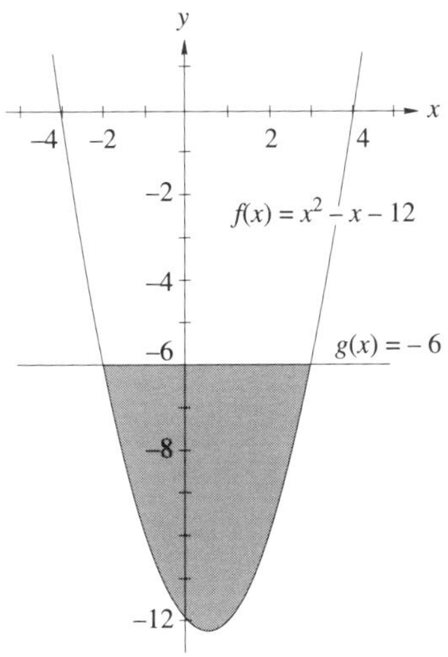

35. $\frac{1}{8}$ **37.** $\frac{9}{16}\approx 0.5625$ **39.** **(a)** 0.776130; **(b)** 0.781752
41. **(a)** 1.613695; **(b)** 1.613656 **43.** **(a)** 1.091812;
(b) 1.086185 **45.** **(a)** 118.5 m; **(b)** $\frac{127}{128}\approx 0.992\,\text{m/s}^2$
47. 30.8 A

Chapter 4 Test

1. 1 **3.** $\frac{1}{5}x^5+3x^3+9x+C$ **5.** $x-2x^{-1}+C$
7. $\frac{2}{3}(10^{3/2}-2^{3/2})\approx 19.196$ **9.** $\frac{125}{6}$ **11.** 0.00107C

ANSWERS FOR CHAPTER 5

Exercise Set 5.1

1. $\bar{y}=1/3$, $f_{\text{rms}}=\sqrt{1/5}=\dfrac{\sqrt{5}}{5}\approx 0.4472$ **3.** $\bar{y}=10/3$,
$h_{\text{rms}}=\sqrt{248/14}\approx 4.0661$ **5.** $\bar{y}=2\frac{38}{15}$,
$f_{\text{rms}}=\sqrt{\frac{13}{2}}\approx 2.5495$ **7.** $\bar{y}=\frac{28}{3}$,
$f_{\text{rms}}=\sqrt{1648/15}\approx 10.4817$ **9.** $\bar{y}=6\frac{2}{3}$,
$f_{\text{rms}}=\sqrt{\frac{226}{5}}\approx 6.7231$ **11.** $\bar{y}=10.7083$, $j_{\text{rms}}=12.8992$
13. 3.7522 cm for 90 days; 5.4755 cm for the year
15. $\bar{s}=240$ ft; $\bar{v}=-75.8947$ ft/s **17.** **(a)** 68° F; **(b)** 71° F
19. 2.6547 A **21.** 23.0309 A **23.** 27136 W
25. **(a)** 1.3927 V; **(b)** 19.40 W **27.** 0.02 C **29.** 2537.5 J
31. 0.6667 V

Exercise Set 5.2

1. 661.33π **3.** $64\pi/3$ **5.** 30π **7.** 8π
9. $16\pi/3$ **11.** $768\pi/7$ **13.** $\dfrac{512\pi}{5}$ **15.** $2\pi/35$
17. $\dfrac{10\pi}{3}$ **19.** $32\sqrt{2\pi}/3$ **21.** $V=\frac{4}{3}\pi r^3$
23. **(a)** $x^2=\frac{2}{3}y$; **(b)** 1.8π **25.** $16{,}913{,}712\pi\,\text{ft}^3$
$\approx 53{,}135{,}993\,\text{ft}^3$ **27.** **(a)** $\frac{800}{3}\pi\approx 837.758\,\text{m}^3$;
(b) $5760\pi\approx 18095.574\,\text{m}^3$

Exercise Set 5.3

1. 8π **3.** 32π **5.** 625π **7.** $768\pi/7$
9. $64\pi/5$ **11.** $\dfrac{8\pi}{35}$ **13.** $10\pi/7\approx 1.42857\pi$
15. $176\pi/15\approx 11.7333\pi$ **17.** $\dfrac{27\pi}{2}$ **19.** $28\pi\sqrt{21}$
21. 99.7 in.^{3}

Exercise Set 5.4

1. 12 **3.** 14/3 **5.** 14/3 **7.** 64.6308π
9. $4\sqrt{2\pi}$ **11.** $32\sqrt{17\pi}\approx 414.5$ **13.** approximately 244.17 m

Exercise Set 5.5

1. $(0.2, 5)$ **3.** $(2.9, 0.2)$ **5.** $\left(\frac{7}{6}, \frac{4}{3}\right)$ **7.** $\left(\frac{1}{74}, \frac{25}{74}\right)$ **9.** $(1.75, 3.25)$ **11.** $(32/7, 4/5)$ **13.** $(253/95, 195/152)$ **15.** $(5/4, 17/5)$ **17.** $\left(-\frac{1}{2}, \frac{8}{5}\right)$ **19.** $(0, 9)$ **21.** $(0.5, 4.4)$ **23.** $(7/4, 0)$ **25.** $\left(0, \frac{3}{5}\right)$ **27.** $(33/28, 0)$ **29.** $(0, 2.75)$ **31.** $(2.4, 0)$ **33.** $(5/9, 0)$ **35.** **(a)** $\left(\frac{5049.5}{199}, \frac{5049.5}{199}\right) \approx (25.374, 25.374)$; **(b)** $\left(\frac{h^2+hw-w^2}{2(2h-w)}, \frac{h^2+hw-w^2}{2(2h-w)}\right)$ **37.** $m = 4040$ g, $\bar{x} = 26.6$ cm from the less dense end.

Exercise Set 5.6

1. $I_x = 0$; $I_y = 93$; $r_0 = \sqrt{93/8} \approx 3.4095$ **3.** $I_x = 52$; $I_y = 19$; $r_0 = \sqrt{71/7} \approx 3.1847$ **5.** $I_y = 8\rho\sqrt{2/15} \approx 0.7542\rho$; $r_y = \sqrt{0.4} \approx 0.6325$ **7.** $I_x = 125\rho$; $r_x = \sqrt{25/3} \approx 2.8868$ **9.** $I_x \approx 0.014286\rho$; $r_x \approx 0.414$ **11.** $I_y = 256\ \text{g}\cdot\text{cm}^2$; $r_y = \sqrt{4.8} \approx 2.1909$ cm **13.** $I_y = 41.6\ \text{g}\cdot\text{cm}^2$; $r_y = \sqrt{156/55}$ cm ≈ 1.6842 cm **15.** $I_y = 4\rho\pi/3$; $r_y = \sqrt{2/3} = 0.8165$ **17.** $I_y = 29.867\pi\rho$; $r_y \approx 1.497$ **19.** $I_x = 2592\pi\rho$; $r_x = 3\sqrt{2} \approx 4.2426$ **21.** $I_y = 512\pi\ \text{g}\cdot\text{cm}^2$; $r_y = \sqrt{7.76} \approx 2.7852$ cm **23.** $I_x = 158.0952\pi\ \text{g}\cdot\text{cm}^2$; $r_y \approx 1.7436$ cm

Exercise Set 5.7

1. $25/8 = 3.125$ ft·lb **3.** 15 J **5.** 4.0 N **7.** 4.8305 ft. **9.** 1125 J **11.** 126 ft·lb. **13.** 127 kJ **15.** 179.76 MJ **17.** 99k J **19.** 80,000 mi-lb. $= 4.224 \times 10^9$ ft·lb. **21.** 1 238 720π J $\approx$ J **23.** 1 286 250 J **25.** 2.866 MJ **27.** 274 400π J **29.** **(a)** 19 600 N/m^2; **(b)** 1 372 000 N **31.** 137 200 N **33.** 39 200 N **35.** 24 578 N

Review Exercises

1. $\bar{y} = 14$, $f_{\text{rms}} = 15.1438$ **3.** $\bar{y} = 16/3$, $h_{\text{rms}} = 6.3666$ **5.** $V = 1488\pi$, $\bar{x} = 3.7742$ **7.** $V = 511\pi/9$, $\bar{x} = 1.8018$ **9.** $V = 16\pi/35 \approx 0.4571\pi$, $\bar{y} = 0.5469$ **11.** $V = 20.25\pi$, $\bar{y} = 4.5$ **13.** $L = 9.0734$, $S = 93.9342\pi$ **15.** $L = 20.1111$, $S = 411.0124\pi$ **17.** 4.5 N·m **19.** 2 A **21.** 35,250 ft·lb **23.** 39 739 392π J

Chapter 5 Test

1. 32 **3.** 227.5π **5.** 14.6238 **7.** 2 250 000π J

Index of Applications

This index categorizes the application problems in this book alphabetically. After each topic, the index lists the text page number, followed by the problem number in parentheses. For example, if you are looking for a problem on acoustical engineering, you would turn to page 117 and look for problem number 35.

Index